Professional Training Program
and Curriculum Standerds in Construction Cost
〔 Sino-British Cooperative Program 〕

工程造价专业人才培养方案及专业课程标准（中英合作办学项目）

主　编　戴明元　鲜　洁　吕　颖
副主编　托尼·夏洛克（英）

北京理工大学出版社
BEIJING INSTITUTE OF TECHNOLOGY PRESS

内 容 提 要

　　本书是四川建筑职业技术学院与英国林肯学院在多年合作办学的基础上编写而成的，主要包括中英文对照工程造价专业（中英合作办学项目）人才培养方案及专业课程标准。人才培养方案及课程标准整合了中英双方工程造价专业课程体系，又保持了各自的专业核心课程，有利于培养熟悉国际工程规范、国际国内工程造价，能组织涉外工程投标报价、工程管理的建设类国际化技术技能型人才。

　　本书可作为高等职业院校中英合作办学项目工程造价专业教学指导性文件，也可供中外合作办学项目制定人才培养方案和课程标准时参考。

版权专有　侵权必究

图书在版编目（CIP）数据

工程造价专业人才培养方案及专业课程标准 / 戴明

元，鲜洁，吕颖主编. --北京：北京理工大学出版社，

2021.8

中英合作办学项目

ISBN 978-7-5763-0235-6

Ⅰ.①工…　Ⅱ.①戴…　②鲜…　③吕…　Ⅲ.①工程造

价－专业人才－人才培养－高等职业教育－教学参考资料

②工程造价－课程标准－高等职业教育－教学参考资料

Ⅳ.①U723.3

中国版本图书馆CIP数据核字（2021）第172959号

出版发行/北京理工大学出版社有限责任公司	
社　　址/北京市海淀区中关村南大街5号	
邮　　编/100081	
电　　话/（010）68914775（总编室）	
（010）82562903（教材售后服务热线）	
（010）68944723（其他图书服务热线）	
网　　址/http://www.bitpress.com.cn	
经　　销/全国各地新华书店	
印　　刷/北京紫瑞利印刷有限公司	
开　　本/787毫米×1092毫米　1/16	
印　　张/32.5	
字　　数/822千字	责任编辑/江　立
版　　次/2021年8月第1版　2021年8月第1次印刷	责任校对/周瑞红
定　　价/162.00元	责任印制/李志强

图书出现印装质量问题，请拨打售后服务热线，本社负责调换

编写说明

工程造价专业（中英合作办学项目）是中国四川建筑职业技术学院与英国林肯学院合作开办的专业。该专业于2012年经四川省教育厅批准、教育部审核备案（证书编号：PDE51UK3A20120568N），同年开始招生，专业代码：540502。本专业课程由中英双方教师共同开发、讲授，学生毕业时获得四川建筑职业技术学院毕业文凭和英国林肯学院国家高级文凭（HND）。

经过几年的实践，中英双方共同编制了一套完整的中英文对照人才培养方案及课程标准，用于指导该专业教学及教学质量控制，供高职院校中外合作办学项目相关专业参考、借鉴。本书的特点在于：

1. 人才培养方案融入了英国学徒制教育理念，体现了中国职业教育特色，兼顾了中外合作办学双方对课程设置、毕业学分的要求，突出职业能力培养，对职业核心能力结构进行了分解。

2. 课程标准从课程概述、课程设计思路、课程目标、课程内容与学时分配、教学实施建议、考核评价六个方面进行了描述，明确了学生应掌握的知识及应达到的能力，形成了较完整的教学指导性文件。

3. 中英文对照有利于中英双方教师及项目学生把握整个课程体系，有效实施教学和教学研究，提升中方教师英语水平、学生英语应用能力，便于中英双方相关机构对教学质量检查和评估。

本书主要包括工程造价专业（中英合作办学项目）人才培养方案及专业课程标准。

本书由四川建筑职业技术学院戴明元、鲜洁、吕颖担任主编，英国林肯学院托尼·夏洛克担任副主编。人才培养方案由戴明元撰写，课程标准由托尼·夏洛克、刘觅等中英双方多名教师共同编写，由吕颖、宗麟、陈南威、谌菊红翻译。四川建筑职业技术学院李育枢主审中文部分，戴明元、鲜洁主审英文部分并提出了宝贵意见。美国教师Shlbi Schadendorf、英国林肯学院教师Matthew King阅读了英文部分并对英文表述提出了修改意见，在此一并表示感谢。

职业院校课程设置随行业发展不断改进，人才培养方案和专业课程标准按国家相关规定和行业要求适时修订，中外合作办学项目亦如此。该人才培养方案和课程标准也需要在使用中根据国家相关要求和行业发展不断进行修改与完善。

本书的编写和翻译是我们对工程造价专业（中英合作办学项目）人才培养方案与课程标准的思考，希望对高职院校中外合作办学项目有所裨益。限于编者水平，编写和翻译中的错误在所难免。诚恳欢迎中外合作办学项目相关专业同人和广大读者批评指正。

编　者

Instructions

The Specialized Sino-British Cooperative Program in Construction Cost is a specialty jointly established by Sichuan College of Architectural Technology (SCAT) and Lincoln College of U.K. The specialty was approved by the Education Department of Sichuan Province and the Ministry of Education in 2012 (Certificate No.: PDE51UK3A20120568N), and began to recruit students in the same year with the major code of 540502. The professional courses are jointly taught by both Chinese and British teachers. After graduation, the students have obtained the diploma of Sichuan College of Architectural Technology and the National Higher Diploma (HND) of Lincoln College.

After several years of practice, SCAT and Lincoln College have jointly worked out a complete set of professional training program and curriculum standard to guide the teaching and teaching quality control of the specialty, and provide reference for the relevant specialties of Sino-foreign cooperative education program in higher vocational colleges. The characteristics of the book are as follows:

1. The professional training program integrates the British apprenticeship education concept, reflects the characteristics of Chinese vocational education, takes into account the requirements of Chinese and foreign cooperative schools for curriculum and graduation credits, highlights the cultivation of vocational ability, and decomposes the structure of professional core competence.

2. The curriculum standards are described from six aspects: curriculum description, curriculum design ideas, curriculum objectives, curriculum content and teaching hours, teaching suggestions and evaluation. The knowledge that students should master and the ability they should achieve are clarified, and a relatively complete teaching guidance document is formed.

3. Chinese and English version is beneficial for Chinese and foreign teachers and project students to grasp the whole curriculum

system, effectively implement teaching and teaching research, improve the English level of Chinese teachers and students' English application ability, and facilitate the inspection and evaluation of teaching quality by the relevant institutions of China and Britain.

This book mainly includes professional training program and curriculum standard of professional courses and professional basic courses for specialized Sino-British cooperative program in construction cost.

This book is edited by Dai Mingyuan, Xian Jie and Lv Ying of Sichuan College of Architectural Technology and Tony Sherlock of Lincoln College. The professional training program is written by Dai Mingyuan. The curriculum standards are jointly compiled by Tony Sherlock (UK), Liu MI and other Chinese and British teachers. Lv Ying, Zong Lin, Chen Nanwei and Chen Juhong translate the book. Li Yushu reviewed the Chinese part, Dai Mingyuan and Xian Jie reviewed the English part, and they put forward valuable opinions. Shelbi Schadendorf, an American teacher, and Matthew King, a teacher from Lincoln College in the U.K., read the English part and put forward some suggestions for revising the English expression. For all the above, we would like to express our thanks.

With the development of the industry, the curriculum of vocational colleges is constantly improved. The professional training program and curriculum standards are timely revised according to the relevant national regulations and industry requirements, and so are the Sino-foreign cooperative education programs. This professional training program and curriculum standards also need to be constantly modified and improved according to the relevant national requirements and industry development.

The compilation and translation of this book is our reflection on the professional training program and curriculum standards of Sino-British Cooperative Program in Construction Cost, hoping to be beneficial to Sino foreign cooperative education program in higher vocational colleges.Limited to the editors' knowledge, mistakes in writing and translation are inevitable. We sincerely welcome the colleagues and readers of the Sino foreign cooperative education program to criticize and correct the mistakes and shortcomings of this book.

Editors

目 录 Contents

工程造价专业人才培养方案
（中英合作办学项目）

专业大类：土木建筑大类　代码：54
专业类别：建设工程管理类　代码：5405
专业名称：工程造价（中英合作办学项目）　代码：540502
培养层次：专科

一、招生对象与学制

招生对象：普通高中毕业生
基本学制：全日制三年

二、培养目标

本专业由中国四川建筑职业技术学院与英国林肯学院合作开办，培养适应社会主义市场经济需要，德、智、体、美等方面全面发展，具有良好诚信品质、敬业精神和责任意识的，具备双语能力、国际化视野，掌握建设工程造价专业的基本理论和专业技能，面向建设单位及涉外建设企事业单位，工程造价咨询、招标代理、工程监理、工程项目管理等机构，工程造价主管等部门，从事建设工程预结算、建设工程招标、建筑工程投标报价、工程造价控制、工程造价审计等工作的高端技术应用型专门人才。可适应的岗位：建设工程及涉外建筑工程预算、投标报价、造价控制、工程结算、造价审计等工作岗位。

三、培养规格

（一）职业核心能力

（1）编制工程预算、计算工程量的能力；
（2）编制工程量清单、具备工程量清单投标报价的能力；
（3）编制招标控制价、编制工程结算的能力；
（4）涉外工程资料管理、涉外工程计量计价的能力；
（5）学习发展的能力，能根据工作和时代发展，不断更新知识和技能。

Professional Training Program for Specialized Sino-British Cooperative Program in Construction Cost

Professional Category: Civil Engineering   Code: 54

Specialty Category: Building and Construction Engineering Management   Code: 5405

Specialty: Construction Cost(Sino-British Cooperative Program)   Code: 540502

Level of Education: Higher Vocational College

1. Target of Enrollment and School System

Target of Enrollment: Graduates of Higher School

School System: Full time education for three years

2. Objectives

This specialty offers courses in cooperation with Sichuan College of Archi tectural Technology Lincoln College, U. K., aiming to cultivate students to meet the requirements of a socialist market economy, to develop students'comprehensive abilities in moral, intellectual, physical and aesthetic fields, to have good quality of integrity, professionalism and responsibility, to have the bilingual ability, international perspective, to have the basic theoretic and professional skills. The graduates of this specialty areapplication-oriented talents in high-end technology who may work with construction cost budget, construction project bidding, quotations for construction project bid, project cost control, project control audit, and they may work in foreign-related construction enterprises and institutions, the organizations of project cost consulting, bidding, supervision, management, and the administrations of project cost supervision, etc.

3. Program Mission

3. 1   Professional Core Competence

Students in this cooperative program should have the professional core competence of

(1) making the project budget and calculating the project quantities;

(2) making the bills of quantities, bidding and quoting for the bills of quantities;

(3) making bidding control price and project settlement;

(4) managing the foreign-related engineering data and calculating the quantities of overseas project;

(5) learning continuously and getting knowledge and skills based on the work environment and development of the times.

职业核心能力结构分解表

序号	能力	内涵要点	相关课程
1	工程量计算	较强的编制工程预算的能力,工程计量的专业知识和基本技能	画法几何、建筑构造、建筑结构基础与识图、建筑识图、工程造价概论、建筑工程预算、钢筋工程量计算、建筑CAD、Revit建模基础、建筑施工工艺等
2	工程量清单投标报价	编制工程量清单的能力,工程量清单投标报价的专业知识和基本技能	工程量清单计价、工程量清单计价实训、工程造价概论等
3	工程计价、预算、结算	编制工程招标控制价的能力,工程预算、决算的专业知识和基本技能	定额原理、建筑与装饰材料、工程造价控制、工程造价软件应用、工程造价软件应用实训、水电安装工程预算、水电安装工程预算实训、工程结算等
4	工程资料管理,涉外工程计量计价	工程资料归档管理的能力,涉外工程计量计价的专业知识和基本技能	建筑工程招投标与合同管理、BIM管理、FIDIC合同、建筑英语、大学英语、剑桥国际英语、计算机基础、英国工程图识读、英国建筑工程施工技术、英国建筑材料、英国工程计量、英国工程计价、英国工程招投标等

(二)基本要求

1. 基本素质

(1)具有强烈的爱国主义精神、社会责任感及良好的思想品德、社会公德和职业道德;

(2)具有求实创新的科学精神、刻苦钻研的实干精神及较强的团队协作意识;

Decomposition Table for the Professional Core Competence

Number	Competence	Related Abilities	Related Courses
1	Calculation of engineering quantities	The ability of preparing project budget and the professional knowledge and basic practical skills in calculating project quantities	Descriptive Geometry, Building Structure, Basic Building Structure and Drawing Reading, Construction Drawing Reading, Introduction to Project Cost, Building Construction Budget, Calculation of Reinforcement Quantity, Construction CAD, Revit Modeling Basis, Building Construction Technology, etc.
2	Quotations of bills of quantities	The ability of preparing bills of quantities and the professional knowledge and basic practical skills in bid and quotation of bills of quantities	Pricing for Bills of Quantities, Practical Training for Bills of Quantities Estimating, Introduction to Project Cost, etc.
3	Project pricing, budget and settlement	The ability of preparing the project bidding control price and the professional knowledge and basic practical skills in project budget and construction settlement	Quota Principle, Building and Decoration Materials, Construction Cost Control, Application of Construction Cost Software, Hydropower Installation Budge, Practical Training for Hydropower Installation Budget, Project Final Account, etc.
4	Construction materials management, foreignrelated project quantities calculation	The ability of archiving and managing of construction data and the professional knowledge and basic practical skills in calculating and pricing quantities in foreign-related project	Bidding and ·Contract Management of Construction Project, BIM Management, Fidic Contract, English for Building and Construction Engineering, College English, Cambridge International English, Fundamentals of Computer Application, Reading &Understanding Drawings in the U. K. , Understanding Construction Technology in the U. K. , Understanding Material Technology in the U. K. , Understanding Measurement Process in Construction in the U. K. , Understanding Construction Estimating in the U. K. , Understanding Construction Tendering in the U. K. , etc.

3.2　Basic Requirements

3.2.1　Basic Qualities

Students should have the quantity of

(1)possessing strong patriotism, social responsibility, and possessing good ideological moralities, social moralities and professional ethics;

(2)possessing the scientific spirit of seeking truth and innovation, possessing a hard-working spirit, and a strong sense teamwork;

（3）具有一定的审美情趣、艺术修养和文化品位,有较高的人文、科学素养;

（4）具有健康的身心素质、健全的人格、坚强的意志和乐观向上的精神风貌。

2. 基本知识

（1）了解国家的基本法律、法规;

（2）掌握必要的基础理论知识:表达与沟通、计算机应用等;

（3）掌握本专业的基础知识和基本理论;具备工程预算、结算等知识;

（4）掌握并能运用本专业的专业知识;具备建筑企事业及涉外工程投标报价、计量计价等专业。

3. 基本能力

（1）具备较强的建筑企事业及涉外工程投标报价、计量计价、预算结算的能力;能够使用工程造价常用专业软件;

（2）能够利用所学的专业知识分析问题、解决问题,具备较强的实践操作能力;

（3）具有获取本专业前沿知识和相关学科知识的自学能力、创新意识和一定的社会活动能力;

（4）具有从事工程造价专业所必需的工程预算、投标报价、造价控制、工程结算、造价审计等能力。

四、职业资格证书

职业资格证书表

序号	证书名称和级别	内涵要点	发证单位
1	全国助理造价工程师	建筑工程造价计价	工程造价管理协会
2	造价工程师	建筑工程造价计价与审计	住建部

五、核心专业课程描述

(一)画法几何

课程学分:2.5;课程总学时:44(其中理论学时:40,实践学时:4);开设学期:第1学期

(3) possessing some aesthetic taste, artistic accomplishment and cultural taste and having higher cultural and scientific qualities;

(4) possessing good mental and physical health, sound personality, strong will and optimistic spirit.

3.2.2 Basic Knowledge

Students will be able to

(1) master laws and basic rules of our country;

(2) master the basic knowledge of some theories, such as some expression and communication skills, computer application, etc. ;

(3) master the basic knowledge and basic theory of this specialty: project budget, project settlement, etc. ;

(4) master the professional knowledge and have the ability to apply some professional knowledge: bidding and quoting for construction projects and foreign-related projects, estimating and pricing for project, etc.

3.2.3 Basic Abilities

Students should have the ability of

(1) possessing strong abilities in bidding and quoting for construction projects and foreign-related projects, in estimating and pricing for project and in budget and settlement of projects; applying most frequently used software of this specialty;

(2) analyzing and solving problems by one's professional knowledge and having strong abilities in practicing;

(3) self-learning of the latest knowledge of this specialty and of the related subject and possessing the sense of innovation and capacity of social activities;

(4) taking job responsibilities related with project budget, bidding and pricing, cost control, project settlement, cost audit, etc.

4. Vocational Qualification Certificates

List of Vocational Qualification Certificates

Number	Certificate Name and Level	Related Abilities	Issuing Unit
1	National Assistant Cost Engineer	The ability of pricing for construction projects	Construction Cost Management Association
2	Cost Engineer	The ability of pricing and auditing for construction projects	Ministry of Housing and Urban-rural Development of People's Republic of China

5. Description of the Core Professional Courses

5.1 Descriptive Geometry

Credits: 2.5   Teaching Hours: 44(For lectures: 40,for practice: 4)

Semester: 1st

主要教学内容:

建立明确的中心投影和平行投影(正投影和斜投影)的概念;掌握点、线、面各种位置的投影特性和作图方法;掌握平面立体和曲面立体的投影特性和作图方法,以及在表面上作点、作线的方法。

教学目标:

《画法几何》是土木建筑工程专业学生的必修课程,是阅读和回执工程图的理论基础,也是其他工科类专业学生应掌握的基本知识。本门课程的教学目标是掌握投影的原理及各几何元素空间相对位置的投影特点,解决空间几个元素的度量和定位问题。

通过本门课程的学习培养学生的空间想象力和空间构思能力,培养学生的读图能力和工程素质,为其他专业课程的学习打下扎实的基础。

参考教材:

李翔、刘觅、凌莉群.《画法几何》,高等教育出版社,2014。

(二)建筑构造

课程学分:2;课程总学时:32(其中理论学时:26,实践学时:6);开设学期:第1学期

主要教学内容:

房屋建筑构造,装饰构造,民用与工业建筑施工图识读。

教学目标:

熟悉房屋建筑构造、装饰构造,能识读民用与工业建筑施工图。

参考教材:

赵西平.《房屋建筑学》,中国建筑工业出版社,2017。

(三)剑桥国际英语

课程学分:2;课程总学时:32(其中理论学时:22,实践学时:10);开设学期:第1学期

主要教学内容:

思辨能力培养,交流能力,团队协作能力,简历制作以及面试技巧。

教学目标:

通过学习本门课程,学生可以在跨文化背景下通过对职业能力进行板块训练,同时掌握板块能力语言词汇,并在设定情景中进行练习运用以达到语言、职业能力双向提升的教学目的。该门课程致力于将中外合作办学项目学生培养成为跨文化背景下语言能力佳、交流能力强、具备良好职业素养并拥有国际竞争力的复合型人才。

Main Contents:

The main contents of this course include: precise knowledge on concepts of central projection and parallel projection (orthogonal projection and oblique projection); projecting characteristics of points, lines, surfaces and their plotting methods; projecting characteristics of plane solids and curved-surface solids and their plotting methods; the methods of plotting point and lines on any surface and so on.

Teaching Objectives:

As well as being a fundamental course for majors of construction cost, *Descriptive Geometry* is a compulsory course for students of civil engineering, and it lays theoretical foundation for reading and acknowledging construction drawing. Teaching objectives of this course are helping students to learn the principles of projection and characteristics of relatively spatial positioning in projecting various geometrical shapes, solving the problems in measuring and locating multiple spatial elements.

By learning this course, students would enhance their spatial visualization and conception, their ability of graphic reading and their intellectual engagement in construction, and it may lay a foundation for other core courses.

Recommended Textbook:

Li Xiang, Liu Mi, Ling liqun. *Descriptive Geometry*. Higher Education Press, 2014.

5.2 Building Structure

Credits: 2   Teaching Hours: 32 (For lectures: 26, for practice: 6)   Semester: 1st

Main Contents:

The main contents of this course include: structures of houses and apartment buildings, decoration structures, reading drawings of civil and industrial buildings.

Teaching Objectives:

By learning this course, students should have specialist knowledge about the structure of houses and apartment buildings as well as decorating structures. They should also be able to read construction drawing of civil and industrial buildings.

Recommended Textbook:

Zhao Xiping. *Building Structure*. China Architecture & Building Press, 2017.

5.3 Cambridge International English

Credits: 2   Teaching Hours: 32 (For lectures: 22, for practice: 10)   Semester: 1st

Main Contents:

The main contents of this course include: critical thinking, communication skills, team work, the writing skills and practice of resume and interview, etc.

Teaching Objectives:

By learning this course, students can train their professional competence in various modules and in the cross-cultural context, master the language vocabulary of different settings, and practice English in set situations to improve both the language and the professional competence. This unit aims to cultivate students of construction cost (Sino-British Program) into interdisciplinary talents with good language ability, strong communication ability, excellent professional quality and international competitiveness in cross-cultural context.

参考教材:

安妮·汤普森.《思辨思维-实践概论》(第三版),罗德里奇出版社,2009。

(四)建筑与装饰材料

课程学分:2;课程总学时:32(其中理论学时:22,实践学时:10);开设学期:第1学期

主要教学内容:

建筑及装饰材料基本性质,常用建筑及装饰材料(石材、水泥、混凝土、钢材、木材、沥青、防水材料及建筑塑料、玻璃、涂料、面砖等)及其制品的种类、名称、规格、质量标准、选用、检验试验方法、保管方法等。

教学目标:

了解常用建筑与装饰材料的基本性质,了解无机胶凝材料、混凝土及砂浆、建筑钢材、木材、防水材料、装饰材料等的种类、作用;熟悉各种建筑与装饰材料的特点、作用等。

参考教材:

李江华、郭玉珍、李柱凯.《建筑材料项目化教程》,华中科技大学出版社,2013。

(五)建筑结构基础与识图

课程学分:4;课程总学时:64(其中理论学时:54,实践学时:10);开设学期:第2学期

主要教学内容:

框架结构、排架结构、钢结构等的梁、板、柱、基础基本知识,结构施工图识读。

教学目标:

了解框架结构、排架结构、钢结构等的梁、板、柱、基础基本知识,能识读结构施工图,掌握平法施工图的识读方法。

参考教材:

杨太生.《建筑结构基础与识图》(第四版),中国建筑工业出版社,2019。

(六)建筑施工工艺

课程学分:3.5;课程总学时:56(其中理论学时:56,实践学时:0);开设学期:第2学期

主要教学内容:

土方工程、桩基工程、砌筑工程、钢筋混凝土工程、结构安装工程、冬雨期施工、大模板施工等等的施工方法、施工工艺与技术。

Recommended Textbook:

Anne Thomson. *Critical Reasoning. A Practical Introduction.* 3rd Edition. Routledge, 2009.

5. 4　Building and Decoration Materials

Credits: 2　　Teaching Hours: 32(For lectures: 22, for practice: 10)　　Semester: 1st

Main Contents:

The main contents of this course include: characteristics of building and decoration materials, the categories, names, qualities, standards, selection, tests and preservation of usual building and decoration materials(stone, cement, concrete, steel, wood, asphalt and waterproof materials, as well as plastics, glass, coating materials and tiles of buildings)and their correlative products.

Teaching Objectives:

By learning this course, students can get the characteristics of common building materials and decorating materials, learn the categories and application of inorganic cement, concrete and mortar, construction steels, timbers, waterproof material and decorating materials. They can also have specialist knowledge about characteristics and applications of various building and decorating materials.

Recommended Textbook:

Li Jianghua, Guo Yuzhen, Li Zhukai. *Project Tutorial of Construction Materials.* Huazhong University of Science and Technology Press, 2013.

5. 5　Basic Building Structure and Drawing Reading

Credits: 4　　Teaching Hours: 64(For lectures: 54, for practice: 10)　　Semester: 2nd

Main Contents:

The main contents of this course include: basic knowledge on beam, plate, pillar and foundation of frame structure, bent structure and steel structure and reading of structural construction drawings.

Teaching Objectives:

By learning this course, students can get the fundamental knowledge on beam, plate, pillar and foundation of frame structure, bent structure and steel structure. Furthermore, students can acquire the ability to read structural construction plans and learn the approaches to read the holistic design of structural construction plans.

Recommended Textbook:

Yang Taisheng. *Basic Building Structure and Drawing Reading.* 4th Edition. China Architecture & Building Press, 2019.

5. 6　Building Construction Technology

Credits: 3. 5　　Teaching Hours: 56(For lectures: 56, for practice: 0)　　Semester: 2nd

Main Contents:

The main contents of this course include: engineering approaches, processes and technologies of earthwork, pile foundations, masonry, steel and concrete, structure and installation, execution in winter and rain, big template construction, etc.

教学目标：

熟悉土方工程、基础工程、砌筑工程、钢筋混凝土工程、预应力混凝土工程、装配式框架结构吊装及滑模施工、防水工程、装饰工程等施工工艺与技术。

参考教材：

李辉,黄敏.《建筑施工技术》(第2版),重庆大学出版社,2017。

(七)建筑识图

课程学分1.5 课程总学时:24(其中理论学时:16,实践学时:8);开设学期:第2学期

主要教学内容：

土木工程建筑施工图识读。

教学目标：

掌握建筑、装饰施工图纸内容构成,培养识读建筑、装饰施工图的能力。

参考教材：

李翔,宋良瑞,张翔.《建筑识图与实务》,高等教育出版社,2014。

(八)工程造价概论

课程学分2;课程总学时:32(其中理论学时:28,实践学时:4);开设学期:第2学期

主要教学内容：

工程造价基本原理、建筑安装工程费用、定额的构成、识读与应用、工程造价计价方法、计价模式等。

教学目标：

理解工程造价基本原理、掌握建筑安装工程费用,熟悉定额的构成,能应用定额,了解工程造价计价方法和计价模式。

参考教材：

袁建新,袁媛.《工程造价概论》(第三版),中国建筑工业出版社,2016。

(九)建筑工程项目管理

课程学分:2;课程总学时:32(其中理论学时:28,实践学时:4);开设学期:第3学期

主要教学内容：

工程项目管理的组织、工程项目计划与控制、工程项目风险管理、工程流水施工、网络技术、工程项目的风险管理等。

教学目标：

熟悉工程项目管理的组织、工程项目计划与控制、工程项目风险管理、工程流水施工、网络技术等内容,能进行单位工程施工组织设计。

Teaching Objectives:

By learning this course, students should have specialist knowledge about engineering processes and technologies of earthwork, foundation, masonry, steel and concrete, prestressed concrete, hoisting and sliding mode of fabricated frameworks, water-proofing structures and decorating structures.

Recommended Textbook:

Li Hui, Huang Min. *Building Construction Technology*. 2nd Edition. Chongqing University Press, 2017.

5.7 Construction Drawing Reading

Credits: 1.5   Teaching Hours: 24   (For lectures: 16, for practice: 8)   Semester: 2nd

Main Contents:

The main contents of this course are the reading of construction drawings in civil engineering.

Teaching Objectives:

By learning this course, students should get the content of construction and decoration drawings, foster their ability in reading construction and decoration drawings.

Recommended Textbook:

Li Xiang, Song Liangrui, Zhang Xiang. *Architectural Drawings Reading and Practice*. Higher Education Press, 2014.

5.8 Introduction to Construction Cost

Credits: 2   Teaching Hours: 32(For lectures: 28, for practice: 4)   Semester: 2nd

Main Contents:

The main contents of this course include: the basic principle of construction cost, the calculation of building installation construction cost, the composition, reading & application of quota, the pricing methods of construction cost and the modes of pricing.

Teaching Objectives:

By learning this course, students should understand the principle of construction cost, master the cost of building installation engineering, be familiar with the composition of quotas and able to apply, understand the pricing methods of construction cost & the modes of pricing.

Recommended Textbook:

Yuan Jianxin, Yuan Yuan. *Overview of Cost Engineering for Projects*. 3rd Edition China Architecture & Building Press, 2016.

5.9 Construction Project Management

Credits: 2   Teaching Hours: 32(For lectures: 28, for practice: 4)   Semester: 3rd

Main Contents:

The main contents of this course include: project management organization, project planning and control, project risk management, construction streamline method, network technology, project risk management, etc.

Teaching Objectives:

By learning this course, students should be familiar with project management organization, project planning and control, project risk management, construction streamline method, network technology etc. and they should be able to carry out construction organization design of unit project.

参考教材：

项建国.《建筑工程项目管理》(第三版)，中国建筑工业出版社,2015。

兰凤林.《工程项目管理实务》(第二版)，大连理工大学出版社,2014。

(十)建设工程招投标与合同管理

课程学分:2;课程总学时:32(其中理论学时:32,实践学时:0);开设学期:第4学期

主要教学内容：

招标投标的基本概念、招标投标的方式及招投标程序、合同的基础知识;各阶段的合同管理、合同实施控制;索赔的起因和条件、索赔值的计算、反索赔等。

教学目标：

熟悉工程招标投标、合同、索赔的概念、分类、程序,掌握招标投标的方式及程序、掌握各阶段合同管理、合同实施控制;能计算费用索赔和工期索赔。

参考教材：

林密.《工程项目招投标与合同管理》(第三版)，中国建筑工业出版社,2013。

(十一)工程经济

课程学分:2;课程总学时: 32(其中理论学时:29,实践学时:3);开设学期:第4学期

主要教学内容：

现金流量与资金时间价值,投资方案经济效果评价,设备更新方案比选,不确定性分析,寿命周期成本分析,价值工程的理论及应用。

教学目标：

能根据财务报表进行投资方案经济效果评价,能进行设备更新方案比选,能进行不确定性分析、寿命周期成本分析,掌握价值工程的理论及应用。

参考教材：

赵彬.《工程技术经济》,高等教育出版社,2010。

(十二)建筑电气设备安装工艺与识图及 Revit 建模基础

课程学分:2;课程总学时:32(其中理论学时: 32,实践学时: 0);开设学期:第4学期

主要教学内容：

学习运用 BIM 技术进行建筑安装工程实体模型的建立,使学生全面和系统地获得 BIM 技术和 Revit 建模软件的相关知识,独立、快速地构建结构模型以及电气模型等。

Recommended Textbooks:

Xiang Jianguo. *Construction Project Management.* 3rd Edition. China Architecture & Building Press, 2015.

Lan Fenglin. *Project Management Practice.* 2nd Edition. Dalian University of Technology Press, 2014.

5.10 Bidding and Contract Management of Construction Project

Credits: 2   Teaching Hours: 32(For lectures: 32, for practice: 0)   Semester: 4th

Main Contents:

The main contents of this course include: the basic concept of bidding, the way of bidding and bidding procedure, the basic knowledge of contract, contract management and contract implementation control at each stage, the cause and condition of claim, the calculation of claim value, counterclaim, etc.

Teaching Objectives:

By learning this course, students should be familiar with the concept, classification and procedure of project bidding, contract & claim, master the approach and procedure of bidding, master contract management & contract implementation control at all stages and they should be able to calculate expense claim and duration claim.

Recommended Textbook:

Lin Mi. *Bidding and Contract Management of Construction Project.* 3rd Edition. China Architecture & Building Press, 2013.

5.11 Engineering Economy

Credits: 2   Teaching Hours: 32(For lectures: 29, for practice: 3)   Semester: 4th

Main Contents:

The main contents of this course include: cash flow and capital time value, economic effect evaluation of investment plan, comparison and selection of equipment upgrade plan, uncertainty analysis, life cycle cost analysis, theory and application of value engineering.

Teaching Objectives:

By learning this course, students should be able to carry on the economic effect evaluation of the investment plan according to the financial statement, be able to carry on comparison & selection the equipment renewal plan, be able to carry on the uncertainty analysis and the life cycle cost analysis, master the value engineering theory and its application.

Recommended Textbook:

Zhao Bin. *Engineering Technology and Economy.* Higher Education Press, 2010.

5.12 Building Electrical Installation, Drawing Reading and Revit Building Model

Credits: 2   Teaching Hours: 32(For lectures: 32, for practice: 0)   Semester: 4th

Main Contents:

The main contents of this course include: building the physical model of construction and installation engineering, basic knowledge of BIM technology and Revit modeling software, building and structural modules, electrical module and the application of the module.

教学目标：

通过学习本课程,学生能全面和系统地获得 BIM 技术和 Revit 建模软件的相关知识,培养学生的科学思想和研究方法,使学生在软件应用、逻辑思维和解决问题的能力等方面都得到基本而系统的训练,为以后工作奠定必要的基础。

参考教材：

陆泽荣、叶雄进.《BIM 建模应用技术》第二版,中国建筑工业出版社,2018。

(十三) 建筑水暖设备安装工艺与识图及 Revit 建模基础

课程学分:2;课程总学时：32;(其中理论学时：32,实践学时：0);开设学期:第 4 学期

主要教学内容：

应用 Revit 软件进行暖通工程、建筑室内给排水工程、消防工程专业的模型建立。

教学目标：

主要任务是学习运用 BIM 技术进行建筑安装工程实体模型的建立,使学生全面和系统地获得 BIM 技术和 Revit 建模软件的相关知识,培养学生的科学思想和研究方法,使学生在软件应用、逻辑思维和解决问题的能力等方面都得到基本而系统的训练,为以后工作奠定必要的基础。

参考教材：

陆泽荣、叶雄进.《BIM 建模应用技术》,中国建筑工业出版社,2018。

(十四) 英国工程图识读

课程学分:4;课程总学时:60(其中理论学时:40,实践学时:20);开设学期:第 2 学期

主要教学内容：

学习英国工程图表述方法、英语专业词汇、施工图绘制的基本要求,识读工程平面图、结构施工图等。

教学目标：

掌握英文识图专业词汇,能识读英国工程平面图、结构施工图。

参考教材：

托尼.《英国工程图识读》,英国林肯学院自编教材。

(十五) 建筑 CAD

课程学分:2;课程总学时:32(其中理论学时:16,实践学时:16);开设学期:第 5 学期

Teaching Objectives:

By learning this course, students should comprehensively and systematically acquire relevant knowledge of BIM technology and Revit modeling software, have scientific thoughts and research methods, and get basic and systematic training in software application, logical thinking and problem-solving ability and lay a necessary foundation for future work.

Recommended Textbook:

Lu Zerong, Ye Xiongjin. *BIM Modeling Application Technology*. 2nd Edition. China Architecture & Building Press, 2018.

5. 13　Building Plumbing Installation, Drawing Reading and Revit Modeling Basis

Credits: 2　　Teaching Hours: 32(For lectures: 32, for practice: 0)　　Semester: 4th

Main Contents:

The main contents of this course include: applying BIM technology to construction and installation engineering entity model, building water supply and drainage module, fire control module, and apply the model.

Teaching Objectives:

By learning this course, students should study the use of BIM technology to construction and installation engineering entity model, have comprehensive and systematical knowledge for BIM technology and Revit modeling software related knowledge. This course should cultivate students'scientific thought and research methods, so that students can be trained in software application, logical thinking and problem-solving skills, thus laying the necessary foundation for future work.

Recommended Textbook:

Lu Zerong, Ye Xiongjin. *BIM Modeling Application Technology*. 2nd Edition. China Architecture & Building Press, 2018.

5. 14　Reading & Understanding Drawings in U. K.

Credits: 4　　Teaching Hours: 60(For lectures: 40, for practice: 20)　　Semester: 2nd

Main Contents:

The main contents of this course include: studying the British drawing method in construction, ESP vocabulary, basic requirements for construction drawing and reading the constructional plan, structural working drawings and so on.

Teaching Objectives:

By learning this course, students should master ESP vocabulary, be able to read the British constructional plan and structural construction drawings.

Recommended Textbook:

Tony. *the U. K. Engineering Drawings Reading*(self-compiled textbook), Lincoln College(U. K.)

5. 15　Construction CAD

Credits:4　　Teaching Hours:32(For lectures: 16, for practice: 16)　　Semester: 5th

主要教学内容：

AutoCAD 基础知识,绘图环境及图层管理,绘制平面图形,编辑平面图形,辅助绘图命令与工具尺寸标注,图纸布局与打印输出,建筑工程图绘图。

教学目标：

本课程以目前广泛应用的 AutoCAD 软件作为典型代表,将课堂理论教学与实践教学相结合,详细介绍计算机绘图软件的使用操作原理和方法,通过教师用多媒体讲授基础理论及学生进行计算机操作,使学生熟练掌握计算机辅助设计软件——AutoCAD 绘制工程图样的方法。

参考教材：

唐英敏、吴志刚、李翔.《AutoCAD 建筑绘图教程》(第二版),北京大学出版社,2014。

董祥国.《建筑 CAD 技能实训》,中国建筑工业出版社,2016。

郭慧.《AutoCAD 建筑制图教程》(第三版),北京大学出版社,2018。

(十六)BIM 技术基础

课程学分:1;课程总学时:18(其中理论学时:18,实践学时:0);开设学期:第 4 学期

主要教学内容：

本课程主要从宏观角度讲述 BIM 的基础知识及所用模型和软件,包括 BIM 工程师的素质要求与职业发展、BIM 基础知识、BIM 建模环境及应用软件体系、项目 BIM 实施与应用、BIM 标准与流程。

教学目标：

掌握 BIM 的基本概念及基本常识;了解 BIM 对建筑业带来的价值及实现方式。

参考教材：

李恒、孔娟.《Revit 2015 中文版基础教程》,清华大学出版社,2015。

廖小烽、王君峰.《Revit 2013/2014 建筑设计火星课堂》,人民邮电出版社,2013。

(十七)企业经营管理

课程学分:1;课程总学时:18(其中理论学时:18,实践学时:0);开设学期:第 1 学期

主要教学内容：

建筑企业的战略环境;建筑行业的发展趋势;建筑企业的组织结构和相关制度;建筑企业的职能部门;管理策略应用;建筑工业化对建筑企业的影响。

Main Contents:

The main contents of this course include: basic knowledge of AutoCAD, drawing environment and layer management, drawing plans, editing plans, aided drawing commands and tools, drawing layout and print output and architectural engineering drawings.

Teaching Objectives:

This course takes AutoCAD software which is widely used at present as a typical representative, combines theory and practice in teaching, and introduces the principles and methods of using computer graphics software in detail. Multimedia teaching and computer operation enable students to master the methods of drawing engineering drawings by AutoCAD.

Recommended Textbooks:

Tang Yingmin, Wu Zhigang, Li Xiang. *AutoCAD Architectural Drawing Course.* 2nd Edition. Peking University Press, 2014.

Dong Xiangguo. *Practical Training for Construction CAD.* China Architecture & Building Press, 2016.

GuoHui. *AutoCAD Architectural Drawing Course*, 3rd Edition. Peking University Press, 2018.

5.16 Fundamentals of BIM Technology

Credits: 1　　Teaching Hours: 18(For lectures:18, for practice: 0)　　Semester: 4th

Main Contents:

The main contents of this course include: the concept and development of BIM technology, BIM technology application at home and abroad, BIM technology platform, application standards and requirements of BIM technology, Revit building model construction and the comprehensive application of Revit model.

Teaching Objectives:

By learning this course, students can understand the origin, development and application of BIM technology and the basic operation methods and application advantage of the software of BIM. Through the practical cases, students can understand and master the process and method for accurately building a BIM project model and extracting model information. Through the application of the information model, it will lay the foundation for future work in architectural design, construction, construction cost consulting and so on.

Recommended Textbook:

Li Heng, Kong Juan. *Basic Course of Revit* 2015 *in Chinese.* Tsinghua University Press, 2015.

Liao Xiaofeng, Wang Junfeng, Haoxing. *Classroom of Revit* 2013/2014 *Architectural Design.* People's Post & Telecomunications Publishing House, 2013.

5.17 Construction Enterprise Management

Credits: 1　　Teaching Hours: 18(For lectures: 18, for practice: 0)　　Semester: 1st

Main Contents:

The main contents of this course include: strategies and environment of construction enterprises, developing trends of construction industry, organization and regime of construction enterprises, functional departments of construction enterprises, application of management strategies, and impacts of industrialization of construction industry on construction enterprises.

教学目标:

让学生了解建筑行业的发展环境、发展趋势;了解并分析建筑企业的组织形式;理解制定管理制度的意义;理解建筑企业各个职能部门的工作内容和管理策略;应用管理策略分析建筑企业中的现实问题;了解建筑工业化对建筑企业的影响,以适应工业化下的建筑企业发展需要。

参考教材:

李渠建.《企业管理基础》,高等教育出版社,2013。

(十八)外教课堂学习法

课程学分:1;课程总学时:16(其中理论学时:16,实践学时:0);开设学期:第 2 学期

主要教学内容:

介绍中西方教育体系、思维方式、教学方法、学习方法的不同,学习掌握跨文化沟通技巧,训练学生目标设定、时间管理、团队精神建设以及与外教沟通的能力。

教学目标:

通过教学,让学生(特别是中外合作办学项目的学生)能够提升跨文化沟通能力,了解西方教学法对学习要求的不同,通过自我学习能力和沟通能力的提升,为更好地适应后期大量的外教课程学习打下基础。

参考教材:

伍慧卿.《外教课堂学习策略》(自编讲义)。

(十九)装配式建筑概论

课程学分:1;课程总学时:18(其中理论学时:18,实践学时:0);开设学期:第 3 学期

主要教学内容:

介绍国内外装配式建筑发展历程以及未来发展趋势,装配式施工的相关施工工艺,并将装配式建筑创新内涵从技术领域延伸至管理领域。

教学目标:

了解装配式建筑的发展历程,掌握装配式施工的相关施工工艺,了解装配式建筑的管理特征及价值。

参考教材:

陈群.《装配式建筑概论》,中国建筑工业出版社,2017。

(二十)Revit 建模基础

课程学分:2;课程总学时:32;(其中理论学时:16,实践学时:16);开设学期:第 4 学期。

Teaching Objectives:

By learning this course, students should learn the ecosystem and trend for the development of construction industry, understand and analyze organizational features of construction enterprises, value management regulations, learn the working content and management strategies of functional departments of construction enterprise, analyze the practice of construction enterprises by applying management strategies, understand the impacts of industrialization of construction industry on construction enterprises and adapt to developing needs of construction enterprises in the industrialization age.

Recommended Textbook:

Li Qujian. *Enterprise Management Basis.* Higher Education Press, 2013.

5.18 Learning Strategy in Foreign Teachers'Classrooms

Credits: 1   Teaching Hours: 16(For lectures: 16, for practice: 0)   Semester: 2nd

Main Contents:

The main contents of this course include: introducing the differences between Chinese and Western educational system, thinking modes, teaching methods and learning methods, cross-cultural communication skills and skills in goal setting, time management, team-working spirit and communicative competence with foreign teachers.

Teaching Objectives:

By learning this course, students (especially those engaged in international-joint programs) can improve their cross-cultural communication skills, understand the different learning requirements among western teaching methods, and lay a solid foundation for follow-up myriad courses taught by foreign teachers through the promotion of self-learning ability and communication ability.

Recommended Textbook:

Wu Huiqing. *Learning Strategies for Foreign Teachers Classroom* (Self compiled Lecture Notes).

5.19 Overview of Prefabricated Construction

Credits: 1   Teaching Hours: 18(For lectures: 18, for practice: 0)   Semester: 3rd

Main Contents:

The main contents of this course include: introduction to the history of prefabricated construction homes abroad and its future trend, to the techniques of prefabricated construction, and exploring revolutionary possibilities of managing prefabricated construction technologies.

Teaching Objectives:

By learning this course, students should understand the development and the history of prefabricated construction, learn processes of prefabricated construction, and learn the characteristic and value in management of prefabricated construction.

Recommended Textbook:

Chen Qun. *An Introduction to Prefabricated Construction.* China Architecture & Building Press, 2017.

5.20 Revit Modeling Basis

Credits: 2   Teaching Hours: 32(For lectures: 16, for practice: 16)   Semester: 4th

主要教学内容:

Revit界面情况,项目新建,项目信息设置;标高、轴网创建方法、墙构件创建方法;柱构件创建方法;梁构件创建方法;板构件创建方法;楼梯创建方法;屋顶创建方法;门窗创建方法;族、体量创建方法;家具、场地、明细表等创建方法。

教学目标:

熟悉Revit软件界面操作命令,掌握Revit软件的基本应用方法,能独立完成一般建筑的模型绘制。

参考教材:

胡煜超.《Revit建筑建模与室内设计基础》,机械工业出版社,2017。

(二十一) BIM工程计价软件应用

课程学分:1;课程总学时:18(其中理论学时:14,实践学时:4);开设学期:第5学期

主要教学内容:

工程造价计价原理,BIM计价软件界面介绍,工程量清单编制、招标控制价编制、投标报价、工程结算价编制。

教学目标:

学习完本课程,学生应掌握工程造价计价原理,能熟练利用相关软件编制工程量清单、招标控制价、投标报价、工程结算,并能分析三个阶段工程造价编制方法的异同。同时,形成能独立使用BIM计价软件编制各种造价文件的能力。

参考教材:

四川省建设工程造价管理总站.《四川省建设工程工程量清单计价定额》,中国计划出版社,2015。

(二十二) 建筑工程预算

课程学分:4;课程总学时:64(其中理论学时:38,实践学时:26);开设学期:第3学期

主要教学内容:

建筑工程计价原理,建筑工程定额应用,建筑工程工程量计算,建筑工程费用计算等。

教学目标:

熟悉建筑工程计价原理,能熟练应用建筑工程预算定额,掌握建筑工程工程量计算方法,掌握建筑工程费用的内容及计算方法等,能独立完成建筑工程预算书的编制。

参考教材:

袁建新.《建筑工程预算》(第五版),中国建筑工业出版社,2015。

Main Contents：

The main contents of this course include： Revit Interface situation，project launching，project information settings，creation methods of elevation，axial network and wall components，pillar components creation method，beam components creation method，plate components creation method，staircase creation method，roof creation method，doors and windows creation method，family and volume creation method，creation methods of furniture，venue，specification，etc.

Teaching Objectives：

By learning this course，students should be familiar with Revit software interface operation commands，master the basic application method of Revit software，be able to complete independently the general architecture model drawing.

Recommended Textbook：

Hu Yuchao. *The Basis of Revit Architecture Modeling and Interior Design*，China Machine Press，2017.

5.21　Application of BIM Construction Cost Software

Credits：1　　Teaching Hours：18(For lectures：14，for practice：4)　　Semester：5th

Main Contents：

The main contents of this course include： basic theory of construction cost pricing，introduction to the operation interface of BIM pricing software，the compiling of bills of quantities and bid control price，the compiling of bidding price and settlement price by the pricing software.

Teaching Objectives：

By learning this course，students can strengthen the basic theory of construction cost pricing and they can be proficient in compiling bills of quantities，bid control price，the bidding price and the settlement price. At the same time，they can analyze the similarities and differences in the compiling methods for the three stages and compile different pricing files by the BIM pricing software independently.

Recommended Textbook：

Construction Cost Management Station in Sichuan Province. *Pricing for Bills of Quantities Quota in Construction Engineering of Sichuan Province.* China Planning Press，2015.

5.22　Building Construction Budget

Credits：4　　Teaching Hours：64(For lectures：38，for practice：26)　　Semester：3rd

Main Contents：

Themain contents of this course include： building construction pricing principle，building construction quota application，bill of quantities and pricing，etc.

Teaching Objectives：

By learning this course，students will be able to be familiar with the principle of building construction pricing，be proficient in the application of building construction budget quota；master the calculation method of bill of quantity，master the content and calculation method of building construction cost，complete independently the preparation of building construction budget book.

Recommended Textbook：

Yuan Jianxin. *Building Construction Budget.* 5th Edition. China Architecture & Building Press，2015.

(二十三) 钢筋工程量计算

课程学分:2.5;课程总学时:40(其中理论学时:28,实践学时:12);开设学期:第4学期

主要教学内容:

了解钢筋工程量计算的基本知识、钢筋工程量计算依据和钢筋工程量计算的基本方法,掌握建筑工程的平法识图和钢筋工程量计算的基本方法及技巧,实际工程的钢筋准确算量方法。

教学目标:

通过学习这门课程,学生能熟练使用16G101系列平法图集,学生能在没有教师的直接指导下独立参考《16G101图集》完成实际工程的钢筋准确算量。

参考教材:

王武齐.《钢筋工程量计算》(第二版),中国建筑工业出版社,2018。

(二十四) 工程量清单计价

课程学分:3;课程总学时:48(其中理论学时:32,实践学时:16);开设学期:第4学期

主要教学内容:

工程量清单及工程量清单计价的基本内容,工程量清单及工程量清单计价编制方法。

教学目标:

熟悉工程量清单及工程量清单计价的基本内容,掌握工程量清单及工程量清单计价编制方法,能编制工程量清单,能编制招标控制价,能确定投标价。

参考教材:

袁建新.《工程量清单计价》第四版,中国建筑工业出版社,2015。

(二十五) 工程造价软件应用

课程学分3;课程总学时:48(其中上机学时:32,实践学时:16);开设学期:第5学期

主要教学内容:

工程量计算软件、钢筋工程量计算软件、工程量清单报价软件的应用及操作方法。

教学目标:

掌握工程量计算软件、钢筋工程量计算软件、工程量清单报价软件的应用及操作方法。

5.23　Calculation of Reinforcement Quantity

Credits：2.5　　Teaching Hours：40(For lectures：28,for practice：12)　Semester：4th

Main Contents：

The main contents of this course include：the rules in calculating reinforcement quantity, the usage of drawings by ichnographic representing method, calculating the quantity of reinforcement in different components of building construction.

Teaching Objectives：

By learning this course, students can master the competence of reading drawings by ichnographic representing method of 16G101 series skillfully; students can read and understand the quantity of reinforcement in different components of building construction based on the drawings by ichnographic representing method of 16G101 by themselves.

Recommended Textbook：

Wang Wuqi. *Calculation of Reinforcement Quantity.* 2nd Edition China Architecture & Building Press, 2018.

5.24　Pricing for Bill of Quantities

Credits：3　　Teaching Hours：48(For lectures：32,for practice：16)　Semester：4th

Main Contents：

The main contents of this course include：preparing bill of quantities, compiling tender sum limit and offer bidding quotations upon bill of quantities according to the "*Pricing Specification of Bill of Quantities in Construction Engineering*" and the "*Calculation Specification of Bill of Quantities for Building and Decoration Construction*".

Teaching Objectives：

By learning this course, students will acquire basic professional qualities, the ideological foundation, thinking style and professional ethics which are necessary for senior professional talents. Students will be able to master the basic knowledge and basic skills of pricing for bill of engineering quantity, and to cultivate the ability to solve practical problems in construction cost occupations and other relevant jobs.

Recommended Textbook：

Yuan Jianxin. *Pricing for Bill of Quantities.* 4th Edition China Architecture Publishing, 2015.

5.25　Application of Construction Cost Software

Credits：3　　Teaching Hours：48(Computer hours：32,for practice：16)　Semester：5th

Main Contents：

The main contents of the course include：being familiar with engineering quantity calculation software, reinforcing steel bar engineering quantity calculation software, bill of quantities quotation software application and operation method.

Teaching Objectives：

By learning this course, students will be able to master the application and operation method of engineering quantity calculation software, reinforcing steel bar engineering quantity calculation software, and bill of quantities quotation software.

参考教材:

张晓敏,李社生.《建筑工程造价软件应用:广联达系列软件》,中国建筑工业出版社,2013。

(二十六) 水电安装工程预算

课程学分3;课程总学时:48(其中理论学时:32,实践学时:16);开设学期:第5学期

主要教学内容:

民用建筑给排水、电气照明安装工程定额应用,安装工程量计算,安装工程费用计算等。

教学目标:

能熟练应用民用建筑给排水、电气照明安装工程定额,掌握水电安装工程量计算方法,掌握水电安装工程费用计算,能独立编制水电安装工程预算书。

参考教材:

刘渊、袁媛.《安装工程计量与计价》,东南大学出版社,2019。

(二十七) 定额原理

课程学分1.5;课程总学时:24(其中理论学时:24,实践学时:0);开设学期:第3学期

主要教学内容:

掌握建筑工程定额的种类;掌握施工过程的构成因素、影响因素以及组成明细;掌握技术测定法中测时法、工作日写实法、写实记录法和简易测定法的适用范围、优缺点以及记录时间的具体方法等。

教学目标:

通过课程学习,学生掌握建筑工程定额的种类、施工过程和工作时间研究、定额的各种具体编制方法等知识点,形成编写人工定额、材料定额和机械台班定额的技能,养成熟练应用定额进而独立编写定额的素质。

参考教材:

袁建新.《企业定额编制原理与实务》,中国建筑工业出版社,2003。

(二十八) 工程造价控制

课程学分2;课程总学时:32(其中理论学时:32,实践学时:0);开设学期:第4学期

Recommended Textbooks:

Zhang Xiaoming, Li Shesheng. *Application of Construction Cost Software: Guanglianda Series Software*, China Architecture & Building Press, 2013.

5.26 Project Budget of Water and Electricity Installation

Credits: 3   Teaching Hours: 48(For lectures: 32, for practice: 16)   Semester: 5th

Main Contents:

The main contents of this course includes: basic knowledge of composition of construction project quota and budget, the analysis of composition of construction cost, the calculation method of construction cost, the compilation method of project completion settlement, and the preparation skills of construction drawing budget.

Teaching Objectives:

By learning this course, students will be able to improve their professional competences, method-applied ability and social ability. Further more, students should learn to complete the preparation and compilation of various cost documents with the quota-pricing model.

Recommended Textbook:

Liu Yuan, Yuan Yuan. *Quantities and Pricing of Installation Construction Engineering.* Southeast University Press, 2019.

5.27 Quota Principle

Credits: 1.5   Teaching Hours: 24(For lectures: 24, for practice: 0)   Semester: 3rd

Main Contents:

The main contents of this course include: different types of construction quotas, the component factors, influencing factors and detailed composition of the construction process, the detailed record of work date, realistic recording and the simple measurement in the technical measuring method, as well as the specific methods of recording time.

Teaching Objectives:

By learning this course, students can master the types of construction engineering quotas, construction process and working time research, various specific methods of quota preparation and other knowledge points, which forms the skills of compiling artificial quotas, material quotas and mechanical quotas, and cultivates the quality of applying quotas and compiling it independently.

Recommended Textbook:

Yuan Jianxin. *Principle and Practice of Enterprise Quota Compilation*, China Architecture & Building Press, 2003.

5.28 Construction Cost Control

Credits: 2   Teaching Hours: 32(For lectures: 32, for practice: 0)   Semester: 4th

主要教学内容:

工程造价控制基本理论,决策阶段的工程造价控制,设计阶段的工程造价控制,招标投标阶段的工程造价控制,工程实施阶段的工程造价控制,工程竣工阶段的工程造价控制。

教学目标:

使学生具备基本的职业素质、具有工程管理高级职业人才所必需的思想基础、思维方式及职业素质,能掌握工程造价控制的基本知识和基本技能,基本具备在工程管理相关岗位上解决实际问题的能力。

参考教材:

尹贻林.《工程造价计价与控制》,中国计划出版社,2011。

(二十九)英国建筑工程施工技术

课程学分:4;课程总学时:48(其中理论学时:36,实践学时:12);开设学期:第3学期

主要教学内容:

学习英国建筑施工技术,了解英国传统施工方法及现代施工方法,认知不同的建筑类型——低层建筑、联排别墅、独立住宅和半独立住宅及其施工技术,学习用于描述施工技术的专业词汇等。

教学目标:

掌握施工技术英文专业词汇,英国建筑工程施工技术。

参考教材:

托尼.《英国建筑工程施工技术》,英国林肯学院自编教材。

(三十)英国建筑材料

课程学分:2;课程总学时:24(其中理论学时:24,实践学时:0);开设学期:第3学期

主要教学内容:

介绍英国建筑材料及其基本性质,了解英国常用建筑材料及其制品的种类、名称、规格、质量标准、选用、检验试验方法、保管方法等。

教学目标:

了解英国常用建筑材料的基本性质,了解建筑砌块、木材、防水材料等的种类及其作用,了解木材的防腐、防白蚁处理。熟悉英国各种建筑材料的特点、作用等。

参考教材:

托尼.《英国建筑材料》,英国林肯学院自编教材。

Main Contents:

The main contents of this course include: the basic theory of construction cost control, project cost control in the decision-making stage, design phase of the project cost control, project cost control in the bidding stage, project cost control during the implementation stage and project cost control at the completion stage.

Teaching Objectives:

By learning this course, students will have basic professional quality, the ideological basis, thinking mode and professional quality which are necessary for senior professional talents in engineering management. Furthermore, they will master the basic knowledge and skills of construction cost control, and have the ability to solve practical problems in engineering management related positions.

Recommended Textbook:

Yin Yilin. *Construction Cost Valuation and Control*. China Planning Press, 2010.

5.29 Understanding Construction Technology in the U. K.

Credits: 4   Teaching Hours: 48(For lectures: 36, for practice: 12)   Semester: 3rd

Main Contents:

The main contents of this course include: basic British constructional technology, traditional & modern British building construction methods; different types of low-rise buildings, townhouse, independent residential and semi-independent residential and their building construction technology, ESP vocabulary for building construction technology and so on.

Teaching Objectives:

By learning this course, students will be able to master English professional vocabulary and the British building construction technology.

Recommended Textbook:

Tony. *Building Construction Technology in the U. K.* (self-compiled textbook), Lincoln College (U. K.).

5.30 Material Technology in the U. K.

Credits: 2   Teaching Hours: 24(For lectures: 24, for practice: 0)   Semester: 3rd

Main Contents:

The main contents of this course include: the commonly-used British building materials and their basic properties, the types, names, specifications, quality standards, selection, test methods and storage methods of the commonly-used building materials and their products in the U. K.

Teaching Objectives:

By learning this course, students will be able to understand the basic properties of building materials commonly used in the UK, understand the types and functions of construction blocks, wood, waterproof materials, etc. , understand the anti-corrosion and termite protection of wood, be familiar with the properties and functions of various building materials in the U. K.

Recommended Textbook:

Tony. *Building Materials in the U. K.* (self-compiled textbook), Lincoln College(U. K.).

(三十一)英国建筑工程计量

课程学分:4;课程总学时:60(其中理论学时:48,实践学时:12);开设学期:第4学期

主要教学内容:

英国建筑工程标准计量法,包括间接费、原有建筑设施拆除、土方工程、现浇或预制混凝土、砌体、结构、桁架、防水、门窗、楼梯、装饰装修、家具设备、路面铺设、排污系统、供水系统、供电系统、通信、机械及电气设施等工程量的计算方法。

教学目标:能根据英国建筑工程标准计量法(SMM7)计算工程量。

参考教材:

托尼.《英国建筑工程计量》,英国林肯学院自编教材。

(三十二)英国建筑工程计价

课程学分:4;课程总学时:60(其中理论学时:48,实践学时:12);开设学期:第5学期

主要教学内容:

英国工程量清单及工程量清单计价的基本原则及内容,英国工程量清单及工程量清单计价编制的基本方法。英国及欧洲业主对工程的要求及风险管理,估算师在项目采购、项目评估中的重要作用。

教学目标:

能根据英国工程量清单计价的基本原理计算工程费用。

参考教材:

托尼.《英国建筑工程计价》,英国林肯学院自编教材。

(三十三)英国工程招投标

课程学分:2;课程总学时:32(其中理论学时:24,实践学时:8);开设学期:第5学期

主要教学内容:

英国建筑工程招标投标的基本概念、招标投标的方式及招标投标程序,了解估算师在招标投标、项目评估、承包商的选择中的重要作用,了解雇主、建筑师、估算师在招标决策中的相互作用,招标投标的相关法律法规。

教学目标:

熟悉英国工程招标投标并能应用所学知识从事海外工程招标投标。

5.31 Measurement Processes in Construction in the U. K.

Credits: 4   Teaching Hours: 60(For lectures: 48, for practice: 12)   Semester: 4th

Main Contents:

The main contents of this course include: the standard measurement method of British construction works, including indirect fees, the demolition of original construction facilities, earthwork, cast-in-water or prefabricated concrete, masonry, structure, truss, waterproof, doors and windows, stairs, decoration, furniture equipment, pavement laying, sewage systems, water supply systems, power supply system, communications, machinery and electrical facilities and other engineering quantities calculation methods.

Teaching Objective:

By learning this course, students will be able to calculate the amount of construction works according to the standard measurement method(SMM7)of construction works in the U. K.

Recommended Textbook:

Tony. *Measurement of Construction Works in the U. K.* (self-compiled textbook), Lincoln College (U. K.).

5.32 Construction Estimating in the U. K.

Credits: 4   Teaching Hours: 60(For lectures: 48,for practice: 12)   Semester: 5th

Main Contents:

The main contents of this course include: the basic principles and contents of the bill of quantities and its estimating in the U. K. , the basic compiling method of bill of quantities and its estimating in the U. K. , the engineering requirements and risk management of the U. K. and European property owners, and the important role of estimators in project procurement and project assessment.

Teaching Objectives:

By learning this course, students will be able to calculate the works costs based on the estimating fundamentals of bill of quantities in the U. K.

Recommended Textbook:

Tony. *Estimating of Construction Works in the U. K.* (self-compiled textbook), Lincoln College (U. K.).

5.33 Construction Tendering in the U. K.

Credits: 2   Teaching Hours: 32(For lectures: 24,for practice: 8)   Semester: 5th

Main Contents:

The main contents of this course include: the basic concept of tendering for building project in the U. K. , the approach and procedures of tendering, the important role of estimator in tendering, project evaluation, and contractor selection, the interaction between employers, architects and estimators in tendering decisions, the relevant laws and regulations for tendering in the U. K.

Teaching Objectives:

By learning this course, students will be familiar with the building project tendering in the U. K. and be able to apply the know-how into overseas building project tendering.

参考教材：

托尼.《英国工程招投标》，英国林肯学院自编教材。

34. FIDIC 合同

课程学分：2；课程总学时：32（其中理论学时：32，实践学时：0）；开设学期：第 5 学期

主要教学内容：

使学生掌握 FIDIC 合同条件，特别是施工合同条件的具体内容，了解基于 FIDIC 的合同管理的基本框架，并应用所学开展合同条件编制、工程变更及索赔等合同管理工作。

教学目标：

通过本课程的学习，使学生较为全面地掌握 FIDIC 合同条件的具体规定，掌握合同管理的相关理论知识，基于 FIDIC 的国际 工程招标投标的基本程序、环节和关键节点；掌握索赔的基本理论知识及 FIDIC 合同条件中的索赔条款的内容。

参考教材：

张水波、何伯森.《FIDIC 新版合同条件导读与解析》，中国建筑工业出版社，2003。

35. 建筑英语

学分：2；课程总学时：32（其中理论学时：26，实践学时：6）；开设学期：第 3 学期

主要教学内容：

以建筑施工为主线，内容涉及建筑材料、建筑工艺、项目管理、室内装饰、室外工程、安装工程、园林绿化、施工报告及施工日志写作等。

教学目标：

培养学生在建筑领域使用业务英语的能力，着重培养学生对建筑工程英文资料的阅读能力和翻译能力以及涉外工程现场口译能力和交际能力。

参考教材：

戴明元.《建筑工程英语》，高等教育出版社，2014。

36. 工程结算

课程学分：2；课程总学时：32（其中理论学时：26，实践学时：6）；开设学期：第 5 学期

主要教学内容：

工程结算；签证资料处理，变更造价处理；费用索赔计算。

Recommended Textbook:

Tony. *Understanding Construction Tendering in the U. K.* (self-compiled textbook), Lincoln College (U. K.)

5.34 FIDIC Contract Conditions and Claims

Credits: 2   Teaching Hours: 32(For lectures: 32, for practice: 0)   Semester: 5th

Main Contents:

The main contents of this course include: FIDIC contract conditions, especially the specific content of conditions of contact for construction, the basic framework of contract management based on FIDIC, and the application of carrying out contract management work, such as contract conditions preparation, engineering changes and claims.

Teaching Objectives:

By learning this course. students can master the specific provisions of FIDIC contract conditions, the relevant theory of contract management, the basic procedures, links and key points of international project bidding based on FIDIC, the basic theoretical knowledge of claims and the content of claim clauses in FIDIC contract conditions.

Recommended Textbook:

Zhang Shuibo, He Bosen. *Introduction and Explanation of FIDIC Conditions and claims.* China Architecture & Building Press, 2011.

5.35 English for Building and Construction Engineering

Credits: 2   Teaching Hours: 32(For lectures: 26, for practice: 6)   Semester: 3rd

Main Contents:

The main contents of this course include: building construction as the basis, including the contents of building materials, construction technology, project management, interior decoration, outdoor building construction, installation engineering, landscaping, construction reports and construction log writing, etc.

Teaching Objectives:

By learning this course, students will be able to use ESP in the field of architecture, and students 'reading ability and translation ability in English materials of building construction will be improved, as well as the ability of international on-site interpretation and communicative competence.

Recommended Textbook:

Dai Mingyuan. *English for Building and Construction Engineering*, Higher Education Press, 2014.

5.36 Engineering Settlement

Credits: 2   Teaching Hours: 32(For lectures: 26, for practice: 6)   Semester: 5th

Main Contents:

The main contents for this course include: engineering settlement, documents processing, processing of pricing variation and cost claim calculation.

教学目标:

掌握工程结算编制方法、步骤,掌握变更、索赔的程序,能编制完整的工程结算书。

参考教材:

胡晓娟.《工程结算》,重庆大学出版社,2015。

(三十七)工程造价职业资格实务培训

课程学分:2;课程总学时:32(其中理论学时:26,实践学时:6);开设学期:第5学期

主要教学内容:

课程主要内容包括介绍工程造价职业资格考试制度、职业资格考试内容和要求、职业岗位及能力要求、工程结算的内容和相关规定、工程结算的程序和计算方法、工程结算管理等。

教学目标:

本课程的目标是要求学生在掌握各专业工程的计量与计价上,通过本课程的学习,进一步整合和提升工程造价职业能力,符合职业资格考试内容的要求。

参考教材:

胡晓娟.《工程结算》,重庆大学出版社,2015。

六、实践教学及基本要求

(一)建筑工程预算实训

学分:3;实践周数:2;开设学期:第3学期;实践地点:校内

实践内容:

建筑工程量计算;套用定额,计算建筑工程定额直接费并进行材料数量分析;建筑工程造价计算。

教学目标:

掌握建筑工程预算编制方法、步骤,掌握工程量计算程序,能编制完整的建筑工程预算书。

(二)钢筋工程量计算实训

学分:1;实践周数:1;开设学期:第4学期;实践地点:校内

Teaching Objectives:

By learning this course, students will master the method and steps of project account settlement compilation, master the procedures of change and claim, and will be able to compile a complete engineering settlement book.

Recommended Textbook:

Hu Xiaojuan. *Engineering Settlement*. Chongqing University Press, 2015.

5.37 Professional Qualification Training of Construction Cost

Credits: 2   Teaching Hours: 32(For lectures: 26, for practice: 6)   Semester: 5th

Main Contents:

The main contents of this course include: the system and examination requirement of professional qualification of construction cost, professional post demand and capacity analysis, engineering settlement and relevant regulations, engineering settlement procedures and calculation methods, engineering settlement management, etc.

Teaching Objectives:

By learning this course, students will master the measurement and valuation methods of various professional engineering, further integrate and improve the professional ability of construction cost through the study of this unit, and meet the requirements of professional qualification examination.

Recommended Textbook:

Hu Xiaojuan. *Engineering Settlement*. Chongqing University Press, 2015.

6. Practical Teaching and Basic Requirements

6.1 Practical Training for Construction Project Budget

Credits: 3   Practice Weeks: 2   Semester: 3rd   Practice Location: In campus

Practice Contents:

The main contents of this practice include: calculation of construction quantity, application of the quota, calculating the direct fee of construction engineering quota and analyzing the quantity of materials, as well as calculation of construction project cost.

Objectives:

By this practice, students will be able to master the methods and steps of construction project budget preparation, master the calculation procedure of project quantity, and be able to compile a complete construction project budget.

6.2 Practical Training for Calculation of Reinforcement Quantity

Credits: 1   Practice Weeks: 1   Semester: 4th   Practice Location: In campus

实践内容:

了解钢筋工程量计算的基本知识、钢筋工程量计算依据和钢筋工程量计算的基本方法,掌握建筑工程的平法识图和钢筋工程量计算的基本方法及技巧。

教学目标:

通过该课程的学习,学生应该学会使用平法图集,并具备一定的自学能力,能够根据相关标准图集读懂建筑工程施工图纸并能准确计算出建筑工程各类构件的钢筋工程量。

(三)工程量清单计价实训

学分:2;实践周数:2;开设学期:第4学期;实践地点:校内

实践内容:

工程量清单编制;工程量清单报价编制。

教学目标:

掌握工程量清单编制方法;掌握工程量清单报价编制方法;能独立完成工程量清单和工程量清单报价的编制。

(四)水电安装工程预算实训

学分:1;实践周数:1;开设学期:第5学期;实践地点:校内

实践内容:

民用建筑水给排水与电气照明工程量计算;套用定额,计算安装工程定额直接费并进行未计价材料数量分析与未计价材料费用计算;安装工程造价计算。

教学目标:

掌握水电安装工程预算编制方法、步骤,掌握工程量计算程序,能编制完整的水电安装工程预算书。

(五)工程造价软件应用实训

学分:1;实践周数:1;开设学期:5,安排在校内进行。

实践内容:

利用软件完成某工程的工程量计算和工程造价计算。

Practice Contents:

The main contents of this course include: mastering the rules in calculating reinforcement quantity, the usage of drawings by ichnographic representing method, skillfully calculating the reinforcement quantity of different components.

Objectives:

By this practice, students will be able to read drawings by ichnographic representing method and have the ability to read construction engineering drawings and accurately calculate the quantity of reinforcement in different components of building construction based on drawings by ichnographic representing method.

6.3 Practical Training for Bills of Quantities

Credits: 2   Practice Weeks: 2   Semester: 4th   Practice Location: In campus

Practice Contents:

The main contents of this practice include: compiling the bill of quantities and preparing quotation of bill of quantities.

Objectives:

By this practice, students will be able to master the method of compiling the bill of quantities, master the quotation method of bill of quantities and be able to prepare bill of quantities and bill of quantities quotation independently.

6.4 Practical Training for Project Budget of Water and Electricity Installation

Credits: 1   Practice Weeks: 1   Semester: 5th   Practice Location: In campus

Practice Contents:

The main contents of this practice include: civil building water supply and drainage and electrical lighting engineering quantity calculation, application of the quota, calculating the direct cost of installation engineering quota, analyzing the quantity of non-priced materials and calculating the cost of non-priced materials, as well as the calculation of installation project cost.

Objectives:

By this practice, students will be able to master the methods and steps of preparing the budget of hydropower installation project, master the calculation procedure of project quantity, and be able to prepare the complete budget of hydropower installation project.

6.5 Construction Cost Software Application Training

Credits: 1   Practice Weeks: 1   Semester: 5th   Practice Location: In campus

Practice Contents:

The main contents of this practice include: calculating the quantity and the cost of a project with specific software.

教学目标:

能熟练操作算量软件和计价软件。

(六)毕业设计

学分:8;实践周数:8;开设学期:第6学期;实践地点:校内

实践内容:

综合运用建筑施工项目管理知识,能够完成一套完整的单项工程的施工图预算编制,计算单项工程建筑工程造价;能编制单位工程施工组织设计;根据前面两项内容进行招标文件和投标文件的编制,完成投标报价的全过程。

教学目标:

综合运用工程造价的知识和施工组织设计的知识,计算单项工程建筑工程造价和编制单位工程施工组织设计,并能够进行投标报价,编制投标文件。

(七)顶岗实习

学分:8;实践周数:8;开设学期:第6学期;实践地点:校外

实践内容:

建筑施工项目管理相关岗位(工程招标、投标报价、内业资料管理、工程施工管理、质量管理、安全管理等)工作的实习。

教学目标:

熟悉并适应建筑施工项目管理相关岗位(工程招标、投标报价、内业资料管理、工程施工管理、质量管理、安全管理等)的工作。

七、毕业学分基本要求

毕业学分基本要求表

课内教学学分	理论教学	必修课学分	102
		限选课学分	15
		任选课学分	4
	实践教学	毕业设计及毕业实习学分	16
		其他实践学分	8
课外教学学分			5
合计			150

Teaching Objectives:

By this practice, students will be familiar with calculation software and pricing software.

6.6 Graduation Design Project

Credits: 8   Practice Weeks: 8   Semester: 6th   Practice Location: In campus

Practical Contents:

The main contents of this practice include: comprehensive usage of construction project management knowledge to complete a complete set of construction drawing budget preparation of a single project, calculating the cost of a single project construction project, compiling the organization design of unit project construction, preparing bidding documents and finishing the whole process of bidding quotation.

Teaching Objectives:

By this practice, students will be able to comprehensively apply the knowledge of project cost and construction organization design, calculate the cost of a single project construction project and compile the organization design of unit project construction, and be able to make bidding quotation and prepare bidding documents.

6.7 Post Practice

Credits: 8   Practice Weeks: 8   Semester: 6th   Practice Location: Off campus

Practice Contents:

The main contents of this practice include: internship in related positions of construction project management (project bidding, bidding quotation, internal data management, project construction management, quality management, safety management, etc.).

Objectives:

By this practice, students will be familiar with and adapt to the construction project management related positions (project bidding, bidding quotation, internal data management, project construction management, quality management, safety management, etc.).

7. Basic requirements for graduation credits

Basic requirements for graduation credit

In-class teaching credits	Theory teaching	Required course credits	102
		Elective course credits	15
		Optional course credits	4
	Practice teaching	Graduation design project and graduation practice credits	16
		Other practice credits	8
Extracurricular teaching credits			5
total			150

八、课程体系

(一) 课程逻辑关系图

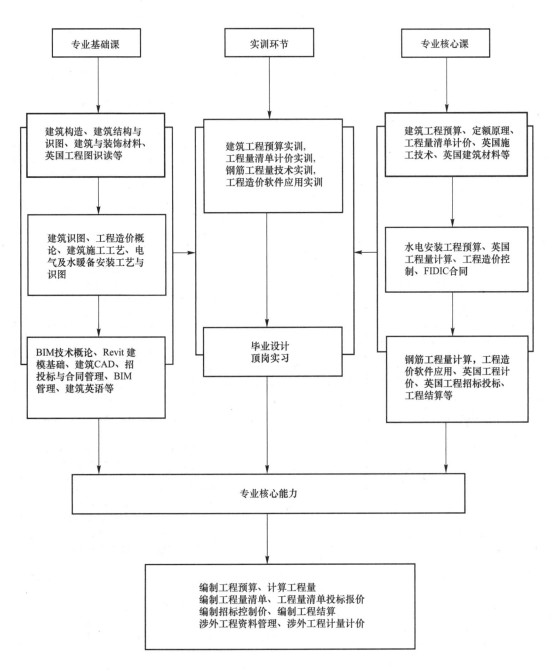

8. Course system

8.1 Logical diagram of the course

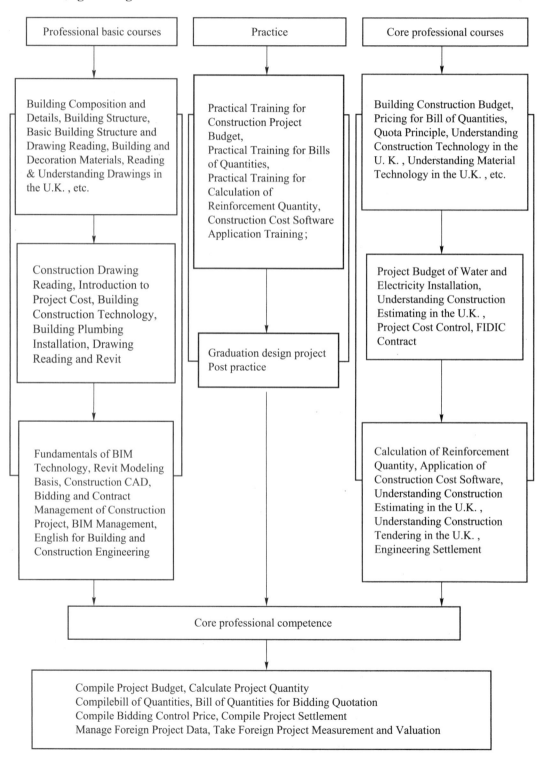

Professional basic courses	Practice	Core professional courses
Building Composition and Details, Building Structure, Basic Building Structure and Drawing Reading, Building and Decoration Materials, Reading & Understanding Drawings in the U.K. , etc.	Practical Training for Construction Project Budget, Practical Training for Bills of Quantities, Practical Training for Calculation of Reinforcement Quantity, Construction Cost Software Application Training;	Building Construction Budget, Pricing for Bill of Quantities, Quota Principle, Understanding Construction Technology in the U. K. , Understanding Material Technology in the U.K. , etc.
Construction Drawing Reading, Introduction to Project Cost, Building Construction Technology, Building Plumbing Installation, Drawing Reading and Revit	Graduation design project Post practice	Project Budget of Water and Electricity Installation, Understanding Construction Estimating in the U.K. , Project Cost Control, FIDIC Contract
Fundamentals of BIM Technology, Revit Modeling Basis, Construction CAD, Bidding and Contract Management of Construction Project, BIM Management, English for Building and Construction Engineering		Calculation of Reinforcement Quantity, Application of Construction Cost Software, Understanding Construction Estimating in the U.K. , Understanding Construction Tendering in the U.K. , Engineering Settlement

Core professional competence

Compile Project Budget, Calculate Project Quantity
Compilebill of Quantities, Bill of Quantities for Bidding Quotation
Compile Bidding Control Price, Compile Project Settlement
Manage Foreign Project Data, Take Foreign Project Measurement and Valuation

（二）课程设置

课程设置细化表（核心专业课程）

课程类别		课程名称	学分分配			开课学期					
课程属性	课程性质		总学分	理论学分	实践学分	第1学期	第2学期	第3学期	第4学期	第5学期	第6学期
专业基础课	必修课	画法几何	2.5	2.5		☆					
		建筑构造	2	2		☆					
		剑桥国际英语	2	2		☆					
		建筑与装饰材料	2	2		☆					
		建筑结构基础与识图	4	4			☆				
		建筑施工工艺	3.5	3.5			☆				
		建筑识图	1.5	1.5			☆				
		工程造价概论	2	2			☆				
		建筑工程项目管理	2	2				☆			
		建筑工程招投标与合同管理	2	2					☆		
		工程经济	2	2					☆		
		建筑电气设备安装工艺与识图及Revit建模基础	2		2				☆		
		建筑水暖设备安装工艺与识图及Revit建模基础	2		2				☆		
		英国工程图识读	4	4			☆				
		建筑CAD	2		2					☆	

8.2　Course Setting

Detailed table of course setting(core professional courses)

Course category		Course name	Credit allocation			Term					
Course property	Course character		Total credit	Theory teaching credit	Practice teaching credit	The first semester	The second semester	The third semester	The fourth semester	The fifth semester	The sixth semester
Professio-nal basic course	Required course	Descriptive Geometry	2.5	2.5		☆					
		Building Structure	2	2		☆					
		Cambridge International English	2	2		☆					
		Building and Decoration Materials	2	2		☆					
		Basic Building Structure and Drawing Reading	4	4			☆				
		Building Construction Technology	3.5	3.5			☆				
		Construction Drawing Reading	1.5	1.5			☆				
		Introduction to Project Cost	2	2			☆				
		Construction Project Management	2	2				☆			
		Bidding and Contract Management of Construction Project	2	2					☆		
		Engineering Economy	2	2					☆		
		Building Electrical Installation, Drawing Reading and Revit	2		2				☆		
		Building Plumbing Installation, Drawing Reading and Revit	2		2				☆		
		Reading & Understanding Drawings in the U. K.	4	4			☆				
		Construction CAD	2		2					☆	

课程类别		课程名称	学分分配			开课学期					
课程属性	课程性质		总学分	理论学分	实践学分	第1学期	第2学期	第3学期	第4学期	第5学期	第6学期
专业基础课	限选课	BIM 技术基础	1	1				☆			
		企业经营管理	1	1		☆					
		外教课堂学习法	1	1			☆				
		装配式建筑概论	1	1				☆			
		Revit 建模基础	2	2					☆		
		BIM 工程计价软件应用	1	1						☆	
专业课	必修课	建筑工程预算	4	4				☆			
		建筑工程预算实训	3		3			★			
		钢筋工程量计算	2.5	2.5					☆		
		钢筋工程量计算实训	1		1				★		
		工程量清单计价	3	3					☆		
		工程量清单计价实训	2		2				★		
		工程造价软件应用	3		3					☆	
		工程造价软件应用实训	1		1					★	

Professional Training Program for Specialized Sino-British Cooperative Program in Construction Cost

continued

Course category		Course name	Credit allocation			Term					
Course property	Course character		Total credit	Theory teaching credit	Practice teaching credit	The first semester	The second semester	The third semester	The fourth semester	The fifth semester	The sixth semester
Profess-ional basic course	Elective course	Fundamentals of BIM Technology	1	1				☆			
		Construction Enterprise Management	1	1		☆					
		Learning Strategy in Foreign Teachers' Classrooms	1	1			☆				
		Overview of Prefabricated Construction	1	1				☆			
		Revit Modeling basis	2	2					☆		
		Application of BIM Construction Cost Software	1	1						☆	
Profess-ional course	Required course	Building Construction Budget	4	4				☆			
		Practical Training for Construction Project Budget	3		3			★			
		Calculation of Reinforcement Quantity	2.5	2.5					☆		
		Practical Calculation Training of Steel Reinforcement	1		1				★		
		Pricing for Bills of Quantities	3	3					☆		
		Practical Training for Bills of Quantities	2		2				★		
		Application of Construction Cost Software	3		3					☆	
		Construction Cost Software Application Training	1		1					★	

课程类别		课程名称	学分分配			开课学期					
课程属性	课程性质		总学分	理论学分	实践学分	第1学期	第2学期	第3学期	第4学期	第5学期	第6学期
专业课	必修课	水电安装工程预算	3	3						☆	
		水电安装工程预算实训	1							★	
		定额原理	1.5	1.5				☆			
		工程造价控制	2	2					☆		
		英国建筑工程施工技术	4	4				☆			
		英国建筑材料	2	2				☆			
		英国建筑工程计量	4	4					☆		
		英国建筑工程计价	4	4						☆	
		英国工程招投标	2	2						☆	
		毕业设计(含答辩)	8		8						★
		顶岗实习	8		8						★
	限选课	FIDIC合同	2	2						☆	
		建筑英语	2	2				☆			
		工程结算	2	2						☆	

continued

| Course category | | Course name | Credit allocation | | | Term | | | | | |
Course property	Course character		Total credit	Theory teaching credit	Practice teaching credit	The first semester	The second semester	The third semester	The fourth semester	The fifth semester	The sixth semester
Profess-ional course	Required course	Project Budget of Water and Electricity Installation	3	3						☆	
		Practical Training for Project Budget of Water and Electricity Installation	1							★	
		Quota Principle	1. 5	1. 5				☆			
		Construction Cost Control	2	2					☆		
		Construction Technology in the U. K.	4	4				☆			
		Material Technology in the U. K.	2	2				☆			
		Measurement Processes in Construction in the U. K.	4	4					☆		
		Construction Estimating in the U. K.	4	4						☆	
		Construction Tendering in the U. K.	2	2						☆	
		Graduation Design Project	8		8						★
		Post Practice	8		8						★
	Elective course	Fidic Contract	2	2						☆	
		English for Building and Construction Engineering	2	2				☆			
		Engineering Settlement	2	2						☆	

续表

课程类别		课程名称	学分分配			开课学期					
课程属性	课程性质		总学分	理论学分	实践学分	第1学期	第2学期	第3学期	第4学期	第5学期	第6学期
专业课	限选课	工程造价职业资格实务培训	2	2						☆	

注:"☆"表示课内理论教学(含课程内实践);"★"表示课内集中周实践教学

九、创新意识和创业能力的培养

(1)采用启发式、讨论式、讨论参与式、探讨式等教学方法提高学生独立思考、综合分析的能力。

(2)通过"职业生涯与发展规划""创新创业与就业指导"等课程教学、开展创业教育讲座等培养学生的创新精神和创业意识。

(3)通过指导教师带领学生前往合作企业进行实习实训、实践锻炼,社会调查,学生创办企业等活动培养学生的创新创业能力。

(4)鼓励学生参与课外科技创新活动获取专利、实践作品、创办企业、论文等成果。

十、人才培养方案制(修)订报告

(一)人才培养方案制(修)订依据

依据川建院教务[2019年春秋]才培养方案修订通知而制定。

(二)人才培养方案制(修)订工作思路

(1)充分考虑中外合作办学项目的特点,做好引进资源的优化与整合。

(2)兼顾中外双方课程体系,满足该专业学生国内国外就业需要。

(3)紧扣国家"一带一路"倡议,有利于建筑业走出去。

(4)人才培养方案要有利于学生学习发展能力。

(三)人才培养方案制(修)订工作程序

(1)学习领会教务处人才培养方案通知精神。

continued

Course category		Course name	Credit allocation			Term					
Course property	Course character		Total credit	Theory teaching credit	Practice teaching credit	The first semester	The second semester	The third semester	The fourth semester	The fifth semester	The sixth semester
Profess-ional course	Elective course	Professional Qualification Training of Construction Cost	2	2						☆	

Note: "☆"Represents in-class theoretical teaching(including in-class practice) ; "★"Represents the weekly practice teaching in class.

9. Cultivation of innovative consciousness and entrepreneurial ability

(1) All through the teaching process, various teaching methods are adopted to improve students'ability of independent thinking and comprehensive analysis, such as heuristic mode, discussion, participation, cooperation and so on.

(2) Through the courses such as " career and development planning ", " innovation and entrepreneurship and employment guidance" and entrepreneurship education lectures, students'innovative spirit and entrepreneurial consciousness are developed.

(3) Students'innovation and entrepreneurship abilities are developed through activities such as internship and practical training, social investigation, and student-owned enterprises, etc.

(4) All through the teaching process, students are encouraged to participate in extracurricular scientific and technological innovation activities to obtain patents, practical works experience, establishing enterprises, papers and other achievements.

10. Professional training program report

10. 1　The basis for making professional training program

The professional training program is based on[spring and autumn of 2019] training plan revision notice from Teaching Affairs Office of Sichuan College of Architectural Technology.

10. 2　Working concepts of making professional training program

(1) The making of this professional training program takes full account of the characteristics of Sino-British cooperative program and it tries to optimize and integrate the resources successfully.

(2) The making of this professional training program takes Chinese and British course system into account and it meets the needs of students'employment in domestic and overseas market.

(3) The making of this professional training program closely links with the propose of our country's "Belt & Road Initiative", and students of this specialty promote the development of construction industry in overseas market.

(4) The making of the professional training programs should be conducive to the overall development of students'ability.

10. 3　The procedure of making the professional training program

(1) Learn the requirements from meetings about the making of professional training program.

（2）征求工程管理系、林肯学院及相关领导、教师、学生的意见。

（3）起草人才培养方案初稿。

（4）征求校内外专家的意见。

（5）国际学院教学工作指导委员会审议。

（6）学院教务处、学院教学工作指导委员会、学院院长审核。

（四）人才培养方案制（修）订人员

序号	姓名	工作单位	职称	职务
	戴明元	四川建筑职业技术学院	教授	
	侯兰	四川建筑职业技术学院	副教授	教研室主任
	杨兴培	四川华西海外投资建设有限公司	高级工程师	董事长
	托尼	英国林肯学院	讲师	
	马修	英国林肯学院	讲师	

（2）Seek advice from the department of engineering management, Lincoln college and relevant leaders, teachers and students.

（3）Make out a draft for further discussion and revision.

（4）Solicit opinions from experts on and off campus.

（5）Seek scrutinizing by the teaching guidance committee of International School of Technical Education.

（6）Let it examined and verified by the teaching affairs office, teaching guidance committee and the president of SCAT.

10.4　Professional training program editing(revision) staff

No.	Name	Work unit	Title	Position
	Dai Mingyuan	Sichuan College of Architectural Technology	Professor	
	Hou Lan	Sichuan College of Architectural Technology	Associate Professor	Department Director
	Yang Xingpei	Sichuan Huaxi Overseas Investment and Construction Co. ,Ltd	Senior Engineer	Chairman
	Tony	Lincoln college	Lecturer	
	Matthew	Lincoln college	Lecturer	

校外专家意见	单位:坦桑尼亚海南国际股份有限公司 学历/学位:大学本科 职称:高级经济师 职务:董事长 　　本人仔细审阅了贵校关于中英合作办学工程造价专业人才培养方案,从人才培养、课程设置、学分分配、毕业要求都做了细致、周到的安排,非常全面。培养目标定位准确,课程设置合理,符合建设单位工程造价人才需求。本人认为该方案可以实施。 专家签名: 　　年　月　日
教务处意见	 处长签名: 　　年　月　日
学校教学工作 委员会意见	 主任签名: 　　年　月　日
学校院长意见	

Opinions of expert outside the college	Company: Hainan international Ltd. Co. in Tanzania Education/degree: undergraduate Title: senior economist Position: chairman I have carefully reviewed the professional training program of specialized Sino-British cooperative program in construction cost. The professional training program, curriculum setting, credit allocation and graduation requirements are specific and practical. The training objective is accurate, the curriculum setting is reasonable, and the training program conforms to the demand for the specialty of construction cost. I think this plan can be implemented. Signature of expert: Date:
Opinions of teaching affairs	Signature of the director: Date:
Opinions of the school teaching guidance committee	Signature of the director: Date:
Opinion of the president	

四川建筑职业技术学院

课 程 标 准

课程名称：　　　　　画法几何

课程代码：　　　　　900024

课程学分：　　　　　2.5

基本学时：　　　　　44

适用专业：　工程造价（中英合作办学项目）

执 笔 人：　　　　　刘　觅

编制单位：　　　　制图教研室

审　　核：　　土木工程系（院）（签章）

批　　准：　　　　教务处（盖章）

Sichuan College of Architectural Technology

Curriculum Standard

Curriculum Name: _____ Descriptive Geometry _____

Curriculum Code: _____ 900024 _____

Credits: _____ 2. 5 _____

Teaching Hours: _____ 44 _____

Applicable Specialty: _____ Construction Cost _____

_____ (Sino-British Cooperative Program) _____

Complied by: _____ Liu Mi _____

On Behalf of: Teaching & Research Section of Construction Drafting

Reviewed by: _____ The Department of Civil Engineering _____

Approved by: _____ Teaching Affairs Office _____

《画法几何》课程标准

一、课程概述

《画法几何》是研究在平面上用图形表示形体和解决空间几何问题的理论与方法的学科,是绘制和阅读工程图样的投影理论基础。本课程在专业中属于专业基础课,也是学院院管课程之一,是学院的绝大部分专业都要学习的核心课程,是工程类相关专业的制图课程的基础先修课程,是从事工程设计、施工、管理等技术人员必备的基本素质技能。本课程适合高职 3 年制工程造价(中英合作办学项目)专业在第 1 学期开设,后续课程是《建筑制图》《建筑 CAD》等。

二、课程设计思路

本课程标准依据《四川建筑职业技术学院人才培养方案》制定。《画法几何》主要研究投影法的基本理论和作图方法以及利用这些方法去解决空间几何元素间的关系,强调空间想象能力和发散性思维。教学方法以课堂讲授为主,练习实践与上习题课结合,强调学生的逻辑思维能力,重点培养学生的空间想象能力和空间分析能力。

课程教学设计原则:

(1)思维切换性:解决学生学习过程中在空间和投影平面图形间不断切换的学习问题。

(2)步骤有效性:教学上强调思维和操作的步骤,让学生通过简单的步骤操作达到空间思维能力和表达能力的培养。

(3)循序渐进性:按照先易后难、先简后繁的形式,让学生循序渐进地实现空间点线面体向投影面投影的空间想象思维转换。

(4)重点深化:选择用常用形体三维模型、典型题目图解过程实现空间形体与平面投影图相互转换的思维强化。

三、课程目标

通过本课程,学生可以掌握用投影方法图示空间形体和图解空间几何问题的基本理论与方法,培养空间思维能力和作图识图的基本能力,培养认真细致的工作习惯。

Descriptive Geometry **Curriculum Standard**

1. Curriculum Description

Descriptive Geometry studies theories and methods of graphically representing shapes and hence the solution for spatial geometry propositions. It helps developing students'fundamental knowledge of projection in reading and preparing constructions drawings. The curriculum is one of specialized fundamental curricula, and one of those administrated directly by the college. *Descriptive Geometry* is also a core curriculum for most other specialties in the college and a prerequisite of drafting curricula for engineering and relevant specialties. Knowledge of *Descriptive Geometry* is the fundamental quality of technicians, designers, or managers in engineering. The curriculum is offered in the first term for the specialty of Construction Cost (Sino-British Cooperative Program), followed by courses like *Construction Drawing Reading* , *Construction CAD* and so on.

2. Curriculum Design Ideas

The curriculum standard is compiled in accordance with *the Professional Training Program of SCAT.*

Descriptive Geometry emphasizes spatial imagination and divergent thinking with a focus on fundamental principles and drawing preparation of projection theories and on the application of them in understanding the relationship between spatial geometric elements. The teaching mode of this curriculum is a combination of classroom instruction, hand-on practice and exercises classes, with classroom instruction being the major part. The teaching process will stress students'logical thinking, and try to foster their spatial imagination and spatial analysis.

Designing principles of the curriculum standard:

(1) A switch of mindset: Students'problem of continuously switching between shapes and their projection will be dealt with in the learning process.

(2) Efficiency of steps: By a focus on steps of thinking and operation, students'spatial imagination and verbal presentation are enhanced in the simple step-by-step learning process.

(3) Progressiveness: In a teaching process from easy and simple ideas to difficulties and complexity, students will be progressively turned from seeing only points, lines and planes to seeing projected shapes.

(4) Enhancement: The mental switch between spatial shapes and their planar projections is enhanced by demonstrating three-dimensional models of usual shapes and illustrating solution to classic problems.

3. Curriculum Objectives

This course helps students master fundamental theories and methods of representing spatial shapes with planar projection and solving spatial geometry problems with graphs. It also tries to cultivate students'abilities in spatial thinking and drawing reading, and to foster working habits of being careful and accurate.

(一)知识目标

(1)了解工程制图有关国家标准的一般规定及图样画法。

(2)掌握投影法(主要是正投影法)的基本理论及其应用。

(3)掌握空间形体图示、度量和定位、图解空间形体问题的方法。

(4)掌握绘制和阅读空间形体图样的方法。

(二)素质目标

(1)培养空间几何问题的图解能力,对空间形状与相关位置的逻辑和形象思维能力。

(2)培养空间想象能力和空间分析能力。

(3)培养认真负责的工作态度和严谨细致的工作作风。

(4)培养自学能力、分析问题和解决问题能力、创造力和审美能力。

(三)能力目标

(1)学生能独立识读常见空间形体的三面正投影图,标示出形体的形状尺寸和空间相对位置关系。

(2)学生能通过图解法完成空间形体投影图的分析与补充。

(3)学生在没有教师直接指导下能独立按照《房屋建筑制图统一标准》要求完成常见工程形体三面正投影图绘制。

四、课程内容与学时分配

教学内容与学时安排表

序号	教学情境/任务/项目/单元	教学目标	知识点	学时	实践项目	学时
1	绪论	建立中心投影和平行投影(正投影和斜投影)的明确概念	1. 投影及其特性	4		
			2. 工程上常用的几种图示法			
			3. 正投影法基本原理(★)			

3. 1　Knowledge Objectives

After learning this course, students will be able to

(1)understand common rules and preparing methods in national standards related to engineering drawings.

(2)master the basic theories of projection(mainly orthography)and their application.

(3)master representation, measurement and orientation of spatial shapes, as well as the illustrated solution of spatial geometric problems.

(4)master methods of preparing and reading drawings of spatial shapes.

3. 2　Quality Objectives

After learning this course, students will have the quality of

(1)applying descriptive geometry solution in spatial geometric problems, and logic thinking and visualization of spatial shapes and their orientations.

(2)developing spatial imagination and spatial analysis capabilities.

(3)developing serious and responsible work attitudes and rigorous and meticulous work style.

(4)enhancing self-learning, analysis capabilities, problem-solving, creativity and the sense of beauty.

3. 3　Ability Objectives

After learning this course, students will have the abilities of

(1) reading independently the three-sided orthographic projection of common spatial shapes and assigning dimensions to those shapes and indicating their relative orientations.

(2) applying descriptive geometry solution to complete analysis and supplement of the spatial projection map.

(3) independently drawing the three-sided orthographic projection of common engineering shapes under the requirement of *the Standards of Construction Drawings for Houses*.

4. Curriculum Contents and Teaching Hours

Table of Teaching Contents and Teaching Hours

Item	Topic	Teaching Objectives	Lesson Contents and Teaching Hours			
			Lesson Contents	Hrs	Student Activities	Hrs
1	Introduction	Establish accurate concepts of central projection and parallel projection (Orthography and Oblique projection)	1. Projection and its characteristics	4		
			2. Graphic representations common to Construction Projects			
			3. Basic principles of orthography(★)			

序号	教学情境/任务/项目/单元	教学目标	教学内容及学时分配			
			知识点	学时	实践项目	学时
2	制图基本知识	了解工程制图有关国家标准的一般规定及图样画法	4. 绘图工具与仪器使用方法	2	1. 平面图形画法	2
			5. 制图的基本规定			
			6. 尺寸标注			
			7. 几何作图			
			8. 尺规作图的方法与步骤			
3	点的投影	掌握点的投影特性和作图方法	9. 点的两面投影	4		
			10. 点的三面投影			
			11. 两点的相对位置			
4	直线的投影	掌握直线的投影特性和作图方法；掌握直线倾角和线段长度的求法；掌握直线上点的投影特性及在平面上作点、作直线的方法；掌握两平行、相交、交叉直线以及一边平行于投影面的直角的投影特性	12. 直线的投影	6		
			13. 直线对投影面的相对位置			
			14. 直线上的点（★）			
			15. 线段的实长与倾角（★）			
			16. 两直线的相对位置			
5	平面的投影	掌握点的投影特性和作图方法；掌握直线与平面及两平面之间的平行、相交及垂直的投影特性和作图方法	17. 平面的投影	8		
			18. 平面对投影面的相对位置			
			19. 平面上的直线和点（★）			
			20. 直线与平面及两平面的相对位置			

continued

Item	Topic	Teaching Objectives	Lesson Contents and Teaching Hours			
			Lesson Contents	Hrs	Student Activities	Hrs
2	Fundamental of drafting	Understand common rules and preparing methods in national standards related to engineering drawings	4. Operation of graphing devices and tools	2	1. Plot two dimensional figures	2
			5. Common rules in drawing preparation			
			6. Assignment of dimensions			
			7. Geometrical plotting			
			8. Method and steps of drawing with rulers			
3	Projection of points	Learn projecting features and plotting methods of points	9. Two-sided projection of points	4		
			10. Three-sided projection of points			
			11. Relative orientation of two points			
4	Projection of straight lines	Learn projecting features and plotting methods of straight lines; Master ways to figure out straight lines'inclination and their length; Learn projecting features of points on straight lines, as well as ways to plot points or straight lines on a plane; Learn projecting features of straight lines parallel to, intersecting or crossing each other and of vertical angels of whom one side is parallel to the projecting plane	12. Projection of straight lines	6		
			13. Relative orientations of a st-raight line to the projecting plane			
			14. Points on straight lines(★)			
			15. The real dimension and in-clination of line segments(★)			
			16. Relative orientation of two straight lines			
5	Projection of planes	Learn projecting features and plotting methods of points; Learn projecting features and plotting methods of a plane parallel to, intersecting or perpendicular to a straight line or another plane	17. Projection of planes	8		
			18. Relative orientation of a pla-ne to the projecting plane			
			19. Straight lines and points on a plane(★)			
			20. Relative orientation of a straight line to a plane, and that of two planes			

续表

序号	教学情境/任务/项目/单元	教学目标	教学内容及学时分配			
			知识点	学时	实践项目	学时
6	基本形体的投影	掌握平面立体和曲面立体的投影特性和作图方法,以及在表面上作点、作线的方法	21. 平面基本形体的投影 22. 曲面基本形体的投影 23. 形体表面上的点与线(★)	6		
7	组合形体的投影	能分析平面立体与立体的截交线、相贯线的性质,掌握平面与平面立体和曲面立体的截交线的作图方法	24. 组合形体的投影(★) 25. 截交线 26. 相贯线 27. 组合形体视图的读法	6		
8	工程形体的表达方法	掌握形体剖断面图的画法;掌握形体轴测投影图的画法	28. 工程形体尺寸标注 29. 剖面图(★) 30. 断面图(★) 31. 简化画法 32. 轴测投影的基本知识 33. 正轴测图(★) 34. 斜轴测图	4	2. 工程形体的画法	2
	合计			40		4

备注:重要知识点标注★

五、教学实施建议

(一)教学参考资料

李翔、刘觅、凌莉群.《画法几何》,高等教育出版社,2014。

李翔、刘觅、凌莉群.《画法几何习题集》,高等教育出版社,2014。

continued

Item	Topic	Teaching Objectives	Lesson Contents and Teaching Hours			
			Lesson Contents	Hrs	Student Activities	Hrs
6	Projection of basic geometric objects	Learn projecting features and plotting methods of objects with planar or curved surfaces, as well as methods of plotting points or lines on those surfaces	21. Projection of basic objects with planar surfaces	6		
			22. Projection of basic objects with curved surfaces			
			23. Points and lines on surfaces of objects(★)			
7	Projection of combined objects	Be able to analyze the section line and the intersecting line of a object with planar surfaces when combined with another; Master the plotting methods of the section line of a object with planar surfaces when combined with one with curved surfaces	24. Projection of combined objects(★)	6		
			25. Section lines			
			26. Intersecting lines			
			27. Ways to read the views of combined objects			
8	Representation of engineering objects	Master plotting methods of sectional and cross sectional drawings of objects. Master plotting methods of axonometric projection of objects	28. Assign dimension to engineering objects	4	2. Plot engineering objects	2
			29. Sectional drawings(★)			
			30. Cross sectional drawings(★)			
			31. Simplified drawings			
			32. Fundamentals of axonometric projection			
			33. Axonometric Orthography(★)			
			34. Axonometric oblique projection			
Total				40		4
Notes: Mark important points with ★						

5. Teaching Suggestions

5.1 Teaching Resources

Li Xiang, Liu Mi, Ling liqun. *Descriptive Geometry*. Higher Education Press,2014.

Li Xiang, Liu Mi, Ling liqun. *Exercise Book for Descriptive Geometry*. Higher Education Press, 2014.

邱小林、周亦人、刘觅.《画法几何与土木工程制图》,华中科技大学出版社,2015。

刘觅、周亦人.《画法几何与土木工程制图习题集》,华中科技大学出版社,2015。

(二)教师素质要求

本课程授课教师应具有本课程或相关课程的教师资格,具有工程类专业背景,有良好的课堂管理组织和识图、绘图、解图能力。

(三)教学场地、设施要求

多媒体教室或绘图教室。

六、考核评价

本课程宜以闭卷考试方式进行课程考核,成绩构成比例按《四川建筑职业技术学院学生学业考核办法》执行,各教学项目考核成绩比例见下表。

序号	考核内容	成绩构成比例/%
1	绪论	10
2	制图基本知识	5
3	点的投影	10
4	直线的投影	15
5	平面的投影	15
6	基本形体的投影	10
7	组合形体的投影	15
8	工程形体的表达方法	20
	合计	100

Qiu Xiaolin, Zhou Yiren, Liu Mi. *Descriptive Geometry and Civil Engineering Drafting.* Huazhong University of Science and Technology Press, 2015.

Liu Mi, Zhou Yiren. *Exercise Book for Descriptive Geometry and Civil Engineering Drafting.* Huazhong University of Science and Technology Press, 2015.

5.2 Teachers'Qualification

Teachers of this course should be pedagogically qualified and well educated in this course or relevant curricula. They should be specialized in engineering, with excellent skills in classroom management and organization and the ability to read, prepare, and analyze engineering drawings.

5.3 Teaching Facilities

Multimedia classrooms or Drafting rooms.

6. Evaluation/Examination

The course should be assessed in closed-book examinations. Assignment of scores is based on the *Regulation of Academic Evaluation for Students in SCAT.* The percentage of each topic in the total score is as follows:

Item	Topics to be assessed	Percentage
1	Introduction	10
2	Fundamental of drawing preparation	5
3	Projection of points	10
4	Projection of straight lines	15
5	Projection of planes	15
6	Projection of basic geometric objects	10
7	Projection of combined objects	15
8	Representation of engineering objects	20
Total		100

四 川 建 筑 职 业 技 术 学 院

课 程 标 准

课程名称：　　　　建筑构造

课程代码：　　　　010556

课程学分：　　　　　2

基本学时：　　　　　32

适用专业：　工程造价(中英合作办学项目)

执 笔 人：　　　　刘　　觅

编制单位：　　　制图教研室

审　　核：　　土木工程系(院)(签章)

批　　准：　　　教务处(盖章)

Sichuan College of Architectural Technology

Curriculum Standard

Curriculum Name: Building Structure

Curriculum Code: 010556

Credits: 2

Teaching Hours: 32

Applicable Specialty: Construction Cost

(Sino-British Cooperative Program)

Complied by: Liu Mi

On Behalf of: Teaching &Research Section of Construction Drafting

Reviewed by: The Department of Civil Engineering

Approved by: Teaching Affairs Office

《建筑构造》课程标准

一、课程概述

《建筑构造》是研究建筑空间组合与建筑构造理论和方法的课程,主要包括民用建筑构造(墙与基础、楼层和地层、楼梯、屋顶、门和窗)、单层厂房构造(外墙和门窗、屋顶、天窗),其主要目标是使学生了解民用与工业建筑设计的基本理论和方法,基本掌握一般民用和工业建筑构造的理论和方法。本课程是工程造价(中英合作办学项目)专业的专业基础课,适宜在第2学期开设。本课程的先修课程为《画法几何》《建筑识读》《建筑与装饰材料》,后续课程为《建筑结构基础》《建筑施工工艺》等。

二、课程设计思路

本课程标准依据《工程造价专业(中英合作办学项目)专业人才培养方案》制定。本课程作为一门内容广泛的综合性学科,涉及建筑功能、建筑艺术、环境规划、工程技术、工程经济等诸多方面的问题。同时,这些问题之间又因共存于一个系统中而相互关联、相互制约、相互影响。随着人类物质生活水平的不断提高以及社会整体技术力量的提高,特别是工程技术水平的不断发展,该系统中的各个层面都会不断发生变化,它们之间的相关关系也会随之发生变化。因此,在学习这门课程的过程中,应当具有系统的眼光和发展的眼光。教学中以教师讲授和案例展示为主,同时需结合建筑模型馆进行现场观摩学习。

三、课程目标

《建筑构造》课程目标是了解和研究建筑设计的思路与过程,了解民用与工业建筑设计的基本理论和方法,掌握一般民用与工业建筑构造的理论和方法。

(一)知识目标

(1)掌握房屋构造的基本理论,了解房屋各组成部分的构成和细部构造要求,掌握不同构造的理论基础。

(2)能够根据房屋的使用要求和材料供应情况及施工技术条件选择合理的构造方案进行构造设计、绘制施工图和熟练地识读施工图。

Building Structure **Curriculum Standard**

1. Curriculum Description

Building Structure is a course studying theories and methods of spatial combination and building structures. The major contents are residential building structures(walls and foundations, floors and strata, stairs, roofs, doors and windows), and single-story industrial building structures(outer walls, doors and windows, roofs, and skylights). The main purpose of the course is to help students learn fundamental theories and methods of residential and industrial building design, and of usual residential and industrial building structures. The course is delivered in the second term as a required fundamental one for the specialty of Construction Cost(Sino-British Cooperative Program). The delivered courses are *Descriptive Geometry*, *Construction Drawing Reading*, *Building and Decoration Materials*, and the following courses are *Building Construction Technology*, *Basic Building Structure* and so on.

2. Curriculum Design Ideas

The curriculum standard is based on the professional training program for the specialty of construction cost(Sino-British Cooperative Program). It covers comprehensive contents, such as architectural function, architectural art, environmental planning, engineering, technology, and economy. All those contents are interrelated and coexist in one system within which, each of those foresaid contents is changing accordingly as overall technology, especially engineering technology develops and material living standards of human beings are improved. Changes happening to them bring about changes in their relationships. Therefore, this course should be studied and viewed with insights into those changes and the foresaid system. Teaching process of the course features classroom instruction and case study, as well as visits to and observation in construction model rooms.

3. Curriculum Objectives

The course is to introduce and study the idea and the process of architectural design, and to learn fundamental theories and methods of residential and industrial building design. It will help students master theories and methods applied in usual residential and industrial building construction.

3. 1　Knowledge Objectives

After learning this course, students will be able to

(1) master fundamental theories of building structure; learn different compositions of buildings and their detailed structural requirements; master fundamental theories of different structures.

(2) choose proper architectonic plan and design accordingly under the functional requirement of buildings and conditions of material supply and construction technology; prepare and read constructional drawings skillfully.

（3）运用建筑设计的理论和方法进行一般建筑的初步设计,从中了解建筑设计的步骤和方法,并完成初步设计所要求的建筑平面、立面、剖面设计图。

(二)素质目标

（1）培养辩证思维能力。

（2）培养认真负责的工作态度和严谨细致的工作作风。

（3）培养自学能力、分析问题和解决问题能力、创造力和审美能力。

(三)能力目标

（1）学生能在没有教师的直接辅导下独立完成一般建筑的初步设计。

（2）学生能在没有教师的直接辅导下独立完成建筑基本组成部分的详图设计。

四、课程内容与学时分配

教学内容与学时安排表

序号	教学情境/任务/项目/单元	教学目标	教学内容及学时分配			
			知识点	学时	实践项目	学时
1	民用建筑构造概述	掌握建筑的构成要素、构造组成、建筑的分类和等级、影响构造的因素和设计原则; 了解模数数列的应用和定位轴线的标定,有关常用专业名词	1. 建筑的构成要素和构造组成(★)	4	1. 建筑的定位轴线设计	2
			2. 建筑的分类和等级划分			
			3. 影响构造的因素和设计原则(★)			
			4. 建筑模数协调标准			
2	基础与地下室的构造	掌握地基与基础的关系和影响基础埋深的因素; 掌握基础类型和地下室防水防潮	5. 地基与基础的基本概念	4		
			6. 基础的埋深及其影响因素(★)			
			7. 基础的类型与构造(★)			
			8. 地下室的防潮和防水构造(★)			

（3）learn architectural designing methods and steps in conducting the preliminary design of usual buildings in the guidance of architectural designing theories and methods; prepare the plan, elevation and section drawings as is required by the preliminary design.

3. 2   Quality Objectives

After learning this course, students will have the quality of

（1）critical thinking.

（2）seriousness and responsibility in work and rigorous work styles.

（3）competence in self-study, problem analyzing and solving, innovation and aesthetics.

3. 3   Ability Objectives

After learning this course, students will have the ability of

（1）completing preliminary design of usual buildings with no direct tutoring.

（2）completing detail design of basic building components with no direct tutoring.

4. Curriculum Contents and Teaching hours

Table of Teaching Contents and Teaching Hours

Item	Topic	Teaching Objectives	Lesson Contents and Teaching Hours			
			Lesson Contents	Hrs	Student Activities	Hrs
1	Introduction to civil architectural structures	Learn the constituents of architecture, the structure of it, the types and grades of buildings, factors impacting structure and the designing principles; Learn to apply modular sequence, and to demarcate coordinate grid lines of building; and learn relevant terms	1. Constituents and structures of architectures（ ★ ）	4	1. Design coordinate grid lines of building	2
			2. Types and grades of buildings			
			3. Factors impacting structure and design principles（ ★ ）			
			4. Standards of construction modular coordination			
2	Structures of foundation and basement	Learn the relation between subgrade and foundation, and the factors impacting buried depth of foundations; Learn the types of foundation, and damp resistance and waterproofing of basement	5. Concepts of subgrade and foundation	4		
			6. Buried depth and factors impacting it（ ★ ）			
			7. Types and structures of foundations（ ★ ）			
			8. Damp-resistant and waterproofing structures of basements（ ★ ）			

序号	教学情境/任务/项目/单元	教学目标	教学内容及学时分配			
			知识点	学时	实践项目	学时
3	墙体构造	掌握砖墙和砌块墙的细部构造； 了解隔墙构造、墙体的装修构造	9. 墙体类型及设计要求 10. 墙体的一般构造和细部构造（★） 11. 隔墙构造 12. 墙体装饰装修构造	4	2. 墙身节点详图设计	1
4	楼地层构造	掌握钢筋混凝土楼板层的构造原理和结构布置特点； 熟悉各种常用地面及顶棚的构造做法； 了解阳台和雨篷的构造原理和做法	13. 楼地层的基本知识 14. 钢筋混凝土楼板构造（★） 15. 地坪层的构造 16. 阳台与雨篷构造	2		
5	楼梯构造	掌握楼梯的组成、尺度、现浇钢筋混凝土楼梯的构造和楼梯的细部构造； 了解台阶和坡道的构造	17. 楼梯的组成及形式 18. 楼梯的尺度（★） 19. 现浇楼梯的构造（★） 20. 踏步和栏杆扶手构造 21. 室外台阶和坡道 22. 电梯和自动扶梯	4	3. 楼梯详图设计	1
6	屋顶构造	掌握平屋顶和坡屋顶的构造； 了解屋顶的细部构造、保温、隔热	23. 屋盖的形式及设计要求（★） 24. 屋盖的排水设计 25. 卷材防水屋面构造（★） 26. 刚性防水屋面构造 27. 涂膜防水屋面构造 28. 瓦屋面的简单构造 29. 屋盖的保温和隔热构造 30. 吊顶棚的构造	3	4. 屋顶平面图的设计	1

continued

Item	Topic	Teaching Objectives	Lesson Contents and Teaching Hours			
			Lesson Contents	Hrs	Student Activities	Hrs
3	The structures of wall	Learn detail structures of brick wall and block wall; Learn the structure of partition walls and decorative structure of wall	9. Types and designing requirements of walls	4	2. Node detail design of walls	1
			10. General structure and detail structure of walls(★)			
			11. The structure of partition walls			
			12. Decorative structures of wall			
4	The structure of floor	Master structural principles and characteristic of reinforced concrete floor; Learn to build structures of usual floors and ceiling; Learn structural principles and construction of balconies and canopies	13. Fundamentals of floor	2		
			14. The structure of reinforced concrete floor(★)			
			15. The structure of ground floor			
			16. The structure of balcony and canopy			
5	The structure of stair	Learn the components and the dimension of stairs, the structure of cast-in-situ reinforced concrete stair and detail structures of stair; Learn the structures of step and ramp	17. Components and types of stair	4	3. Detail design of stairs	1
			18. The dimension of stair(★)			
			19. The structure of cast-in-situ stair(★)			
			20. The structures of tread and handrail			
			21. Outdoor steps and ramps			
			22. Elevators and escalators			
6	The structure of roof	Learn the structures of flat roof and pitched roof; Learn detail structure and thermal insulation of roof	23. Types and design require－ments of roofs(★)	3	4. The design of roof plan	1
			24. Drainage design of roofs			
			25. The structure of membrane roofing(★)			
			26. The structure of stiff waterproofing roof			
			27. The structure of coated waterproofing roof			
			28. Simple structure of tile roof			
			29. Thermal insulation structure of roof			
			30. The structure of secondary roof			

序号	教学情境/任务/项目/单元	教学目标	教学内容及学时分配			
			知识点	学时	实践项目	学时
7	门窗构造	掌握窗和门与墙体的连接构造；理解窗与门的构造理论及正确选用与识读标准图的方法，了解遮阳构件的构造	31. 门窗的形式和尺寸(★)	1		
			32. 常见门窗构造			
8	变形缝的构造	掌握建筑变形缝的类型、作用与构造	33. 变形缝的作用、类型及构造(★)	1		
9	工业建筑设计概述	掌握工业建筑的构造要求，单层厂房的组成与类型	34. 工业建筑特点、类型及设计要求	3	5. 单层厂房的构造设计	1
			35. 单层厂房的组成与类型(★)			
			36. 单层厂房的设计(★)			
			37. 单层厂房的构造			
	合计			26		6

备注：重要知识点标注★

五、教学实施建议

(一)教学参考资料

赵西平.《房屋建筑学》(第二版),中国建筑工业出版社,2017。

赵研.《房屋建筑学》(第二版),高等教育出版社,2013。

吴启凤.《房屋构造》(第2版),西南交通大学出版社,2009。

卓维松.《房屋建筑构造》,中国海洋大学出版社,2011。

(二)教师素质要求

本课程的授课教师应具有本课程或相关课程的教师资格,具有工程类专业背景,有良好的课堂管理组织和建筑工程图识图、绘图能力。

(三)教学场地、设施要求

多媒体教室、建筑模型馆。

continued

Item	Topic	Teaching Objectives	Lesson Contents and Teaching Hours			
			Lesson Contents	Hrs	Student Activities	Hrs
7	The structure of door and window	Learn the structure connecting door and window to walls; Understand structural theories and choices of door and window and the ways to read standard drawings; Learn the structure of sunshade members	31. Types and dimensions of doors and windows(★)	1	—	—
			32. Usual structures of doors and windows			
8	The structure of deformation joint	Learn the types, function and structure of constructional deformation joints	33. The function, types and structure of deformation joints(★)	1		
9	Introduction to the design of industrial buildings	Master the structural requirements of industrial buildings and the components and types of single-story factory	34. Features, types and designing requirements of industrial buildings	3	5. Structural design of single-story factory	1
			35. Components and types of single-story factory(★)			
			36. The design of single-story factory(★)			
			37. The structure of single-story factory			
Total				26		6
Notes: Mark the important points with ★						

5. Teaching Suggestions

5.1 Teaching Resources

Zhao Xiping. *Building Structure*. 2nd Edition. China Architecture & Building Press, 2017.

Zhao Yan. *Building Structure*. 2nd Edition. Higher Education Press, 2013.

Wu Qifeng. *Building Compositions*. 2nd Edition. Southwest Jiaotong University Press, 2009.

Zuo Weisong. *Architectural Composition of Buildings*. China Ocean University Press, 2011.

5.2 Teachers'Qualification

Teachers of the course should be pedagogically ready and well educated in the specialty or relevant specialties. They should be specialized in engineering, with excellent skills in classroom management and organization and abilities in preparing and reading building construction drawings.

5.3 Teaching Places & Facility Requirement

Multimedia classrooms and construction model rooms.

六、考核评价

本课程宜以闭卷考试方式进行课程考核,成绩构成比例按《四川建筑职业技术学院学生学业考核办法》执行,各教学项目考核成绩比例见下表:

序号	考核内容	成绩构成比例/%
1	民用建筑构造概述	10
2	基础与地下室的构造	15
3	墙体构造	15
4	楼地层构造	10
5	楼梯构造	15
6	屋顶构造	10
7	门窗构造	5
8	变形缝的构造	5
9	工业建筑设计概述	15
	合计	100

6. Evaluation/Examination

This course should be assessed in closed-book examinations whose scoring follows *Regulation of Academic Evaluation for Students in SCAT*. Score distribution among topics of this course is as follow:

Item	Topics to be assessed	Percentage
1	Introduction to civil architectural structures.	10
2	Structures of foundations and basements	15
3	Structures of wall	15
4	Floor structure	10
5	The structure of stair	15
6	The structure of roof	10
7	The structure of door and window	5
8	The structure of deformation joint	5
9	Introduction to the design of industrial building	15
	Total	100

四川建筑职业技术学院

课 程 标 准

课程名称： 剑桥国际英语(一)

课程代码： 210016

课程学分： 2

基本学时： 32

适用专业： 中外合作办学专业

执 笔 人： 王姣姣

编制单位： 中澳合作办学教研室

审　　核： 国际技术教育学院(签章)

批　　准： 教务处(盖章)

Sichuan College of Architectural Technology

Curriculum Standard

Curriculum Name: Cambridge International English I

Curriculum Code: 210016

Credits: 2

Teaching Hours: 32

Applicable Specialty: Construction Cost

(Sino-British Cooperative Program)

Compiled by: Wang Jiaojiao

On Behalf of: Teaching & Research Section of Sino-Australia Program

Reviewed by: The Department of ISTE

Approved by: Teaching Affairs Office

《剑桥国际英语(一)》课程标准

一、课程概述

本课程是主要针对中外合作办学项目学生开设,旨在提升项目学生英语应用能力的同时帮助学生建立未来职业能力素养。学生通过学习本门课程,可以在跨文化背景下通过对职业能力进行板块训练,同时掌握板块能力语言词汇,并在设定情景中进行练习运用以达到语言、职业能力双向提升的教学目的。该门课程致力于将中外合作办学项目学生培养成为跨文化背景下语言能力佳、交流能力强、具备良好职业素养并拥有国际竞争力的复合型人才。

二、课程设计思路

通过对跨文化背景下需具备的基本职业能力进行板块划分,各个板块按照头脑风暴、理论讲解、课堂讨论、案例分析、模块训练等步骤进行教学任务设计。教学过程中引导学生对实际情景建立英文思维习惯,以英语为载体配合实际情景进行训练,在强化学生运用英语思维解决实际问题的同时,从而达到提升项目学生职业能力与素养的目标。

三、课程目标

(一)知识目标

(1)了解跨文化背景下文化冲突的概念;
(2)掌握跨文化背景下的交流技能;
(3)了解思辨思维的重要性;
(4)掌握团队协作的方法与技巧;
(5)掌握面试基本技巧。

(二)素质目标

(1)具备跨文化沟通的能力;
(2)具备个人、团队协作完成任务的能力;
(3)具备良好跨文化沟通能力;
(4)具备思辨思维能力。

Cambridge International English I
Curriculum Standard

1. Curriculum Description

This course is compulsory for the specialty of construction cost students (Sino-British Cooperative Program). It aims to improve the students'English application ability and help them build up their future professional competence. Through this course, students can train their professional competence in various modules and in the cross-cultural context, master the language vocabulary of different settings, and practice English in set situations to achieve both the language and the professional competence improvement. This course aims to cultivate students of Sino-British Cooperative Program into interdisciplinary talents with good language ability, strong communication ability, excellent professional quality and international competitiveness in cross-cultural context.

2. Curriculum Design Ideas

This course is designed in accordance with the cognitive discipline of students'modular learning. Each module encompasses steps activities of brainstorming, theoretical explanation, classroom discussion, case analysis and modular training. In the process of teaching, students should be encouraged and guided to build up the habit of thinking in English, using English as the carrier to complete different simulated tasks and to solve practical problems, and achieve the goal of improving their professional ability and competence as well.

3. Curriculum Objectives

3. 1   Knowledge Objectives

After learning this course, students will be able to

(1) understand the concept of cultural conflict in a cross-cultural context.

(2) master communication skills in a cross-cultural context.

(3) understand the importance of critical thinking.

(4) master the methods and skills of teamwork.

(5) master the basic skills for interview.

3. 2   Quality Objectives

After learning this course, students will have the quality of

(1) communicating across cultures.

(2) working with others to complete tasks.

(3) good cross-cultural communication skills.

(4) critical thinking.

(三)能力目标

(1)熟悉各模块应使用的技能。

(2)能够使用英文思维进行情景分析、解决问题。

(3)能够运用英文进行团队协作完成任务。

(4)能够应对英文职场面试。

四、课程内容与学时分配

教学内容与学时安排表

序号	教学情境/任务/项目/单元	教学目标	教学内容及学时分配			
			知识点	学时	实践项目	学时
1	课程引入	正确客观认识自己及课程大体内容框架; 了解课堂要求与考核标准; 了解 Gap year 的概念	1. 使用正确英文介绍并定义自己	4	小组讨论:尝试对自己未来五年发展进行设想与规划	2
			2. Gap Year 的概念			
			3. 对个人未来职业规划进行设想			
2	思辨能力	了解思辨思维的概念及重要性; 学会分辨事实、观点和偏好; 了解思辨能力的基本方法	4. 掌握思辨能力的定义	4	英文辩论会	2
			5. 运用思辨能力对情景进行分析			
			6. 掌握事实、观点、偏好的差别并能正确区分			
			7. 运用思辨思维分析文章、进行写作			
3	交流能力	了解交流技巧的重要性; 在谈判中运用交流技巧; 认识不同的交流方式	8. 了解交流对象、目的	4	课堂小组演讲展示	2
			9. 选取合适的交流方式			
			10. 掌握交流的有效性			

3. 3 Ability Objectives

After learning this course, students will have the ability of

(1) using skills acquired in each module;

(2) using English thinking for situational analysis and problem solving;

(3) communicating in English and cooperate with other team members to accomplish tasks;

(4) knowing how to succeed in job interviews in English.

4. Curriculum Contents and Teaching Hours

Table of Teaching Contents and Teaching Hours

Item	Topic	Teaching Objectives	Teaching Contents and Teaching Hours			
			Lesson Contents	Hrs	Student Activities	Hrs
1	Introduction	Correctly and objectively understand oneself and the content framework of the course; Understanding classroom requirements and assessment criteria; Understanding the concept of Gap year	1. Define and introduce oneself in English	4	Group discussion: try to make 5-year plans for future and talk about it	2
			2. The concept of gap year			
			3. Make career plans for future			
2	Critical thinking	Understanding the concept and importance of critical thinking; Learn to distinguish facts, opinions and preferences; The basic ways to understand critical thinking ability	4. Grasp the definition of critical thinking	4	English debate	2
			5. Use critical thinking to analyze settings			
			6. Understand and distinguish facts, opinions and preferences			
			7. Apply critical thinking in analyze reading materials and writing			
3	Communication skills	Understand the importance of communication skills; Use communication skills in negotiation; Understand different ways of communication	8. Understand the object and purpose of communication	4	Group presentation	2
			9. Choose the right way to communicate			
			10. Master the effectiveness of Communication			

序号	教学情境/任务/项目/单元	教学目标	教学内容及学时分配			
			知识点	学时	实践项目	学时
4	团队协作	了解团队协作的重要性; 了解个人特点对于团队的贡献意义; 掌握团队协作能力以提升工作效率	11. 了解团队协作的意义	4	团队完成造桥、造塔任务实践	2
			12. 学会使用SWOT分析法对自己进行客观评价			
			13. 如何提升团队工作效率			
5	制作工作简历	了解简历的基本格式与作用; 掌握如何制作一份合格的简历; 了解简历对于应聘工作的重要性	14. 简历中应该避免的缩写、用词等	4	全英文简历制作	2
			15. 简历遇到的通病			
			16. 如何构建自己的简历			
6	面试技巧	了解不同面试类型与风格; 知道第一印象的重要性; 了解为什么需要不同的面试	17. 传统与行为面试的内容	2		
			18. 掌握面试中常见的问题			
			19. 面试准备			
合计				22		10
备注						

五、教学实施建议

(一)教学参考资料

安妮·汤普森.《思辨思维－实践概论》(第三版),罗德里奇出版社,2009。

罗杰·费雪、威廉姆·于锐.《无须让步达成共识:让对方在谈判中说是》(修订版),企鹅图书出版社,2011。

亚瑟·蒂·罗森博格.《如何写出适合所有情况的漂亮简历和求职信》(第五版),亚当传媒出版社,2007。

普拉蒂·梅斯纳、沃夫·梅斯纳.《赢得适合的工作——面试制胜计划》,泛麦克米伦印第安出版社,2015。

(二)教师素质要求

任课教师要求管理相关专业或英语专业。

continued

Item	Topic	Teaching Objectives	Teaching Contents and Teaching Hours			
			Lesson Content	Hrs	Student Activities	Hrs
4	Teamwork	Understand the importance of teamwork; Understand the contribution of personal styles to teams; Develop the ability of teamwork to improve work efficiency	11. Understand the significance of teamwork	4	Completing bridge and tower building tasks for teamwork practice	2
			12. Learn to use SWOT analysis to evaluate oneself objectively			
			13. How to improve teamwork efficiency			
5	Make C. V. (resume)	Understand the basic format and function of resume; Master how to make a qualified resume; Understand the importance of resume for job application	14. Abbreviations and words to be avoided in resumes	4	Make a perfect English C. V.	2
			15. Common faults in resumes			
			16. How to make your resume			
6	Interview	Understanding different interview types and styles; The importance of knowing the first impression; Understand why different interviews are needed	17. Tradition and Behavior Interview	2		
			18. Master common questions in interviews			
			19. Interview preparation			
Total				22		10
Note						

5. Teaching Suggestions

5.1 Teaching Resources

Anne Thomson. *Critical Reasoning-A Practical Introduction.* 3rd Edition. Routledge,2009.

Roger Fisher, William Ury. *Getting to Yes: Negotiating Agreement Without Giving In.* Revised Edition. Penguin Books, 2011.

Arthur D. Rosenberg. *How to Write Outstanding Resumes and Cover Letters for Every Situation.* 5th Edition. Adams Media, 2007.

Pratibha Messner, Wolfgang Messner. *Winning the Right Job—A Blueprint to Acing the Interview.* Pan Macmillan India, 2015.

5.2 Teachers'Qualification

The teachers should have English and/or management-related major background and a teacher qualification certificate.

(三)教学场地、设施要求

多媒体教室、活动教室

六、考核评价

考核方式:模拟面试考核。

成绩构成:期末考试70%,实践项目30%。

5.3　Teaching Facilities

This course can be delivered in multimedia classrooms and activity classroom.

6. Evaluation

The Method of Assessment: Students are required to pass the simulated job interview.

Assessment Requirement: The final grade score is the total of the learning process assessment (account for 30%) and the final test score(account for 70%).

四 川 建 筑 职 业 技 术 学 院

课 程 标 准

课程名称： 建筑与装饰材料

课程代码： 900037

课程学分： 2

基本学时： 32

适用专业： 工程造价(中英合作办学项目)

执 笔 人： 傅 岩

编制单位： 建筑材料检测教研室

审 核： 材料工程系(签章)

批 准： 教务处(盖章)

Sichuan College of Architectural Technology

Curriculum Standard

Curriculum Name: Building and Decorative Materials

Curriculum Code: 900037

Credits: 2

Teaching Hours: 32

Applicable Specialty: Construction Cost

(Sino-British Cooperative Program)

Compiled by: Fu Yan

On Behalf of: Teaching & Research Section of Building Material Testing

Reviewed by: The Department of Material Engineering

Approved by: Teaching Affairs Office

《建筑与装饰材料》课程标准

一、课程概述

本课程在工程造价(中英合作办学项目)、建设工程管理(中英合作办学项目)专业中属于专业基础课;在第 2 学期开设。

二、课程设计思路

根据本专业培养从事工程造价的应用型人才,课程涉及了相关岗位必备的知识:建筑与装饰材料。教学的目标是使学生掌握各种建筑及装饰材料的概念、性质,技术标准、生产及应用,为后续课程打下基础。以岗位具体工作要求来确定教学内容,包括建筑材料的基本性质、气硬性胶凝材料、水泥、混凝土等。

三、课程目标

通过本课程的学习,学生能掌握各种建筑及装饰材料的概念、性质,技术标准、生产及应用,包括建筑材料的基本性质、气硬性胶凝材料、水泥、混凝土等,培养学生分析及解决问题的能力,提高学生的专业素质,为继续深造打下基础。

(一)知识目标

掌握各种建筑及装饰材料的概念、性质,技术标准、生产及应用;包括建筑材料的基本性质、气硬性胶凝材料、水泥、混凝土等。

(二)素质目标

培养学生艰苦朴素、吃苦耐劳、诚信可靠的优良作风。

Building and Decorative Materials Curriculum Standard

1. Curriculum Description

This course is a professional and fundamental one for the specialties of construction cost and engineering management(Sino-British Cooperative Programs). It is delivered in the second semester.

2. Curriculum Design Ideas

This specialty aims at cultivating students to meet the requirements of works in construction cost and provides the necessary knowledge for them in building and decorative materials. The teaching objective of the course is to help students master the concepts, properties, technical stands, production and application of building and decorative materials, which can lay a great foundation for their future work. The selection of the teaching content is based on the requirement of the positions related with construction cost, so the basic properties of building materials, air-hardening binding materials, cement, concrete and so on are included in the course.

3. Curriculum Objectives

Through this course, students should master the concepts, properties, technical standards, production and application of various construction and decorative materials, so as to lay the foundation for further studies. The teaching contents are selected and determined according to the specific job requirements of construction industry, including the basic properties of building materials, air-hardening binding materials, cement, concrete, etc. Students should learn to develop analysis and problem-solving abilities and improve their professional qualifications.

3.1 Knowledge Objectives

After learning this course, students will be able to

(1) master the concepts, properties, technical standards, production and application of various construction and decorative materials;

(2) master the basic properties of building materials, air-hardening binding materials, cement, concrete, etc.

3.2 Quality Objectives

After learning this course, students will have the quality of bearing hardship, taking simple living style, hard working, honesty and integrity.

(三)能力目标

(1)培养学生分析及解决建筑及装饰材料在使用中出现问题的能力；

(2)提高学生的专业素质，为继续深造打下基础。

四、课程内容与学时分配

教学内容与学时安排表

序号	教学情境/任务/项目/单元	教学目标	教学内容及学时分配			
			知识点	学时	实践项目	学时
1	绪论	1. 掌握建筑材料的定义、分类、技术标准； 2. 认识建筑材料在建筑工程中的地位	1. 建筑材料的定义、分类、技术标准(★)； 2. 建筑材料在建筑工程中的地位； 3. 我国建筑材料及建材工业的发展	1	—	—
2	建筑材料的基本性质	1. 掌握材料与质量有关的性质； 2. 掌握材料的吸湿性、耐水性、抗渗性、吸水性、抗冻性的概念及其表示方法； 3. 掌握材料的导热性、比热容、保温隔热性能； 4. 知道材料的填率与空隙率； 5. 材料的力学性质	4. 建筑材料的实际密度、表观密度与堆积密度的概念、含义、计算公式(★)； 5. 孔隙率、密实度的概念、含义、计算公式(★)； 6. 材料的填充率与空隙率； 7. 材料与水有关的性质； 8. 材料和热性质； 9. 材料的力学性质	3		
3	气硬性胶凝材料	1. 掌握石灰的熟化； 2. 掌握石膏的特点、用途、储存； 3. 熟悉石灰的品种	10. 石灰的生产、熟化、硬化与特点、应用与保管(★)； 11. 建筑石膏的生产、凝结硬化、技术要求、特性	2		

3.3   Ability Objectives

After learning this course, students will have the ability of

(1) analyzing and solve problems in using the building and decorative materials;

(2) improving professional qualification and lay a solid foundation for further studies.

4. Curriculum Contents and Teaching Hours

Table of Teaching Contents and Teaching Hours

Item	Topic	Teaching Objectives	Teaching Contents and Teaching Hours			
			Lesson Contents	Hrs	Student Activities	Hrs
1	Introduction	1. Grasp the definition, classification and technical standard of building materials; 2. Understand the role of building materials in construction engineering	1. The definition, classification and technical standard of building materials(★); 2. The role of building materials in construction engineering; 3. The development of building materials and the industry in China	1		
2	The basic properties of building materials	1. Master the qualityrelated property of the material; 2. Master the concepts and expressive means of hygroscopicity, water resistance, impermeability, water absorbency, and frost resistance; 3. Master the material's thermal conductivity, heat capacity and thermal insulation properties; 4. Understand the material's voids filling capa-city and voidage; 5. Master the mechanical properties of materials	4. Concepts, meanings and calculation formulas of actual density, apparent density and bulk density of building materials(★); 5. Concepts, meanings and calculation formulas of porosity and compactness(★); 6. The material's voids filling capacity and voidage; 7. Water-related property of the material; 8. Heat-related property of the material; 9. The mechanical properties of materials	3		
3	Air-hardening binding material	1. Master the curing of lime; 2. Understand the characteristics, uses and storage of gypsum; 3. Be familiar with lime varieties	10. Lime production, curing, hardening and its properties, application and storage(★); 11. Production, condensation, hardening, technical requirements and properties of building gypsum	2		

序号	教学情境/任务/项目/单元	教学目标	教学内容及学时分配			
			知识点	学时	实践项目	学时
4	水泥	1. 掌握硅酸盐水泥的定义、凝结硬化、结构； 2. 掌握硅酸盐水泥的主要技术性质中凝结时间、标准稠度用水量、体积安定性的定义、含义、检验方法及国家标准规定； 3. 掌握水泥强度及强度等级的划分及合格品质量等级的划分； 4. 掌握掺混合材料的硅酸盐水泥的定义及分类； 5. 掌握掺混合材料的目的、混合材料的种类； 6. 掺混合材料硅酸盐水泥的技术性质及特点； 7. 熟悉硅酸盐水泥中，游离氧化钙、三氧化硫、碱含量的技术要求	12. 水泥的概括、分类(★)； 13. 硅酸盐水泥的生产、矿物组成 及其特点、硅酸盐水泥的凝结硬化(★)； 14. 水泥石的结构(★)； 15. 硅酸盐水泥技术标准； 16. 水泥强度(★)； 17. 掺混合材料的硅酸盐水泥(★)	6		

continued

Item	Topic	Teaching Objectives	Teaching Contents and Teaching Hours			
			Lesson Contents	Hrs	Student Activities	Hrs
4	Cement	1. Master the definition of silicate cement, its condensation, hardening and structure; 2. Master the main technical properties of silicate cement, esp. the definition, meaning, testing methods and national standards of condensation time, standard consistency of water consumption, and volume stability; 3. Master the division of cement strength and the strength grade, as well as the grade division of qualified product; 4. Master the definition and classification of mixedmaterial silicate cement; 5. Master the purpose of mixing materials and the types of mixed materials; 6. The technical properties and features of mixedmaterial silicate cement; 7. Familiar with the technical requirements of free calcium oxide, sulfur trioxide, alkali content of silicate cement	12. The definition and classification of cement(★); 13. The production, mineral composition and features, condensation and hardening of silicate cement(★); 14. Structure of cement pastes (★); 15. The technical standard of silicate cement; 16. The cement strength(★); 17. Silicate cement with mixed materials(★)	6		

序号	教学情境/任务/项目/单元	教学目标	教学内容及学时分配			
			知识点	学时	实践项目	学时
5	混凝土	1. 掌握混凝土的定义、分类及特点; 2. 掌握混凝土组成材料分类及技术要求; 3. 掌握混凝土拌合物的主要技术性质——和易性; 4. 掌握硬化混凝土的主要技术性质——强度、耐久性; 5. 掌握普通混凝土的初步配合比设计	18. 混凝土的定义、分类及特点(★); 19. 混凝土组成材料分类及技术要求(★); 20. 混凝土拌合物的和易性(★); 21. 硬化混凝土的强度、耐久性(★); 22. 普通混凝土的初步配合比设计	10		
6	水泥标准稠度用水量及强度试件制作	掌握试验的目的、仪器设备的使用、检测步骤、结果计算及评定			水泥标准稠度用水量及强度试件制作	2
7	水泥强度试验				水泥强度试验	2
8	普通混凝土用砂试验				普通混凝土用砂试验	2
9	普通混凝土拌合物性质测定				普通混凝土拌合物性质测定	2

continued

Item	Topic	Teaching Objectives	Teaching Contents and Teaching Hours			
			Lesson Contents	Hrs	Student Activities	Hrs
5	Concrete	1. Master the definition, classification and properties of concrete; 2. Master the materials classification and technical requirements in the concrete composition; 3. Master the main technical property ofconcrete mixture: workability; 4. Master the main technical properties of harde-ned concrete: strength and durability; 5. Master the preliminary mix design of ordinary concrete	18. Definition, classification and properties of concrete(★); 19. Classification and technical requirements of the composition materials(★); 20. The workability of concrete mixture(★); 21. Strength and durability of hardened concrete; 22. The preliminary mix design of ordinary concrete	10		
6	Water consumption in cement standard consistency and strength test parts production				Water consumption in cement standard consistency and strength test parts production	2
7	Cement strength test	Grasp the purpose of the test, the use of instruments and equipment, testing procedures, calculation and evaluation of the results			Cement strength test	2
8	Sand test for ordinary concrete				Sand test for ordinary concrete	2
9	Determination of properties of ordinary concrete mixtures				Determination of properties of ordinary concrete mixtures	2

序号	教学情境/任务/项目/单元	教学目标	教学内容及学时分配			
			知识点	学时	实践项目	学时
10	混凝土强度试验	掌握试验的目的、仪器设备的使用、检测步骤、结果计算及评定			混凝土强度试验	2
合计				22		10
备注:重要知识点标识★						

五、教学实施建议

(一)教学参考资料

李江华、郭玉珍、李柱凯.《建筑材料项目化教程》,华中科技大学出版社,2013。

(二)教师素质要求

任课教师必须有高校教师资格、材料专业背景,具有一定的教学能力。

(三)教学场地、设施要求

具有多媒体设施的教学场地。

六、考核评价

(一)考核方式

闭卷。

(二)考核说明

平时成绩评定占总成绩30%,主要考查出勤、课堂发言、作业等;期末终结性考核占总成绩70%。

continued

Item	Topic	Teaching Objectives	Teaching Contents and Teaching Hours			
			Lesson Contents	Hrs	Student Activities	Hrs
10	Strength test of concrete	Grasp the purpose of the test, the use of instruments and equipment, testing procedures, calculation and evaluation of the results.			Strength test of concrete	2
Total				22		10
Note: Mark the important points with ★						

5. Teaching Suggestions

5.1　Teaching Resources

Li Jianghua, Guo Yuzhen, Li Zhukai. *Project Tutorial of Building Materials.* Huazhong University of Science and Technology Press, 2013.

5.2　Teachers'Qualification

Teachers must have teacher qualification certificate, with specialty background and adequate teaching experience.

5.3　Teaching Facilities

This course can be delivered in multimedia classrooms.

6. Evaluation

6.1　The Method of Assessment

Closed-book examination is required.

6.2　Assessment Requirement

The final grade score is the total of the learning process assessment(including the usual attendance, in-class performance and homework, accounting for 30%) and the final-term examination(accounting for 70%).

四川建筑职业技术学院

课程标准

课程名称：　　　　　建筑结构基础与识图

课程代码：　　　　　010077

课程学分：　　　　　4

基本学时：　　　　　64

适用专业：　　　　　工程造价（中英合作办学项目）

执　笔　人：　　　　　张爱莲

编制单位：　　　　　结构教研室

审　　核：　　　　　土木系（院）（签章）

批　　准：　　　　　教务处（盖章）

Sichuan College of Architectural Technology

Curriculum Standard

Curriculum Name: Basic Building Structure and Drawing Reading

Curriculum Code: 010077

Credits: 4

Teaching Hours: 64

Applicable Specialty: Construction Cost

(Sino-British Cooperative Program)

Complied by: Zhang Ailian

On Behalf of: Teaching & Researching Section of Structures

Reviewed by: The Department of Civil Engineering

Approved by: Teaching Affairs Office

《建筑结构基础与识图》课程标准

一、课程概述

《建筑结构基础与识图》是工程造价(中英合作办学项目)专业的公共课程,主要讲述钢筋、混凝土材料、砌体材料的基本力学性能、设计方法、各类混凝土构件(拉、压、弯、剪、扭)及基本结构的受力性能及其计算方法、混凝土构件的配筋构造、砌体结构的受力性能及其计算方法;平法施工图识读。该部分内容的学习,目的是让学生掌握混凝土结构设计的基本原理;掌握梁板结构、砌体结构的设计方法、学会平法施工图的识读。本课程有助于学生建立起建筑结构的基本概念,为学习后续课程打下坚实的基础。

本课程先修课程:建筑制图、建筑与装饰材料、理论力学、材料力学等。

本门课程的后续课程:建筑设备工程与识图、钢结构施工、建筑施工组织等。

二、课程设计思路

本课程采用"知识+实例+实践"的教学模式,打破传统单一的知识传授教学模式。学生通过具体的工程实例,掌握建筑结构中各种构件的内力计算方法、配筋计算方法和配筋图的绘制方法,从设计角度对建筑工程结构施工图有一个全面深入的了解,在学习过程中,不断练习构件配筋图的传统表达方式,让学生建立丰富的空间想象力,能够熟练掌握不同构件钢筋的空间定位,每一种钢筋的作用和构造要求,为以后学习钢筋放样,钢筋下料长度的计算奠定坚实的理论基础。

Basic Building Structure and Drawing Reading Curriculum Standard

1. Curriculum Description

Basic Building Structure and Drawing Reading is a non-major course offered to the specialized Sino-British cooperative program in construction cost. Contents of the course include the mechanical properties and the design of reinforcement, concrete materials and masonry materials; the loading capacities and their calculations of various concrete members (being pulled, pressed, bended, sheared and twisted) and their basic structures; the reinforcement structures of concrete members; the loading capacities and their calculations of masonry structures; and the reading of construction drawings prepared by ichnographic representing methods. In the study, students will master the basic theories in concert structure design and the designing methods of beam-slab structures and masonry structures. They will also be able to read construction drawings prepared by ichnographic representing methods. By studying this course, students will establish a conceptual system of architectural structures, laying foundation for courses followed up.

The delivered courses are *Architectural Drawing*, *Building and Decoration Materials*, *Theoretical Mechanics*, *Mechanics of Materials* and so on.

Courses followed this are *Building Services and Drawing Reading*, *Steel Structure*, *Steel Structure Construction*, *Construction Organization* and so on.

2. Curriculum Design Ideas

As a breach from traditional knowledge-centered teaching approaches, the course adopts the mode of integrating knowledge, cases and practice in the teaching process. In case studies of specified constructions, students can master the calculation of inner forces and reinforcement of various construction members and the preparation of reinforcement drawings, attaining an in-depth understanding to construction drawings of building structures from the designer's perspective. In the learning process, students should repeatedly practice traditional representation of reinforcement drawings of members, enriching their spatial imagination and familiarizing themselves with the reinforcement location of different members and the function and structural requirements of reinforcement. In doing so, they would lay a theoretic foundation for learning reinforcement setting out and cutting length calculation of reinforcing bars later.

三、课程目标

了解结构设计规范、规程及相关标准图集;对简单的建筑工程,能够进行结构设计及施工图的绘制;对已有建筑物,可以进行结构构件的承载力复核;能够读懂简单的钢筋混凝土和砌体结构的施工图。

(一)知识目标

(1)掌握建筑结构常用材料的种类和材性;

(2)掌握建筑结构及结构构件的构造知识,包括抗震构造知识;

(3)掌握现浇钢筋混凝土肋形楼盖的设计方法;

(4)钢筋混凝土柱、梁、板、剪力墙平法施工的识读能力及钢筋翻样能力。

(二)素质目标

(1)具有进行一般建筑结构构件(受弯、轴向受力构件)截面设计与承载力复核的能力;

(2)具有分析和处理实际施工过程中遇到的一般结构问题的能力;

(3)具有正确识读建筑结构施工图的能力。

(三)能力目标

(1)专业能力目标:了解建筑结构的基本概念;对于简单的建筑工程,能够进行结构设计及施工图的绘制。

(2)方法能力目标:对已有建筑物,可以进行结构构件的承载力复核;能读懂结构施工图。

(3)社会能力目标:具有较强的口头与书面表达能力、沟通协调能力;具有团队精神和协作精神;具有良好的心理素质和克服困难的能力;能与建筑设计企业、建筑施工企业建立良好、持久的关系;具有工作责任感。

3. Curriculum Objectives

The course aims for students to learn requirements, regulations and standard atlas of structural design; to be able to conduct structural design and construction drawings preparation for simple building construction; to review bearing capacities of structural members for existing buildings; to be able to read and understand construction drawings of simple reinforced concrete and masonry structures.

3.1 Knowledge Objectives

After learning this course, students will be able to

(1) learn the types and mechanical properties of usual materials in building structure;

(2) learn about building structures and structural members, including aseismic structures;

(3) master designing approaches of cast-in-situ reinforced concrete slab floors;

(4) acquire the ability to read ichnographic representing of reinforced concrete columns, beams, slabs and shear walls, and the ability to set out reinforcement.

3.2 Quality Objectives

After learning this course, students will have the quality of

(1) designing the section of usual structural members (bending and axially loaded members) and reviewing their bearing capacity;

(2) analyzing and solving usual structural problems in real construction;

(3) reading construction drawings of building structures.

3.3 Ability Objectives

After learning this course, students will have the ability

(1) in the specialty: basic concepts of building structure; competence in designing the structure of simple buildings and preparing construction drawings for them.

(2) methodology: the ability in reviewing bearing capacity of structural members of existing buildings; the ability to read construction drawings of structures.

(3) sociality: excellent oral and writing expression; apt communication and negotiation; teamwork and cooperation; steady psychology against difficulties; the ability to establish sustainable relationships with both architecture designing firms and construction companies; and a sense of duty in the work.

四、课程内容与学时分配

教学内容与学时安排表

序号	教学情境/任务/项目/单元	教学目标	教学内容及学时分配			
			知识点	学时	实践项目	学时
1	绪论	1. 掌握建筑结构的概念； 2. 了解建筑结构的发展与应用状况； 3. 掌握本课程的内容、学习目标及学习要求	1. 建筑结构的概念	1	1. 通过多媒体课件展示各种建筑结构的分类	1
			2. 本课程的内容、学习目标及学习要求	1	2. 本课程的学习方法及学习目标以及可能遇到的困难	
2	建筑结构计算基本原则	1. 了解荷载分类和荷载代表值的概念； 2. 掌握建筑结构极限状态及使用表达式	3. 荷载分类和荷载代表值的概念	2	3. 讨论荷载的分类方式	1
			4. 建筑结构极限状态及使用表达式	2	4. 讨论建筑结构极限状态及使用表达式，计算建筑结构荷载	
3	建筑结构材料	1. 掌握建筑钢材的品种和规格，建筑钢材的力学性能； 2. 掌握混凝土的强度，混凝土的设计指标； 3. 了解砌体材料种类及强度等级	5. 建筑钢材的品种和规格，建筑钢材的力学性能	1	5. 讨论钢材的品种和规格，掌握建筑钢材的力学性能	1
			6. 混凝土的强度，混凝土的设计指标	1	6. 讨论混凝土的强度指标	
			7. 砌体材料种类及强度等级	1	7. 讨论砌体材料种类及强度等级	

4. Curriculum Contents and Teaching Hours

Table of Teaching Contents and Teaching Hours

Item	Topic	Teaching Objectives	Lesson Contents and Teaching Hours			
			Lesson Contents	Hrs	Student Activities	Hrs
1	Introduction	1. Understand the concept of building structure; 2. Learn the development and application of building structure; 3. Understand contents, learning objectives and requirements of this unit	1. The concept of building structure	1	1. Multimedia presentation of varied types of building structure	1
			2. Contents, learning outcomes and requirements of this unit	1	2. Learning methods, goals and possible difficulties of this unit	
2	Basic rules for the calculation of building structure	1. Learn the types of loads and the concept of representative value of loads; 2. Master the limit state of building structure and the expression of its application	3. Types of loads and the concept of representative value of loads	2	3. Discussion on ways to distinguish different types of loads	1
			4. The limit state of building structure and the expression of its application	2	4. Discussion on the limit state of building structure and the expression of its application; calculation loads of building structures	
3	Materials of building structures	1. Learn the types, dimensions and mechanic property of architectural steel; 2. Learn the strength and designing index of concrete; 3. Learn the types and strength grades of masonry envelope materials	5. The types, dimensions and mechanic property of architectural steel	1	5. Discussion on the types, dimensions and mechanic property of architectural steel	1
			6. The strength and designing index of concrete	1	6. Discussion on the strength index of concrete	
			7. Types and the strength grades of masonry materials.	1	7. Discussion on types and the strength grades of masonry materials	

序号	教学情境/任务/项目/单元	教学目标	教学内容及学时分配			
			知识点	学时	实践项目	学时
4	钢筋混凝土受弯构件	1. 了解受弯构件的构造要求; 2. 掌握单筋矩形截面正截面承载力计算公式,了解单筋 T 形截面正截面承载力计算公式,了解双筋截面的正截面承载力计算公式; 3. 掌握斜截面承载力计算公式; 4. 了解变形及裂缝宽度验算方法	8. 受弯构件的构造要求	2	8. 讨论受弯构件钢筋的种类及作用	1
			9. 斜截面承载力计算	2	9. 讨论受弯构件斜截面的三种破坏形态	
			10. 变形及裂缝宽度验算方法	1	10. 讨论变形及裂缝宽度验算方法	
5	钢筋混凝土受压构件	1. 了解受压构件的构造要求; 2. 掌握轴心受压构件正截面承载力计算方法; 3. 掌握偏心受压构件正截面承载力计算方法	11. 受压构件的构造要求	2	11. 讨论受压构件钢筋的种类及作用	1
			12. 轴心受压构件正截面承载力计算	2	12. 轴心受压构件正截面承载力计算练习	
			13. 偏心受压构件正截面承载力计算	2	13. 讨论偏心受压构件的计算方法	
6	钢筋混凝土受扭构件	1. 掌握受扭构件的受力特点和配筋构造; 2. 了解钢筋混凝土受扭构件承载力计算	14. 受扭构件的受力特点和配筋构造	1	14. 讨论受扭构件的受力特点和配筋构造	1
			15. 了解钢筋混凝土受扭构件承载力计算	1	15. 了解钢筋混凝土受扭构件承载力计算	

continued

Item	Topic	Teaching Objectives	Lesson Contents and Teaching Hours			
			Lesson Contents	Hrs	Student Activities	Hrs
4	Reinforced concrete bending member	1. Learn the structural requirements of bending member; 2. Master the formula of the bearing capacity of single-reinforced rectangular cross section, of single-reinforced T-shape cross section, and of double reinforced cross section; 3. Master methods to calculate the bearing capacity of oblique section; 4. Understand the methods to check the deformation and crack width	8. Structural requirements of bending member	2	8. Discussion on types and functions of steel in bending members	1
			9. Calculating the bearing capacity of oblique section	2	9. Discussion on three kinds of failures happening to oblique sections of bending member	
			10. Validating calculation of deformation and crack width	1	10. Discussion on validating calculation of deformation and crack width	
5	Reinforced concrete compression member	1. Learn the structural requirements of compression member; 2. Master methods to calculate the bearing capacity of cross section of axial compression members; 3. Master methods to calculate the bearing capacity of cross section of eccentric compression members	11. Constructional requirements of compression members	2	11. Discussion on the types and functions of compression steel members	1
			12. Calculating the bearing capacity of cross section of axial compression members	2	12. Calculation practice on the bearing capacity of cross section of axial compression members	
			13. Calculating the bearing capacity of cross section of eccentric compression members	2	13. Discussion on calculation of cross section of eccentric compression members	
6	Reinforced concrete torsion member	1. Learn the mechanical characteristics and reinforcement structure of torsion member; 2. Learn methods to calculate the bearing capacity of reinforced concrete torsion members	14. The mechanical characteristics and reinforcement structure of torsion member	1	14. Discussion on loading features and reinforcement structure of torsion member	1
			15. Learning methods to calculate the bearing capacity of reinforced concrete torsion members	1	15. Learn how to calculate the bearing capacity of reinforced concrete torsion members	

序号	教学情境/任务/项目/单元	教学目标	教学内容及学时分配			
			知识点	学时	实践项目	学时
7	预应力混凝土构件	1. 了解预应力混凝土的基本概念； 2. 掌握预应力混凝土的施工方法； 3. 了解张拉控制应力与预应力损失； 4. 了解预应力混凝土构件的构造要求	16. 预应力混凝土的基本概念	1	16. 讨论预应力混凝土的基本原理	1
			17. 预应力混凝土的施工方法	1	17. 讨论预应力混凝土的施工方法的施工工艺顺序	
			18. 张拉控制应力与预应力损失	1	18. 讨论如何减小预应力的损失	
			19. 预应力混凝土构件的构造要求	1	19. 讨论预应力混凝土构件的构造要求	
8	钢筋混凝土楼（屋）盖	1. 了解现浇钢筋混凝土楼盖的分类和受力特点，掌握单向板肋形楼盖的构造和内力计算方法； 2. 了解双向板楼盖的受力特点； 3. 了解装配式楼盖的构造要求； 4. 掌握钢筋混凝土楼梯的分类、受力特点和构造要求	20. 现浇钢筋混凝土楼盖的分类和受力特点	1	20. 单向板肋形楼盖的构造和内力计算练习	1
			21. 双向板楼盖的受力特点	2	21. 讨论单向板肋形楼盖和双向板楼盖的区别	
			22. 装配式楼盖的构造要求	2	22. 讨论装配式楼盖和现浇式楼盖的区别	
			23. 钢筋混凝土楼梯的分类、受力特点和构造要求	2	23. 讨论各种类型楼梯的优缺点以及适用范围	

continued

Item	Topic	Teaching Objectives	Lesson Contents and Teaching Hours			
			Lesson Contents	Hrs	Student Activities	Hrs
7	Prestressed concrete member	1. Learn the concepts of prestressed concrete; 2. Master methods of prestressed concrete construction; 3. Learn controlled stretching stress and prestressing lose; 4. Learn structural requirements of prestressed concr-ete members	16. Fundamental concepts of prestressed concrete	1	16. Discussion on the fundamental principles of prestressed concrete	1
			17. Building operation of prestressed concrete construction	1	17. Discussion on ste-pvaried building operations of prestressed concrete	
			18. Controlled stretching stress and prestressing lose	1	18. Discussion on ways to avoid prestressing lose	
			19. Structural requirements of prestressed concrete members.	1	19. Discussion on structural requirements of prestressed concrete members	
8	Reinforced concrete floors	1. Learn the types and loading features of reinforced concrete floors; 2. Learn the structure of beamsupported one way slab floor and master the calculation of its inter force; 3. Learn the structural requirements of assembled floor; 4. Learn the types ,loading features and structural requirements of reinforced concrete stairs	20. The types and loading features of castins-itu reinforced concrete floor	1	20. Calculation practice on the structure of beam-supported one way slab floor and its internal forces.	1
			21. Loading features of beam-supported two-way slab floor	2	21. Discussion on the difference between beam-supported one way slab floor and beamsupported two way slab floor	
			22. Structural requirements of assembled floor	2	22. Discussion on the difference between assembled floor and cast-in-situ floor	
			23. Types, loading features and structural requirements of reinforced concrete stair	2	23. Discussion on the pros and cons of varied stair types, as well as their application	

序号	教学情境/任务/项目/单元	教学目标	教学内容及学时分配			
			知识点	学时	实践项目	学时
9	多层及高层钢筋混凝土房屋	1. 了解钢筋混凝土结构常用结构体系; 2. 熟悉框架结构的形式,结构布置和受力特点以及框架结构的构造要求; 3. 了解剪力墙结构受力特点及构造要求; 4. 了解框架-剪力墙结构的受力特点和构造要求	24. 钢筋混凝土结构常用结构体系	2	24. 讨论钢筋混凝土结构常用结构体系的优缺点	1
			25. 框架结构的形式及构造要求	2	25. 讨论框架结构的优缺点及适用范围	
			26. 剪力墙结构受力特点及构造要求	2	26. 讨论剪力墙结构的优缺点及适用范围	
			27. 框架-剪力墙结构的受力特点和构造要求	2	27. 讨论框架-剪力墙结构的优缺点及适用范围	
10	砌体结构	1. 了解砌体结构构件的计算方法; 2. 熟悉砌体结构房屋的构造要求; 3. 了解砌体结构中的过梁、墙梁、挑梁和雨篷的构造要求和受力特点	28. 砌体结构构件的计算方法	2	28. 讨论影响砌体结构承载力的因素	1
			29. 砌体结构房屋的构造要求	2	29. 讨论砌体结构房屋的构造要求	
			30. 过梁、墙梁、挑梁和雨篷	3	30. 讨论过梁、墙梁、挑梁和雨篷各自的作用	
11	平法施工图识读	掌握柱、梁、剪力墙和板的平法施工图制图规则及标准构造详图	31. 柱、梁、剪力墙和板的平法施工图制图规则及标准构造详图	6		
合计				54		10
备注						

continued

Item	Topic	Teaching Objectives	Lesson Contents and Teaching Hours			
			Lesson Contents	Hrs	Student Activities	Hrs
9	Multistory and highrise reinforced concrete building	1. Learn usual structural systems of reinforced concrete structures; 2. Understand forms, structure and loading features of frame structures and their structural requirements; 3. Learn loading features and structural requirements of shear wall structures; 4. Learn loading features and structural requirements of frame-shear wall structures	24. Usual structural systems of reinforced concrete structures	2	24. Discussion on pros and cons of common structural systems of reinforced concrete structures	1
			25. Forms and structural requirements of frame structures	2	25. Discussion on the pros and cons of frame structures and their application	
			26. Loading features and structural requirement of shear wall structures	2	26. Discussion on the pros and cons of shear wall structures and their application	
			27. Loading features and structural requirement of frame-shear wall structures	2	27. Discussion on the pros and cons of frame-shear wall structures and their application	
10	Masonry structures	1. Learn the calculation of masonry structure members; 2. Understand structural requirements for buildings of masonry structure; 3. Learn structural requirements and loading features of lintels, wall beams, cantilever beams and canopies in masonry structures	28. The calculation of masonry structural members	2	28. Discussion on factors impacting bearing capacity of masonry structures	1
			29. Structural requirements for building of masonry structures	2	29. Discussion on structural requirements for building of masonry structures	
			30. Lintels, wall beams, cantilever beams and canopies	3	30. Discussion on respective functions of lintels, wall beams, cantilever beams and canopies	
11	Constructiondrawings prepared by ichnographic representation method	Master rules in preparing ichnographic work drawings of columns, beams, shear walls and slabs and learn their standard structural detail drawings	31. Rules in preparing ichnographic work drawing of columns, beams, shear walls and slabs and their standard structural detail drawings	6		
Total				54		10
Notes						

五、教学实施建议

(一)教学参考资料

杨太生.《建筑结构基础与识图》第四版,中国建筑工业出版社,2019年。

中国建筑科学研究院.《混凝土结构设计规范(2015年版)》(GB 50010—2010),中国建筑工业出版社,2016。

中国建筑科学研究院.《建筑抗震设计规范(2016年版)》(GB 50011—2010),中国建筑工业出版社,2016。

中国建筑标准设计研究院有限公司.《混凝土结构施工图平面整体表示方法制图规则和构造详图(现浇混凝土框架、剪力墙、梁、板)》(16G101-1),中国计划出版社,2016。

其他教辅资料:

建筑工程教育网;

中国建造师网。

(二)教师素质要求

热爱教育事业,具有先进的教学理念和建筑结构设计的工作经历,具备较强的沟通能力和爱岗敬业、为人师表、锐意进取的职业道德;具备施工图识读能力;具备系统的建筑结构的专业知识;具有一定的建筑施工现场经验;具备宽广厚实的专业知识以及实践技能;具备健康的体质、充沛的精力和良好的心理素质。

(三)教学场地、设施要求

多媒体教室,应配有视频播放软件、工程结构设计所必需的软件。

六、考核评价

(一)考核的形式

笔试 + 平时表现。

5. Teaching Suggestions

5.1　Teaching Resources

Yang Taisheng. *Basic Building Structure and Drawing Reading.* 4th Edition. China Architecture & Building Press, 2019.

China Academy of Building Research. *Code for Design of Concrete Structures* (2015) (GB 50010—2010). China Architecture & Building Press, 2016.

China Academy of Building Research. *Code for Seismic Design of Buildings* (2016) (GB 50011—2010). China Architecture & Building Press, 2010.

China Institute of Building Standard Design & Research. *Regulation for Ichnographic Representation in Construction Drawings of Concrete Structure and Constructional Detail Drawings* (16G101-1). China Planning Press, 2016.

Other Resources:

www. jianshe99. com

www. constructor. cn

5.2　Teachers'Qualification

Teachers of this course should be dedicated to education with advanced teaching philosophy and working experience in building structure design. They must be model teachers with excellent communication skills, devotion to teaching career, and aggressive working ethic. They should also be competent in reading constructional drawing with systematic knowledge in building structures and rich experience of building construction site. Extensive and solid professional knowledge and practice are required for the teachers, as well as physical health, high spirit and balanced mind state.

5.3　Teaching Facilities

Multimedia classroom, with video players and engineering structure designing software installed are necessary for this course.

6. Evaluation/Examination

6.1　Approaches of Evaluation

A combination of daily performance and final examination

（二）考核评价表

考核评价表

目标	评价要素	评价标准	评价依据	考核方式	评分	权重/%
知识	基本知识	按教学大纲要求掌握的知识点；运用知识完成书面作业；运用知识分析和解决问题	个人作业 课堂笔记 课堂练习 期末考试	小组互评		5
				教师评定		5
				作业成绩		5
				期末考试：笔试		70
情感与素质	学习态度	遵守课堂纪律、积极参与课堂教学活动、按时完成作业、按要求完成准备	课堂表现记录、考勤表、同学及教师观察、课堂笔记	学生自评		5
				小组自评		
				教师评定		
	沟通协作管理	乐于请教和帮助同学、小组活动协调和谐、协作教师教学管理、做好教室值日工作、按要求做好课前准备和课后整理	小组作业、小组活动记录、自评及互评记录、值日记录、同学及教师观察	学生自评		5
				小组自评		
				教师评定		
	创新精神	有自主学习计划、再作业练习中能提出问题和见解、对教学或管理提出意见和建议、积极参与小组活动方案设计	个人作业、自主学习计划、学习活动、个人口头或书面提议	学生自评		5
				小组自评		
				教师评定		

6. 2　The Chart of Evaluation Details

The chart of evaluation details

Target	Key	Standard	Evidence	Evaluation	Grade	Percent
Knowledge	Fundamentals	Master key contents as is required by the unit standard. Finish writings. Analyze and solve problems	Personal works Notes taken in class Class works The Final examination	Group evaluation		5%
				Teacher's evaluation		5%
				Scores of works		5%
				The Final examination		70%
Spirit and qualities	Learning attitudes	Abide by class rules. Active participation in class activities. Timely handing in of home works. Prepare as is required.	Records of performance in classes Records of attendance. Comments of classmates and the teacher Notes taken in class	Self evaluation		5%
				Group evaluation		
				Teacher's evaluation		
	Communication, cooperation and management	Be willing to help others or be helped. Coordinate group activities. Help the teacher with classroom management. Finish duty works. Prepare for classes and clear up after them	Team works Records of group activities Records of selfe-valuation and group evaluation, Records of duty Comments of classmates and teachers	Self evaluation		5%
				Group evaluation		
				Teacher's evaluation		
	Innovation	Make plans of self-study. Raise questions in and provide insights of home works and practices. Propose suggestions on the unit or class management. Active participation in designing group activities	Personal works Plans of selfstudy Learning activities Oral or written suggestions for the unit	Self evaluation		5%
				Group evaluation		
				Teacher's evaluation		

四 川 建 筑 职 业 技 术 学 院

课 程 标 准

课程名称： 建筑施工工艺

课程代码： 010233

课程学分： 3.5

基本学时： 56

适用专业： 工程造价（中英合作办学项目）

执 笔 人： 孟小鸣

编制单位： 施工教研室

审 核： 系（院）（签章）

批 准： 教务处（盖章）

Sichuan College of Architectural Technology

Curriculum Standard

Curriculum Name: Building Construction Technology

Curriculum Code: 010233

Credits: 3. 5

Teaching Hours: 56

Applicable Specialty: Construction Cost

(Sino-British Cooperative Program)

Complied by: Meng Xiaoming

On Behalf of: Teaching& Researching Section of ConstructionTechnology

Reviewed by: The Department of Civil Engineering

Approved by: Teaching Affairs Office

《建筑施工工艺》课程标准

一、课程概述

建筑施工工艺是工程造价(中英合作办学项目)专业的一门专业课程,是学生学习建筑施工技术、建筑施工组织及其他相关知识的重要环节。其任务是研究建筑工程施工技术的一般规律,建筑工程中各主要工种工程的施工技术及工艺原理以及建筑施工新技术、新工艺的一般规律,使学生掌握建筑施工的基本知识、基本理论和决策方法,具有解决一般建筑施工的能力,为今后从事建筑施工技术及管理及相关工作打下坚实的基础。本课程内容覆盖面广,实践性极强,学生应先学习建筑工程测量、建筑材料、建筑力学、建筑结构、房屋建筑学、施工预算等相关课程,后期进一步学习建筑工程施工组织设计,建筑工程质量事故分析与处理等课程。本课程宜在第2学期开设。

二、课程设计思路

本课程依据职业岗位标准和人才培养目标设置。主要内容包含土方工程、基础工程、砌体结构工程、钢筋混凝土结构工程、预应力混凝土结构工程、结构安装工程、防水工程、装饰装修工程等内容。本课程开发思路是以市场需求为出发点,以职业能力培养为核心,以工作过程为导向,以工作任务为载体,遵循教学基本原则,吸取先进的教学方法。不仅注重大量的理论学习,还通过教学模型、教学多媒体、教学录像、教学资源库、实践教学结合,按照工作过程对知识进行排序,重在培养学生的职业行动能力。

三、课程目标

(一)知识目标

(1)熟悉建筑施工的程序;

Building Construction Technology Curriculum Standard

1. Curriculum Description

Building Construction Technology is one of major-specified courses for Construction Cost(Sino-British Cooperative Program). As a key course for students to learn building construction technology, organization and other relevant field, the curriculum undertakes the task of studying common rules in building construction technology, studying rules, skills and processes of major trades in building construction projects and studying common rules in applying new technology and new processes in building construction. The course will help students master fundamental knowledge, theories and strategies in building construction, and help their ability to conduct usual building construction projects, laying solid foundation for their future jobs relevant to building construction technology and management. The pre-delievered courses are *Building Surveying and Setout*, *Building Materials*, *Architectural Mechanics*, *Architectural Structures*, *Architecture Design*, *Concepts and Details* and *Construction Project Budget*. And after this course, students can learn further curricula like *Organization Design of Building Construction*, *Analysis and Solution to Quality Problems of Building Construction Projects*, and so on. This course is offered in the 2rd semester.

2. Curriculum Design Ideas

The curriculum standard is in accordance with the course description, as well as job description of the profession. Major contents of this course include earthwork, foundation work, masonry structure work, reinforced concrete structure work, prestressed concrete structure work, structure installation, waterproofing work, decoration and finishing, and the like. Starting from needs in job market and following pedagogic rules and advanced teaching methods, the unit takes professional skill as the focus to help students fulfil real duties in the working process of real jobs. Executive skills for students'construction career are highlighted by theoretic studies in large amount and by rearranged knowledge of working processes demonstrated in pedagogic models, multimedia materials, teaching videos, reference databases and hand-on units.

3. Curriculum Objectives

3.1 Knowledge Objectives

After learning this course, students will be able to

(1) get familiarized with construction sequences;

（2）掌握主要工种工程的施工工艺；

（3）熟悉质量检查的内容和方法。

（二）素质目标

（1）能够不断获取新的技能与知识、将习得的技能知识在各种学习和工作实际场合迁移和应用；

（2）能够合理地处理社会关系、人际关系；

（3）具有团队协作、诚实守信、职业道德的优良品质；

（4）具有良好的身体素质和心理素质。

（三）能力目标

（1）能组织简单建筑工程的施工；

（2）能解决简单施工问题；

（3）能进行建筑工程施工的质量验收。

四、课程内容与学时分配

教学内容与学时安排表

序号	教学情境/任务/项目/单元	教学目标	教学内容及学时分配			
			知识点	学时	实践项目	学时
1	施工基本知识	理解建筑施工的基本概念	1. 基本建设程序，施工招标投标	2		
			2. 施工规范体系			
			3. 分部分项工程的划分，建筑施工程序			
2	土方工程施工	了解土方工程施工的特点，能进行土方工程量的计算；掌握土方开挖与回填技术；熟悉土壁支护的类型；了解土壁的稳定分析；了解施工降水与排水；熟悉常用的施工机械；了解地基加固的方法	4. 土方的种类，工程性质，施工特点	10		
			5. 土方工程计算（★）			
			6. 土方开挖与验槽			
			7. 土壁支护与稳定			
			8. 降水与排水；回填与压实（★）			

(2) master construction technology of major trades in construction projects;

(3) get familiarized with the contents and approaches of quality inspection.

3.2 Quality Objectives

After learning this course, students will have the quality of

(1) acquiring new skills and knowledge and applying them in various studying and working occasions;

(2) establishing proper social and interpersonal relationships;

(3) cooperation, teamwork, integrity and work ethics;

(4) good physical health and balanced mind state.

3.3 Ability Objectives

After learning this course, students will have the ability of

(1) organizing construction projects of simple buildings;

(2) solving simple problems in construction process;

(3) carrying out inspection for quality acceptance of constructional projects.

4. Curriculum Contents and Teaching Hours

Table of Teaching Contents and Teaching Hours

Item	Topic	Teaching Objectives	Lesson Contents	Hrs	Student Activities	Hrs
1	Fundamentals of building construction	Learn fundamental concepts in building construction.	1. Usual sequence of construction and bidding for construction projects	2		
			2. The system of construction regulations			
			3. The division of elements and sub-elements of construction project and construction sequences			
2	Earthwork	Learn features of earthwork and volume calculation of earthwork; Master skills in excavation and backfilling. Get familiarized with types of soil wall support; Learn to analyze stabilization of soil walls; Learn water depression and drainage in construction; Get familiarized with usual construction machinery. Learn ways to reinforce foundation.	4. Types of earthwork, character of construction and features of operation	10		
			5. Calculation of volume of earthwork (★)			
			6. Excavation and inspection of foundation subsoil			
			7. Soil wall support and stabilization			
			8. Water depression and drainage, as well as backfilling and compacting (★)			

序号	教学情境/任务/项目/单元	教学目标	教学内容及学时分配			
			知识点	学时	实践项目	学时
3	地基与基础工程施工	了解各种地基处理方法、原理及施工工艺;了解钢筋混凝土预制桩的预制、起吊、运输及堆放方法;熟悉浅埋式基础的分类和施工要点;熟悉预制桩、灌注桩质量检验标准,掌握锤击法施工的全过程和施工要点(打桩设备、打桩顺序、方法和质量控制);掌握干作业成孔灌注桩、人工挖孔灌注桩、泥浆护壁灌注桩和套管成孔灌注桩的施工工艺、要点和质量控制方法	9. 地基处理与加固	6		
			10. 浅基础工程施工			
			11. 深基础工程施工(★)			
4	砌体结构施工	熟悉脚手架及施工机械;了解砌筑准备工作,掌握基础、主体施工方法与要求;熟悉施工安全技术;了解常见质量问题与施工及其处理;进行砌体结构施工方案编制和砌筑实训	12. 脚手架及施工机械	8		
			13. 砌筑材料准备			
			14. 基础分部工程施工方法与要求(★)			
			15. 主体分部工程施工方法与要求(★)			
			16. 施工安全技术			
			17. 常见质量问题与事故及其处理			

continued

Item	Topic	Teaching Objectives	Lesson Contents and Teaching Hours			
			Lesson Contents	Hrs	Student Activities	Hrs
3	Groundwork and foundation work	Learn varied methods to treat the ground, as well as theories and technology in groundwork; Learn prefabrication, hoisting, delivery and storage of premoulded reinforced concrete piles; Get familiarized with types and key operation of shallow foundation; Get familiarized with quality inspection standards of precast piles and filling piles; Learn the whole process and key operation of hammer driving(the driving plant, the sequence, methods and quality control); Master the operation, keys and quality control of grouting piles with holes formed by dry work, digging or pipe casing and grouting piles with slurrysupported soil wall	9. The treatment and strengthen of ground	6		
			10. Construction of shallow foundation			
			11. Construction of deep foundation(★)			
4	Masonry structure work	Get familiarized with scaffold and construction machinery; Learn masonry preparation, construction technology and requirements of foundations and main buildings; Get familiarized with safety technologyin construction projects; Learn usual quality problems and solutions in construction projects; Compile construction plans for masonry structures and practice mason skills	12. Scaffold and construction machinery	8		
			13. Preparation of masonry materials			
			14. Construction technology and requirements in the construction elements of foundation(★)			
			15. Construction technology and requirements in construction elements of main building			
			16. Safety technology in construction projects			
			17. Usual quality problems, accidents and their solutions			

序号	教学情境/任务/项目/单元	教学目标	教学内容及学时分配			
			知识点	学时	实践项目	学时
5	钢筋混凝土结构施工	了解模板的类型;熟悉模板的设计要求,掌握模板安装与拆除方法;了解钢筋现场检验要求;熟悉钢筋加工过程,掌握钢筋施工方法,掌握混凝土过程施工过程和施工方法;熟悉模板、钢筋和混凝土质量控制及标准;了解混凝土结构过程施工安全技术,能编制现浇多层钢筋混凝土结构施工方案;进行模板、钢筋实训	18. 模板过程施工	10	参观模型馆	
			19. 钢筋过程施工(★)			
			20. 混凝土过程施工(★)			
			21. 混凝土过程质量要求和安全技术			
6	预应力混凝土工程施工	了解先张法、后张法施工工艺与要求;了解无黏结预应力混凝土施工工艺和预应力混凝土结构施工;了解常见质量问题	22. 先张法(★)	4		
			23. 后张法:有黏结预应力混凝土施工(★)			
			24. 无黏结预应力混凝土结构施工(★)			
			25. 常见质量问题及处理			
			26. 施工安全技术			

continued

Item	Topic	Teaching Objectives	Lesson Contents and Teaching Hours			
			Lesson Contents	Hrs	Student Activities	Hrs
5	Reinforced concrete structures work	Learn types of moulds; Get familiarized with requirements of mould design; Master the instalation and dismantling of moulding boards; Learn requirements for insite tests of steel; Understand the processing of steel and master skills in steel work; Master the process and technology of concrete work; Get familiarized with quality standards and quality control of moulds, steel and concrete; Learn safety technology in concrete structure process; Compile construction schemes of multistory reinforced cast-in-situ concrete structures; Practice on moulding and steel work	18. Process of moulding work	10	Visit model rooms	
			19. Process of reinforcement work (★)			
			20. Process of concrete work (★)			
			21. Quality requirements and safety technology for concrete work			
6	Prestressed concrete construction works	Learn building operation and requirements of pretensioning and post-tensioning; Learn building operation of nonadhesive prestressed concrete and the construction of prestressed concrete structure; Learn about usual quality problems	22. Pretensioning(★)	4		
			23. Posttensioning: construction of adhesive prestressed concrete (★)			
			24. Construction of non-adhesive prestressed concrete(★)			
			25. Usual quality problems and solutions			
			26. Constructional safety technology			

序号	教学情境/任务/项目/单元	教学目标	教学内容及学时分配			
			知识点	学时	实践项目	学时
7	结构安装工程施工	熟悉混凝土结构安装的周边工作;掌握安装施工工艺和结构安装方法;了解起重机械选择;了解安装安全技术;了解常见质量问题及其处理;进行混凝土结构施工施工方案的编制	27. 起重设备的类型和选择	4		
			28. 构件的吊装(★)			
			29. 单层工业厂房的安装(★)			
			30. 多层混凝土结构的安装			
8	防水工程施工	熟悉屋面防水的类型、构造、掌握卷材防水屋面的施工工艺与方法;熟悉常见质量问题及其处理;了解地下防水方案;掌握地下防水各种施工方法及其施工工艺;了解堵漏技术,进行防水工程施工方案编制	31. 屋面防水的类型、构造、施工工艺与方法,常见施工质量问题及其处理(★)	6		
			32. 地下防水方案、施工工艺与方法(★)			
			33. 常见施工质量问题及其处理;堵漏技术,防水工程施工方案编制			
9	装饰工程施工	装饰工程施工	34. 抹灰工程施工(★)	6		
			35. 门窗工程施工(★)			
			36. 饰面工程施工			
			37. 楼地面工程施工			
			38. 涂料工程施工			
			39. 幕墙与吊顶、隔墙工程施工			
			40. 常见质量问题与处理			
合计			56			
备注:重要知识点标注★						

continued

Item	Topic	Teaching Objectives	Lesson Contents and Teaching Hours			
			Lesson Contents	Hrs	Student Activities	Hrs
7	Structure and installation	Get familiarized with works concerned in concrete structure installation; Master installation operation and structure installation. Learn about hoisting choices; Learn safety technology in installation. Learn usual quality problems and solutions; Compile construction scheme of concrete structures	27. Types of hoisting choices	4		
			28. Hoisting of members(★)			
			29. Installation of single story factories			
			30. Installation of multistory concrete structure			
8	Waterproofing process	Learn types and structures of waterproofing roofs; Learn operation and methods of applying waterproofing membrane to roofs; Get familiarized with usual quality problems and their solutions; Master operations of underground waterproofing processes; Learn leaking stoppage and compile construction scheme of waterproofing projects	31. Types, structures, construction processes and methods of waterproofing roofs; Usual quality problems and solutions in the construction processes(★)	6		
			32. The construction plan, processes and methods of underground waterproofing projects(★)			
			33. Usual quality problems and their solutions. Leaking stoppage. The compiling of waterproofing construction schemes			
9	Operation in decoration process	Operation in decoration process	34. Operation in masonry process(★)	6		
			35. Operation in door and window process(★)			
			36. Operation in finishing process			
			37. Operation in floor process			
			38. Operation in coating process			
			39. Operation of screen, ceiling and partition			
			40. Usual quality problems and their solutions			
	Total		56			
	Notes: Mark the important points with ★					

五、教学实施建议

(一)教学参考资料

李辉、黄敏.《建筑施工技术》(第二版),重庆大学出版社,2017。

赵育红.《建筑施工技术》(第四版),电力出版社,2015。

其他教辅资料:

江正荣.《建筑施工计算手册》(第四版),中国建筑工业出版社,2018。

(二)教师素质要求

由教学和工程实践经验丰富的双师素质的教师担任教学工作。

(三)教学场地、设施要求

可在多媒体教室进行教学,同时可以在工程现场以及本院的建筑模型体验馆开展教学。

六、考核评价

成绩构成比例原则按《四川建筑职业技术学院学生学业考核办法》执行。

5. Teaching Suggestions

5.1 Teaching Resources

Li Hui, Huang Min. *Building Construction Technology.* 2nd Edition. Chongqing University Press, 2017.

Zhao Yuhong. *Building Construction Technology.* China Electric Power Press, 2015.

Other resources:

Jiang Zhengrong. *Calculation Manual of Building Construction.* 4th Edition. China Architecture & Building Press, 2018.

5.2 Teachers' Qualification

Teachers of the course should be richly experienced in both educational and engineering practice.

5.3 Teaching Facilities

The course can be conducted in both multimedia classrooms and construction sites, as well as construction model rooms on campus.

6. Evaluation/Examination

Score assignment in the examinations follows the *Regulation of Academic Evaluation for Students in SCAT.*

四 川 建 筑 职 业 技 术 学 院

课 程 标 准

课程名称： 建筑识图

课程代码： 010557

课程学分： 1.5

基本学时： 24

适用专业： 工程造价（中英合作办学项目）

执 笔 人： 刘 觅

编制单位： 制图教研室

审 核： 土木工程系（院）（签章）

批 准： 教务处（盖章）

Sichuan College of Architectural Technology

Curriculum Standard

Curriculum Name: Understanding Construction Drawings

Curriculum Code: 010557

Credits: 1. 5

Teaching Hours: 24

Applicable Specialty: Construction Cost

(Sino-British Cooperative Program)

Complied by: Liu Mi

Edited by: Teaching & Research Section of Construction Drafting

Reviewed by: The Department of Civil Engineering

Approved by: Teaching Affairs Office

《建筑识图》课程标准

一、课程概述

工程图纸是工程师的通用语言。《建筑识图》课程主要学习建筑工程图样的基本原理和原则,学习《房屋建筑工程统一制图标准》《建筑制图标准》《总图制图标准》和《建筑结构制图标准》的相关规定和要求,系统介绍绘制和阅读建筑工程图样的方法,是一门实践性很强的专业基础课程。本课程适用高职3年制工程造价(中英合作办学项目)专业在第1学期开设,其先修课程为《画法几何》,后续课程包括《建筑结构》《建筑 CAD》和《建筑施工技术》等核心专业基础课程和专业课程。

二、课程设计思路

本课程标准依据《四川建筑职业技术学院人才培养方案》制定。《建筑识图》其教学的主要作用是培养学生运用制图工具,按照国家制图标准绘制和阅读建筑工程图样的能力,培养学生工程形体想象、逻辑思维和平面图示的能力,旨在为后续专业课程的学习奠定基础。在教学活动中,以教师讲授和图示为主要手段,同时应该为学生动手练习留出足够的学时。

三、课程目标

学生通过各项识图任务的完成,熟悉一般建筑施工图的图示特点及内容,掌握建筑施工图一般的作图步骤,掌握一般建筑施工图的阅读方法步骤,识别建筑施工图常用的符号、建筑材料、总平面图、建筑构造及配件图例;熟练识读钢筋混凝土梁、板、柱的配筋图的内容与图示方法、熟练识读基础图和楼层结构布置图;能找出图纸自身的缺陷和错误;能检查施工图纸的功能设计是否满足建设单位的要求等。

Understanding Construction Drawings
Curriculum Standard

1. Curriculum Description

Engineering drawing is the way engineers communicate with each other. In the course of *Understanding Construction Drawings*, fundamental theories and principles relevant to construction drawings are the main contents, as well as the provisions and requirements of *Unified standards for building drawings*, *Standard for architectural drawings*, *Standard for general layout drawings* and *Standards for structural drawings*. As one of the major-specified fundamental courses, *Understanding Construction Drawings* systematically introduces methods of preparing and reading construction drawings, and puts emphases on hand-on practice. This course is offered in the first term for students enrolling in the specialty of Construction Cost (Sino-British Cooperative Program), a 3-year vocational program. The delivered courses are *Descriptive Geometry*. What follow up are major-specified core courses and other courses like *Building Structure*, *Construction CAD*, *Building Construction Technology*, etc.

2. Curriculum Design Ideas

The curriculum standard is based on *The Professional Training Program for the Specialty of Construction Cost*. Major function of the course is to foster students'ability to prepare construction drawings with proper tools in accordance with national construction drawing standards and to read those drawings. It will also boost students'imagination of constructional objects, their logical thinking and graphic illustration, laying foundation for more major-specified units. In the teaching process, teacher's instruction and illustration in classrooms is the basis, as well as leaving enough time for students'hand-on practice.

3. Curriculum Objectives

The objective of this course is to help students, through various reading tasks of drawings, learn the graphical features and contents of common construction drawings, as well as the steps of preparing and reading them. After learning this course, students should be able to identify usual annotations on construction drawings, building materials, master plans and marginal data on drawings of building composition and fittings. They should also be efficient in preparing and reading reinforcement drawings of reinforced concrete beams, slabs and columns, and in reading foundation drawings and structural floor plans. They will be able to figure out defects and errors on working drawings as well as to inspect functional design of construction drawings against requirements of the construction client.

(一) 知识目标

(1) 掌握基本建筑制图标准;

(2) 了解建筑各部分的基本构造;

(3) 掌握施工图的形成方式及阅读方法;

(4) 掌握建筑施工图一般的绘图方法。

(二) 素质目标

(1) 培养辩证思维的能力;

(2) 具有严谨的工作作风和敬业爱岗的工作态度;

(3) 遵纪守法,自觉遵守职业道德和行业规范。

(三) 能力目标

(1) 能独立识读常见工程类型的建筑施工图;

(2) 能在没有教师直接指导的情况下,找出图纸自身的缺陷和错误;

(3) 学生能在小组协同讨论的情况下,检查施工图纸的功能设计是否满足建设单位的要求;

(4) 学生能在没有教师直接指导的情况下,配合其他工种进行施工图会审。

四、课程内容与学时分配

教学内容与学时安排表

序号	教学情境/任务/项目/单元	教学目标	教学内容及学时分配			
			知识点	学时	实践项目	学时
1	制图的基本知识与技能	掌握制图工具和仪器的使用方法;了解绘图的一般步骤及要求;熟练掌握有关国家制图标准的基本规定	1. 制图工具和仪器的使用方法	2	1. 线型图例练习	2
			2. 绘图的一般步骤及要求			
			3. 有关国家制图标准的基本规定(★)			

3. 1 Knowledge Objectives

After learning this course, students will be able to

(1) Understand standards applied in preparing construction drawings;

(2) Learn structures of basic components of a building;

(3) Master methods of compiling and reading construction drawings;

(4) Master usual methods of preparing construction drawings.

3. 2 Quality Objectives

After learning this course, students will have the quality of

(1) critical thinking;

(2) rigorous work styles and work attitudes out of respect and dedication;

(3) compliance with laws and regulations, and conscious observation of work ethics and industry norms.

3. 3 Ability Objectives

After learning this course, students will have the ability of

(1) reading construction drawings of usual projects independently;

(2) figuring out the defects and errors of construction drawings without direct tutoring;

(3) inspecting functional design of construction drawings against requirements of the construction clients in team work;

(4) cooperating with other technicians to make a joint check up on the blue prints for a project independently.

4. Curriculum Contents and Teaching Hours

Table of Teaching Contents and Teaching Hours

| Item | Topic | Teaching Objectives | Lesson Contents and Teaching Hours | | | |
			Lesson Contents	Hrs	Student Activities	Hrs
1	Fundamentals and skills in preparing construction drawings	Master the operation of drafting tools and devices; Learn the steps and requirements of construction drawings preparation; Master basic rules of relevant national drafting standards	1. Operation of drafting tools and devices	2	Practice on lines and legends	2
			2. Steps and requirements of drafting			
			3. Basic rules of relevant national drafting standards(★)			

续表

序号	教学情境/任务/项目/单元	教学目标	教学内容及学时分配			
			知识点	学时	实践项目	学时
2	建筑施工图识读	掌握建施图(施工图首页、平面图、立面图、剖面图、建筑详图)的内容和识读方法	4. 房屋的组成及其作用 5. 施工图的产生及其分类 6. 施工图的内容和图示特点 7. 建筑施工图中常用的符号(★) 8. 建筑总平面图的内容和识读方法(★) 9. 建筑平面图的内容和识读方法(★) 10. 建筑立面图的内容和识读方法(★) 11. 建筑剖面图的内容和识读方法(★) 12. 建筑详图的内容和识读方法	8	2. 建筑施工图综合训练	4
3	结构施工图识读	掌握结施图(结构设计说明、基础图、楼层及屋面结构布置图、构件详图)的内容和识读方法	13. 常见的结构形式 14. 结构施工图的内容和图示特点 15. 常用结构构件代号及常用图例(★) 16. 钢筋混凝土构件详图(★) 17. 平面整体表示方法简介	6	3. 结构施工图综合训练	2
	合计			16		8
备注:重要知识点标识★						

五、教学实施建议

(一)教学参考资料

李翔、宋良瑞、张翔.《建筑识图与实务》,高等教育出版社,2014。

continued

Item	Topic	Teaching Objectives	Lesson Contents and Teaching Hours			
			Lesson Contents	Hrs	Student Activities	Hrs
2	Construction drawing reading	Master the contents of construction drawings (Including the cover page, plans, elevations, sections and details) and the way to read them	4. The composition and function of houses	8	2. Comprehensive practice on construction drawings	4
			5. Preparation and classification of construction drawings			
			6. Contents and graphic marks of construction drawings			
			7. Common marks on construction drawings(★)			
			8. Contents of master plans and the way to read them(★)			
			9. Contents of plans and the way to read them(★)			
			10. Contents of elevations and the way to read them(★)			
			11. Contents of sections and the way to read them(★)			
			12. Contents of details and the way to read them			
3	Readstructural working drawings.	Master contents of structural working drawings (Including the explanation of the design, foundation plan, structural designs of floors and roofs and details of members) and the way to read them	13. Common structures	6	Comprehensive practice on structural working drawings	2
			14. Contents and graphic marks of structural working drawings			
			15. Common codes and legends of structural members(★)			
			16. Details of reinforced concrete members(★)			
			17. Introduction to ichnographic representation method			
Total				16		8
Notes: Mark the important points with ★						

5. Teaching Suggestions

5.1 Teaching Resources

Li Xiang, Song Liangrui, Zhang Xiang. *Architectural Drawings Reading and Practice.* Higher Education Press, 2014.

李翔.《施工图识读与会审》,高等教育出版社,2010。

吴启凤.《建筑工程制图》(第2版),西南交通大学出版社,2012。

(二)教师素质要求

本课程授课教师应具有本课程或相关课程的教师资格,具有工程类专业背景,有良好的课堂管理组织和建筑工程图识图、绘图能力。

(三)教学场地、设施要求

多媒体教室或绘图教室。

六、考核评价

本课程宜以闭卷考试方式进行课程考核,成绩构成比例按《四川建筑职业技术学院学生学业考核办法》执行,各教学项目考核成绩比例见下表:

序号	考核内容	成绩构成比例/%
1	单元一　制图的基本知识与技能	25
2	单元二　建筑施工图识读	50
3	单元三　结构施工图识读	25
合　计		100

Li Xiang. *Read and Makejoint check-ups on Construction Drawings.* Higher Education Press, 2010.

Wu Qifeng. *Prepare Construction Drawings.* 2nd Edition. Southwest Jiaotong University Press, 2012.

5. 2 Teachers'Qualification

Teachers of this course should be pedagogically ready in the specialty or relevant specialties. They should be professionals in construction and engineering, with excellence in class organization and management and competence in reading and preparing construction drawings.

5. 3 Teaching Places & Facility Requirement

Multimedia classrooms or drafting rooms

6. Evaluation/Examination

This course should be evaluated in closed-book examinations whose scoring follows *Regulation of Academic Evaluation for Students in SCAT.* Score distribution among topics of this course is as follows:

Item	Topics to be evaluated	Percentage
1	Topic 1 Fundamentals and skills in drafting	25
2	Topic 2 Read construction drawings	50
3	Topic 3 Read structural working drawings	25
Total		100

四川建筑职业技术学院

课 程 标 准

课程名称：　　　　工程造价　　　　

课程代码：　　　　150516　　　　

课程学分：　　　　2　　　　

基本学时：　　　　32　　　　

适用专业：　工程造价(中英合作办学项目)

执 笔 人：　　　　侯 兰　　　　

编制单位：　　建筑工程造价教研室　　

审　　核：　　工程管理系(院)(签章)　

批　　准：　　　教务处(盖章)

Sichuan College of Architectural Technology

Curriculum Standard

Curriculum Name: Introduction to Construction Cost

Curriculum Code: 150516

Credits: 2

Teaching Hours: 32

Applicable Specialty: Construction Cost

(Sino-British Cooperative Program)

Compiled by: Hou Lan

On Behalf of: Teaching & Research Section of Construction Cost

Reviewed by: The Department of Engineering Management

Approved by: Teaching Affairs Office

《工程造价概论》课程标准

一、课程概述

本课程是一门理论性相对较强的课程,是工程造价专业相关知识的导入课程。本课程是工程造价专业的专业基础课程,该课程包括工程造价专业导论、工程造价计价原理、工程造价组成、工程造价计价定额分析应用四个部分内容。本课程的先修课程有《建筑施工工艺》《建筑装饰工程材料》等,本课程的后续课程有《建筑工程预算》《工程量清单计价》等。

二、课程设计思路

《工程造价概论》是工程造价专业第一门涉及造价知识的课程,故授课内容包括工程造价专业介绍、专业发展现状等。该门课程内容应搭建工程造价专业理论知识框架,让学生对工程造价计算方法在理论上有整体认识。同时,计价依据是工程造价原理中的重要内容,故该课程包含了工程造价计价定额的分析与使用的内容。该课程主要利用多媒体等公共教学资源来组织教学。

三、课程目标

学完本课程,学生应了解工程造价专业,熟悉工程造价计价方式,掌握工程造价组成、工程造价计价原理、工程计价定额组成及应用方法;形成定额识读和定额应用能力,且能识读和应用其他定额;养成分析问题和解决问题的习惯。

(一)知识目标

(1)了解工程造价专业、工作岗位设置、工作岗位的主要工作内容、工程造价行业现状;

(2)掌握工程造价的概念;

(3)熟悉工程造价两种计价方式的异同;

(4)掌握工程造价组成;

(5)掌握工程造价计价程序和计价原理;

(6)掌握工程造价计价定额的组成;

Introduction to Construction Cost Curriculum Standard

1. Curriculum Description

This course is the theoretical basis of the specialty of construction cost and it is the introductory course to the professional knowledge for the specialty of construction cost. It is the basic professional course, which includes general introduction to construction cost, basic principles of construction cost pricing, construction cost components and construction cost pricing quota analysis and application. The courses delivered are *Building Construction Technology*, *Building and Decoration Materials* and so on. The units followed are *Building Construction Budget*, *Pricing for Bills of Quantities* and so on.

2. Curriculum Design Ideas

This is the first professional course for students to get basic knowledge of construction cost, so it should include the content of the introduction to this specialty, the development history and the promising future for this specialty. The purpose of this course is to give a framework for theoretical knowledge of this specialty and make the students have an overall understanding of the calculation method of construction cost in theory. The basis of pricing is an important part of the construction cost, so it includes construction cost pricing quota analysis and application. This course mainly uses public teaching resources such as multimedia.

3. Curriculum Objectives

Through this course, students will have basic knowledge of the specialty of construction cost, be familiar with the calculation method of construction cost, and master the knowledge of construction cost component, theoretical basis of construction cost pricing, construction cost pricing quota component and application; they also will have the ability to read and apply construction cost pricing quota and other quotas; they will develop the habit of analyzing and solving problems.

3.1 Knowledge Objectives

After learning this course, students will be able to

(1) understand the basic job responsibilities of the specialty of construction cost and the current status of construction cost industry;

(2) master the concept of construction cost;

(3) master the similarities and differences of two pricing modes in construction cost;

(4) master the components of construction cost;

(5) master the process and theory of construction cost pricing;

(6) master the components of construction cost quota;

（7）掌握工程造价计价定额的应用原理。

(二)素质目标

（1）培养学生对专业的热爱、对工程造价岗位的热爱，为培养高尚的职业道德打下基础；

（2）培养学生的组织、协调能力；

（3）培养学生分析问题、解决问题的能力；

（4）培养学生的学习能力。

(三)能力目标

（1）能在没有老师直接指导下完成定额分析工作；

（2）能在没有老师直接指导下完成定额识读，且能举一反三的识读其他定额；

（3）能在没有老师直接指导下完成定额应用，且能举一反三的应用其他定额。

四、课程内容与学时分配

教学内容与学时安排表

序号	教学情境/任务/项目/单元	教学目标	教学内容及学时分配			
			知识点	学时	实践项目	学时
1	工程造价专业概述	了解工程造价专业及行业动态	1. 工程造价专业简介	4		
			2. 工程造价岗位工作简介			
			3. 工程造价就业单位简介			
			4. 工程造价职业资格及考试简介			
			5. 工程造价行业发展简介			

(7) master the application theory of construction cost quota.

3.2 Quality Objectives

After leaning this course, students will have the quality of

(1) continuous passion for this specialty and the vocational posts, and the solid foundation for developing their noble professional ethics;

(2) organization and cooperation;

(3) analyzing problems and solving problems;

(4) continuous learning.

3.3 Ability Objectives

After learning this course, students will have the ability of

(1) finishing the analysis of quota by themselves;

(2) reading the quota by themselves and applying it into the reading of other quotas;

(3) applying the quota by themselves and learning and applying other quotas.

4. Curriculum Contents and Teaching Hours

Table of Teaching Contents and Teaching Hours

Item	Topic	Teaching Objectives	Teaching Contents and Teaching Hours			
			Lesson Contents	Hrs	Student Activities	Hrs
1	Introduction to the major of construction cost	Understand the major of construction cost and the development trend of construction cost industry	1. Introduction to the major of construction cost	4		
			2. Introduction to different jobs for the major of construction cost			
			3. Introduction to the potential working enterprises for this major			
2	The basic principle of construction cost pricing	Be familiar with the basic principle of construction cost	4. Introduction to professional qualification certificate of construction cost and the examination for it	6		
			5. Introduction to the development of construction cost industry			

序号	教学情境/任务/项目/单元	教学目标	教学内容及学时分配			
			知识点	学时	实践项目	学时
2	工程造价计价原理	熟悉工程造价的基本原理	6. 工程造价的含义	6		
			7. 工程造价计价模式（★）			
			8. 两种计价模式的异同			
			9. 工程造价计算的依据简介			
			10. 工程造价计算的程序			
3	工程造价组成	掌握工程造价组成及理论计算方法	11. 工程造价组成（第一种含义、第二种含义）	6	工程造价计算案例分析	4
			12. 工程造价组成——44号文（★）			
			13. 工程造价理论计算方法（★）			
4	工程造价计价依据	掌握工程造价计价定额分析方法；掌握工程造价计价定额的应用方法	14. 工程造价计价依据概述	12		
			15. 工程造价定额组成			
			16. 工程造价定额分析（★）			
			17. 工程造价定额应用（★）			
合计				28		4
备注：重要知识点标识（★）						

五、教学实施建议

（一）教学参考资料

四川省建设工程造价管理总站.《四川省建设工程工程量清单计价定额》,中国计划出版社,2015。

袁建新、袁媛.《工程造价概论》(第三版),中国建筑工业出版社,2016。

continued

Item	Topic	Teaching Objectives	Teaching Contents and Teaching Hours			
			Lesson Contents	Hrs	Student Activities	Hrs
2	The basic principle of construction cost pricing	Be familiar with the basic principle of construction cost	6. The concept of construction cost	6		
			7. The basic mode of construction cost pricing(★)			
			8. The similarities and differences of two construction cost pricing modes			
			9. Introduction to the basis of construction cost pricing			
			10. The process of construction cost pricing			
3	The components of construction cost	Master the components of construction cost and the calculation methods for it in theory	11. The components of construction cost(concept one and two)	6	The analysis of practical construction cost project	4
			12. The components of construction cost(according to code 44)(★)			
			13. The calculation methods for construction cost in theory(★)			
4	The basis for construction cost pricing	Master the analysis methods construction cost pricing quota; master the application methods of construction cost pricing quota	14. Introduction to the basis of construction cost pricing	12		
			15. Introduction to the components of construction cost pricing quota			
			16. The analysis of construction cost quota(★)			
			17. The application methods of construction cost pricing quota(★)			
Total				28		4
Note: Mark the important points with ★						

5. Teaching Suggestions

5.1　Teaching Resources

Construction Cost Management Station in Sichuan Province. *Pricing for Bills of Quantities Quota in Construction Engineering of Sichuan Province*. China Planning Press, 2015.

Yuan Jianxin, Yuan Yuan. *Overview of Cost Engineering for Projects*. 3rd Edition. China Architecture & Building Press, 2016.

(二) 教师素质要求

工程造价专业课老师应具备教师资格证,应为工程造价专业或相关专业人才具有相应的执业资格和一定的社会实践和专业实践能力,有一定的专业教学经验。

(三) 教学场地、设施要求

实施该课程教学,无须特殊环境,一般多媒体教室即可。

六、考核评价

(一) 考核方式

闭卷(可以带定额)。

(二) 考核说明

总评成绩的构成比例原则按《四川建筑职业技术学院学生学业考核办法》执行:平时:期末 = 3:7。

考核重点应覆盖的范围:工程造价基础知识、建筑工程定额分析与应用、工程造价计算原理、建筑安装工程费用构成(相关文件)。

考核的原则要求:以是否掌握工程造价计价原理、工程定额的分析与应用为考核重点,理论和动手能力均需培养。

5. 2 Teachers'Qualification

Teachers for this course should have the teachers'qualification certificate; they should have the education background of construction cost or other relative majors, have the relative qualification certificates and certain social or professional practice and they should have some professional teaching experience.

5. 3 Teaching Facilities

This course can be delivered in common classrooms equipped with multimedia.

6. Evaluation

6. 1 Evaluation Method

Examination without reference materials(except the Quota).

6. 2 Evaluation Instruction

The score of the course is based on the *Regulation of Academic Evaluation for Students in SCAT*. The daily performance accounts for 30% of the total score and the final examination accounts for 70%.

The focus of the evaluation is the basic knowledge of construction cost, the analysis and application of construction engineering quota, the basic principle of construction cost pricing, the components of installation construction cost.

The principle for the evaluation: the main intention is to test whether the students master the knowledge of construction cost theory and the analysis & application of engineering quota or not. The evaluation should be based on students'theoretical and practical ability.

四 川 建 筑 职 业 技 术 学 院

课 程 标 准

课程名称：　　　建筑工程项目管理

课程代码：　　　150224

课程学分：　　　2

基本学时：　　　32

适用专业：　工程造价（中英合作办学项目）

执 笔 人：　　　李婧妮

编制单位：　建筑工程管理教研室

审　　核：　工程管理系（院）（签章）

批　　准：　　教务处（盖章）

Sichuan College of Architectural Technology

Curriculum Standard

Curriculum Name: Construction Project Management

Curriculum Code: 150224

Credits: 2

Teaching Hours: 32

Applicable Specialty: Construction Cost

    (Sino-British Cooperative Program)

Compiled by: Li Jingni

On Behalf of: Teaching & Research Section of Project Management

Reviewed by: The Department of Engineering Management

Approved by: Teaching Affairs Office

《建筑工程项目管理》课程标准

一、课程概述

本课程是为中英合作办学工程造价专业开设的一门专业基础课,在大二上学期开设。本课程应该建立在《识图与构造》《建筑工程施工技术》《招投标与合同管理》《建筑工程预算》等课程的学习基础上,要求学生对法律法规、建筑构造、施工技术和工程造价等知识有一定的了解,并为后续的《建筑工程资料管理》等专业课以及毕业设计和毕业实习打下基础。

二、课程设计思路

本课程教学目标的确定是以建筑工程项目全生命周期为对象,运用项目管理的基本理论及方法,研究建筑工程项目整个运作过程。通过教学,使学生掌握建筑工程项目管理的基本理论和知识,包括建筑工程项目管理基本概念、工程项目招标投标、施工准备、工程项目组织机构、工程项目进度管理、工程项目质量管理、工程项目成本管理等内容,能够在今后的工作中参与项目管理工作。

本课程理论教学过程中要求理论学时为 28 学时,配合案例习题或讨论 4 学时。课程内容强调了工程项目管理作为一种新的管理技术,它考虑了工程项目的多种界面和复杂环境,强调了工程项目的组织形式和动态管理,由此组成的项目管理系统具有计划、组织和控制等职能,本课程以建筑工程项目为中心,综合阐述建筑工程项目从组织、计划、实施到竣工验收等全过程的管理理论和方法。教学资源除了教学资料具备学生学习本课所需要的理论知识,还配备了具体的案例工程进行深入讲解和分析。

Construction Project Management Curriculum Standard

1. Curriculum Description

This course is a compulsory and professional basic one for the specialty of construction cost (Sino-British Cooperative Program) . It is delivered in the third semester after students complete those prerequisite courses: *Drawing Reading and Structure*, *Construction Engineering Technology*, *Bidding and Contract Management*, *Construction Project Budget* and so on. Through the course, students should understand knowledge, laws and regulations of construction structure, construction technology and construction cost, and lay a solid foundation for follow-up professional courses such as *Construction Project Documents Management* as well as graduation design and enterprise internships.

2. Curriculum Design Ideas

The teaching objectives are based on the whole life cycle of the construction project, applying the basic theory and method of project management to the acquisition of the whole operation process of construction project.

Through the course, students should master the basic theory and knowledge of construction project management, including the basic concept of construction project management, project tendering and bidding, construction preparation, project organization, project progress management, project quality management, project cost management, etc.

The course is required of 28 hours theoretical study, with case studies or 4 hours discussion. The teaching contents emphasize the project management as a new management technology, which takes into consideration the various interfaces and complex environments of construction projects, focus on the organization form and dynamic management of the project, and therefore the resulting project management system has functions of planning, organization and control. This course is centered on the construction project, comprehensively expounding the management theory and method of the whole process of construction projects from organizing, planning and implementing to final completion acceptance.

三、课程目标

本课程要求学生掌握建筑工程项目管理的基本规定和要求,具备独立参与建筑工程项目管理的能力,在实践中自觉遵守国家的法律法规、遵守职业道德和项目管理要求的行业规范,将理论与实际工作结合,实现学以致用。

(一)知识目标

(1)了解并掌握建筑工程项目管理的基本概念,形成建筑工程以项目形式进行运作的理念。

(2)全面掌握工程项目招标投标、施工准备、工程项目组织机构、工程项目进度管理、工程项目质量管理、工程项目成本管理等内容,能够进行建筑工程项目全过程管理。

(二)素质目标

(1)提高综合运用专业知识进行建筑工程项目管理的专业素质。

(2)提升与各单位进行协调、合作的沟通能力等综合素质。

(三)能力目标

(1)能在没有教师直接指导下独立按照《建筑工程项目管理规范》等相关专业规范从事建筑工程项目管理中各项基础工作。

(2)能在没有教师直接指导下独立按照《建筑工程项目管理规范》等相关专业规范掌握工程项目实施过程中工期、成本、质量、安全等目标控制的基本方法。

3. Curriculum Objectives

This course requires students to master the basic regulations and requirements of construction project management, and to develop the ability to independently participate in the management of construction projects. In practice, they shall consciously abide by national laws and regulations, comply with professional norms and project management requirements, and be able to apply theoretical knowledge into the actual work.

3.1 Knowledge Objectives

After learning this course, students will be able to

(1) understand and master the basic concept of construction project management, and develop the idea that construction engineering operates as projects.

(2) fully grasp the project bidding, construction preparation, project organization, project progress management, project quality management, project cost management and other content, and be able to conduct the whole process management of construction project.

3.2 Quality Objectives

After learning this course, students will have the quality of

(1) improving the professionalism of comprehensively using knowledge and skills for construction project management.

(2) improving the overall quality of communicative skills such as coordination and cooperation with various departments.

3.3 Ability Objectives

After learning this course, students will have the ability of

(1) completing independently various basic work of construction project management without the direct tutoring in accordance with the relevant professional norms such as the *Construction Engineering Project Management Regulations*.

(2) grasping independently the basic methods of target control such as duration, cost, quality and safety control during the implementation of the construction project without the direct tutoring in accordance with relevant professional norms such as the *Construction Engineering Project Management Regulations*.

四、课程内容与学时分配

教学内容与学时安排表

序号	教学情境/任务/项目/单元	教学目标	教学内容及学时分配			
			知识点	学时	实践项目	学时
1	第一章 工程项目管理概述	掌握项目、建筑工程项目、项目管理、建筑工程项目管理的概念和特征;掌握建筑工程项目管理的内容、方法和基本理论;了解项目的分类、工程项目管理的主体及任务,建筑工程项目建设以及施工的程序	1. 工程项目:项目的概念、特征(★),建筑工程项目的概念、特征、构成(★),建筑产品及施工生产的特点,建筑工程项目的建设程序,建筑工程项目施工程序(★)	3		
			2. 工程项目管理:建筑工程项目管理的内容、主体及分类,项目管理规划			
			3. 工程项目管理模式:平行发包,总包,施工总承包,CM 模式,PM 模式			
2	第二章 施工准备	了解如何进行施工准备,掌握施工准备的内容和施工准备的方法,从而保证施工顺利进行	4. 施工准备工作的分类、要求、内容,工程技术经济资料的准备,资源的准备,施工现场准备,季节性施工准备	1		
3	第三章 工程项目组织机构	了解组织及组织机构的基本概念;掌握工程项目组织机构的具体形式和组织机构设置的原则;了解工程项目管理组织的各种协调;了解项目经理部的作用、组成和对项目经理的要求	5. 工程项目组织:组织及组织机构的基本概念,组织机构设置的思路和原则(★),工程项目组织的形式(★)	3	1. 练习组织机构基本概念、工程项目组织机构的具体形式	1
			6. 项目经理部与项目经理			

4. Curriculum Contents and Teaching Hours

Table of Teaching Contents and Teaching Hours

Item	Topic	Teaching Objectives	Teaching Contents and Teaching Hours			
			Lesson Contents	Hrs	Student Activities	Hrs
1	Chapter 1 Introduction to Construction Project Management	Master the concepts and features of project, construction project, project management, construction project management; Master the content, method and basic theory of construction project management; Understand the classification of the project, the main body and tasks of project management, project construction and construction procedures	1. Construction project: concept and features of the project (★); features, concept, composition of the construction project (★); features of construction products and construction production; construction procedures of construction projects; operation procedures (★) 2. Project management: the content, subject and classification of construction project management; project management planning 3. Project management models: parallel contracting, total package, general construction contracting, CM model, PM model	3		
2	Chapter 2 Construction preparation	Learn how to prepare for construction, master the content of construction preparation and methods, so as to ensure the smooth progress of construction	4. Construction preparation classification, requirements, content, engineering and economic information preparation, resource preparation, construction site preparation, seasonal construction preparation	1		
3	Chapter 3 Construction project organization	Understand the basic concepts of the organization and the organizational structure; Master the specific form of the project organization and the principle of the organization establishment; Understand the various coordination of the project management organization; Understand the role, constitution, and requirements of the project manager	5. Project organization: the basic concept of organization and its structure; ideas and principles for organizational structure settings (★); the form of project organization(★) 6. Project management department and project managers	3	1. Exercise of the basic concept of the organization and the specific form of the project organization	1

序号	教学情境/任务/项目/单元	教学目标	教学内容及学时分配			
			知识点	学时	实践项目	学时
4	第四章 工程项目进度管理	了解项目进度计划的概念和种类,了解进度计划的编制方法;掌握建筑施工生产的三种组织方式;掌握流水施工原理及参数;掌握网络计划技术的原理和方法;熟悉常用的进度比较和控制方法	7. 工程项目进度管理概述:工程项目进度管理的内涵,项目进度计划的概念和种类,项目进度计划的编制,工程项目进度计划的编制方法(★)	14	2. 练习施工生产的三种组织方式、流水施工的组织、网络图的绘制及参数的计算	2
			8. 流水施工(★):建筑施工生产的三种组织方式(依次施工、平行施工、流水施工),流水施工的分类和表示方法,流水施工的主要参数(工艺参数、空间参数、时间参数),流水施工的主要方式及应用			
			9. 网络计划技术及其应用(★):网络图的基本概念,双代号网络图的基本要素及绘制、时间参数的分类及计算,单代号网络图,其他形式网络图(时标网络计划),网络计划的优化(工期优化、资源优化、费用优化)			
			10. 工程项目进度控制:工程项目进度控制的任务,工程项目进度的比较分析方法(★),工程项目进度偏差的分析(★)			

continued

Item	Topic	Teaching Objectives	Teaching Contents and Teaching Hours				
			Lesson Contents	Hrs	Student Activities	Hrs
4	Chapter 4 The process management of construction project quantity of engineering	Understand the concepts and categories of project schedules and how they are developed; Master the three organizational methods of construction production; Master the principle and parameters of flow process construction; Master the principles and methods of network planning technology; Be familiar with the common progress comparison and control methods	7. Overview of project progress management: the connotation of project progress management, the concept and types of project progress schedules, the preparation of project progress schedules, the preparation of project progress schedules(★) 8. Flow process construction (★): three organizational methods of construction production (sequential construction, parallel construction, flow construction), classification and representation of flow construction, main parameters of flow construction (technologic parameters, spatial parameters, time parameters), main methods and applications of flowing construction 9. Network planning technology and its application (★): basic concepts of network diagram, basic elements and drawing of dual-code network diagram, time parameter classification and calculation, single code network map, other forms of network map(time-scale network plan), optimization of network plan (duration optimization, resource optimization, cost optimization) 10. Project progress control: the task of project progress control, the comparative analysis method of project progress (★), the analysis of project progress deviation (★)	14	2. Exercise of the three organizational methods of construction production, the organization of flowing process construction, the drawing of network diagrams and the calculation of parameters	2

序号	教学情境/任务/项目/单元	教学目标	教学内容及学时分配			
			知识点	学时	实践项目	学时
5	第五章 工程项目质量管理	了解工程项目质量的概念和特性,熟悉工程项目质量管理的主要方法、体系和原则;掌握工程项目施工阶段的质量控制内容;熟悉工程验收阶段的质量控制内容,掌握工程质量验收结果的处理原则;了解工程质量事故的分类及特点,熟悉常见的引起工程质量事故的原因及事故处理的依据和程序	11. 工程项目质量管理概述:工程项目质量的含义、特点,工程项目质量管理的主要方法(★),工程项目质量管理体系	3		
			12. 工程项目施工阶段的质量控制:施工阶段质量控制的分类,生产要素的质量控制,施工过程的质量控制,产成品的质量保护			
			13. 工程项目验收阶段的质量控制:施工质量要收要求(★),工程质量验收项目的划分(★),工程质量验收合格的条件,工程质量验收不合格的处理			
			14. 工程质量事故处理			
6	第六章 工程项目成本管理	熟悉施工成本的构成、分类;熟悉施工成本计划的编制依据、步骤、方法;了解施工成本管理的任务、影响因素和措施;了解施工成本核算;掌握成本控制和成本分析方法	15. 施工成本管理概述:施工成本的构成和分类,施工成本管理的任务,施工成本的影响因素和管理措施	3	3. 练习成本分析和成本控制方法的计算	1
			16. 施工项目成本计划的编制			
			17. 施工成本控制(★)			
			18. 施工成本核算			
			19. 施工成本分析(★)			

continued

Item	Topic	Teaching Objectives	Lesson Contents	Hrs	Student Activities	Hrs
5	Chapter 5 Project quality management	Understand the concepts and features of engineering project quality, and familiar with the main methods, systems and principles of project quality management; Master the quality control content of the construction phase of the project; be familiar with the quality control content of the project acceptance stage and master the handling principle of the project quality acceptance results; Understand the classification and features of project quality accidents; familiar with the common causes of those accidents and the basis and procedures for handling accidents	11. Overview of project quality management: the definition and features of project quality, the main methods of project quality management (★), and the project quality management system 12. Quality control in the construction phase of the project: the classification of quality control in the construction phase, the quality control of the production factors, the quality control of the construction process, the quality protection of finished products 13. Quality control of the acceptance phase: construction quality requirements (★), division of quality acceptance items (★), qualified conditions for project acceptance, treatment of unqualified project 14. Handle the project quality accidents	3		
6	Chapter 6 Project cost management	Be familiar with the composition and classification of project costs; Be familiar with the basis, steps and methods for the preparation of the construction cost plan; Understand the tasks, influencing factors and measures of construction cost management; Understand the construction costs accounting; Master methods of cost analysis cost control	15. Overview of construction cost management: construction cost composition and classification, construction cost management tasks, construction cost factors and management measures 16. Construction project cost planning 17. Construction cost control (★) 18. Construction cost accounting 19. construction cost analysis (★)	3	Exercise of the methods of cost planning, cost control and cost analysis	1

序号	教学情境/任务/项目/单元	教学目标	教学内容及学时分配			
			知识点	学时	实践项目	学时
7	第七章 工程收尾管理	熟悉工程收尾工作,竣工验收的条件、标准和程序,项目竣工结算的程序,工程保修与回访;了解项目后评价的作用、内容、程序和评价指标	20. 项目竣工收尾,项目竣工验收,项目竣工结算,项目保修和回访	1		
合计				28		4
备注:重要知识点标注★						

五、教学实施建议

(一)教学参考资料

项建国.《建筑工程项目管理》第三版,中国建筑工业出版社,2015。

兰凤林.《工程项目管理实务》第二版,大连理工大学出版社,2014。

(二)教师素质要求

授课教师需要具备高等学校教师资格和建筑工程管理相关专业背景,能独立并较为完善地进行本课程所有理论课程教学及实践内容指导,对建筑识图、施工技术、项目管理、招标投标与合同管理等课程内容熟悉并能融会贯通。该课程理论性、实践性较强,应理论联系实践、结合现实多讲实例,增强同学的理解和掌握。同时,在授课中应结合最新的法律法规、专业规范等,补充新的内容。实际教学过程中该课程可根据学时增减有关内容。

(三)教学场地、设施要求

教学场地要求在普通教室进行,最好配合施工现场管理实践。

六、考核评价

本课程考核方式采用闭卷考试形式,期末卷面成绩占70%,平时考核成绩占30%。考核重点应覆盖工程项目管理基本概念、组织机构、流水施工、网络计划、质量成本管理等。

continued

Item	Topic	Teaching Objectives	Teaching Contents and Teaching Hours			
			Lesson Contents	Hrs	Student Activities	Hrs
7	Chapter 7 Project acceptance management	Be familiar with the finishing work of the project; the conditions, standards and procedures for the completion and acceptance of the project; the procedure of the completion and settlement of the project; the project warranty and the revisit; Understand the role, content, procedures and evaluation indicators of postproject evaluations	20. The completion of the project, the completion of the project acceptance, the completion of the project settlement, project warranty and revisit	1		
Total				28		4
Note: Mark the important points with ★						

5. Teaching Suggestions

5.1　Teaching Resources

Xiang Jianguo. *Construction Project Management*. 3rd Edition. China Architecture & Building Press, 2015.

Lan Fenglin. *Practice of Construction Project Management*. 2nd Edition. Dalian Polytechnic University Press, 2014.

5.2　Teachers'Qualification

Teachers shall have the qualification of higher education and the relevant professional background of construction engineering management. They shall independently and completely carry out all the theoretical knowledge teaching and practical training guidance of this course. They shall be familiar with the content of construction drawing reading, construction technology, project management, bidding and contract management, etc. Teachers shall combine theoretical knowledge with practical examples to enhance the students'understanding and mastery. At the same time, the latest laws and regulations and professional norms shall be instructed in the course of teaching, together with new content supplemented. In the actual teaching process, the curriculum content can be increased or decreased according to the time of study.

5.3　Teaching Facilities

This course can be delivered in common classrooms, better with the construction site practice.

6. Evaluation

The total score of the course is consisted of the daily performance(accounts for 30%) and the final examination(accounts for 70%). The final is closed-book examination.

The highlights in examination include: the basic concept of construction project management, organizational structure, flow process construction, network planning, quality and cost management and so on.

四 川 建 筑 职 业 技 术 学 院

课 程 标 准

课程名称： 建设工程招投标与合同管理

课程代码： 150033

课程学分： 2

基本学时： 32

适用专业： 工程造价（中英合作办学项目）

执 笔 人： 柳 茂

编制单位： 工程管理教研室

审 核： 工程管理系（院）（签章）

批 准： 教务处（盖章）

Sichuan College of Architectural Technology

Curriculum Standard

Curriculum Name: Bidding and Contract Management of Construction Project

Curriculum Code: 150033

Credits: 2

Teaching Hours: 32

Applicable Specialty: Construction Cost

(Sino-British Cooperative Program)

Complied by: Liu Mao

On Behalf of: Teaching & Research Section of Construction Management

Reviewed by: The Department of Engineering Management

Approved by: Teaching Affairs Office

《建设工程招投标与合同管理》课程标准

一、课程概述

《建设工程招投标与合同管理》是学院工程造价专业(中英合作办学项目)开设的专业必修课,本课程内容涉及建设工程招投标与合同两个方面的内容,是一门专业性很强的课程。

该课程的教学任务:通过本课程的学习,学生以建设工程为角度,掌握建设工程招投标与合同管理的活动规律和规则,从而初步具备解决实际问题的能力。

本课程在大二第2学期开设。之前开设《建设工程项目管理》课程。

二、课程设计思路

(1)本课程目标根据工程造价(中英合作办学项目)专业培养目标而制定。

(2)本课程目标根据知识、能力、方法三个维度设计,三个方面相互渗透,相互联系,螺旋上升,最终全面实现总目标。

(3)本课程教学内容侧重建设工程招标、建设工程投标、建设工程合同管理、建设工程索赔四个章节的讲解。

三、课程目标

(一)知识目标

通过对建设工程招标投标和合同管理的全面讲述,使学生理解并掌握建设工程招标投标的基本理论,以及施工合同的基本内容和施工合同订立、履行、违约、索赔等管理问题。

(二)素质目标

通过对建设工程招标投标与合同管理的讲解,特别是以实际建设工程进行实例分析,培养学生的专业兴趣和工作热情,便于学生就业后做好本职工作。

Bidding and Contract Management of Construction Project Curriculum Standard

1. Curriculum Description

Bidding and Contract Management of Construction Project is a compulsory course for specialized Sino-British Cooperative Program in Construction Cost. It is specified with contents concerning bidding and contracts of construction projects.

The task of this course is to promote student's professional understanding to the natures and rules in bidding and contract management of construction projects, and to enable them to get hands on in solving problems in real construction projects.

The course is offered in the 2nd terms of the 2nd college year. The course delivered prior to this one is *Construction Project Management*.

2. Curriculum Design Ideas

(1) The Curriculum standard is based on the professional training program of construction cost (Sino-British Cooperative Program).

(2) Three dimensions are taken into consideration when compiling this standard: knowledge, skill and methodology, which are interrelated and integrated to achieve the curriculum objectives in an up-spiraling manner.

(3) The course focuses on contents from the following 4 chapters: Invitation for bids of construction projects, Bidding for construction project, Contract management of construction project and Construction project claim.

3. Curriculum Objectives

3.1　Knowledge Objectives

A comprehensive account for the bidding and contract management of construction projects will enable students to understand fundamental theories in bidding for construction project, basic contents of constructional contracts and managing issues like entering into and executing construction contracts, breaching from them and lodging claims.

3.2　Ability Objectives

Studying this course will enable students to compile tender documents, to prepare construction contracts, and to file and calculate construction claims. They will be ready in solving problems concerning biding and contract management of real construction projects.

(三)能力目标

通过学习,学生能够编写招标投标文件、拟订施工合同、进行施工索赔计算,能够初步具备解决建设工程招投标与合同管理实际问题的能力。

四、课程内容与学时分配

教学内容与学时安排表

序号	教学情境/任务/项目/单元	教学目标	知识点	学时
1	绪论	1. 了解我国招标投标制度的历史沿革、建设工程市场概述、建设工程承发包概述; 2. 掌握建设工程招标投标概述	1. 我国招标投标制度的历史沿革 2. 建设工程市场概述 3. 建设工程承发包概述 4. 建设工程招标投标概述(★)	4
2	建设工程施工招标	1. 了解建设工程施工招标; 2. 掌握招标准备阶段的工作、招标阶段的工作、决标成交阶段的工作	5. 建设工程施工招标概述 6. 招标准备阶段的工作(★) 7. 招标阶段的工作(★) 8. 决标成交阶段的工作(★)	8
3	建设工程施工投标	1. 了解建设工程施工投标; 2. 掌握投标文件的编制、投标报价的策略与技巧	9. 建设工程施工投标概述 10. 投标文件的编制(★) 11. 投标报价的策略与技巧(★)	4
4	合同法原理	1. 了解合同法基础知识、合同法、合同的基本内容和主要形式; 2. 掌握合同的订立、合同的效力、合同的履行、合同的变更、转让与终止、违约责任、合同争议的解决	12. 合同法基础知识 13. 合同法概述 14. 合同的基本内容和主要形式 15. 合同的订立(★) 16. 合同的效力(★) 17. 合同的履行(★) 18. 合同的变更、转让与终止 19. 违约责任(★) 20. 合同争议的解决	6

3.3 Quality Objectives

Students'interests in their profession and working enthusiasm shall be ignited in an introduction to biding and contract management of construction projects, especially in the case study of real projects, bringing about their faith in the profession and helping their future employment.

4. Curriculum Contents and Teaching Hours

Table of Teaching Contents and Teaching Hours

Item	Topics	Teaching Objectives	Lesson Contents	Hrs
1	Introduction	1. Learn the history of bidding regulations in China, basics of building construction market and of contracting and contract awarding; 2. Learn basics of bidding for building construction	1. The history of bidding regulations in China 2. An introduction to the building construction market 3. An introduction to contracting and contract awarding 4. An introduction to bidding for building construction(★)	4
2	Bidding for construction projects	1. Learn basics of bidding for construction project; 2. Master ways to prepare invitation for bid, and learn the bidding process and contract awarding	5. An introduction to bidding for construction projects 6. The preparation of invitation for bid(★) 7. The bidding process(★) 8. Contract awarding(★)	8
3	Tender of building construction projects	1. Learn basics of tendering for cons-truction projects; 2. Master ways to compile tender document, and pricing skills and strategies in tender	9. An introduction to tendering for construction projects 10. Compile tender document(★) 11. An introduction to pricing skills and strategies in tender(★)	4
4	Theories in Contract Laws	1. Learn basics of contract laws, the general ideas, the basic contents and major types of contract; 2. Master the skills of entering into, enforcing, executing, changing, transferring and quitting contracts; Understand responsibilities of non-compliance and the settlement of contract dispute	12. Fundamental knowledge of contract law 13. An introduction to contract law 14. Basic contents and major types of contract 15. Enter into contracts(★) 16. Effectiveness of contracts(★) 17. Execute contracts(★) 18. Change, transfer and quit contracts 19. Responsibilities of noncompliance(★) 20. Settle contract disputes	6

序号	教学情境/任务/项目/单元	教学目标	教学内容及学时分配	
			知识点	学时
5	建设工程施工合同管理	1. 了解建设工程施工合同管理、建设工程施工合同的订立； 2. 掌握《建设工程施工合同》示范文本的主要内容、建设工程施工合同的履行、违约责任的承担	21. 建设工程施工合同管理概述 22. 建设工程施工合同的订立 23.《建设工程施工合同》示范文本的主要内容（★） 24. 建设工程施工合同的履行（★） 25. 违约责任的承担（★） 26. 建设工程施工合同争议的解决	6
6	建设工程施工索赔管理	1. 了解索赔、承包商的索赔策略与技巧、反索赔、索赔案例； 2. 掌握索赔值的计算	27. 索赔概述 28. 索赔值的计算（★） 29. 承包商的索赔策略与技巧 30. 反索赔 31. 索赔案例	4
合计				32

备注：重要知识点标注（★）

五、教学实施建议

（一）教学参考资料

江怒.《建设工程招投标与合同管理》,大连理工大学出版社,2015。

其他教辅资料：

殷时奎、虞永强、陈浩文.《建设工程施工合同法律风险及防范》,中国商业出版社,2007。

姜晨光.《土建工程招投标书编制方法与范例》,化学工业出版社,2011。

（二）教师素质要求

（1）有良好的职业道德,爱岗敬业,为人师表,言传身教。

（2）工程管理专业学习经历,了解招标投标程序,熟悉建设工程施工合同,能够进行建设工程索赔案例的分析,专业知识扎实。

（3）教学方法适宜,因材施教。

（三）教学场地、设施要求

本课程教学场地为室内,不需要多媒体。

continued

Item	Topics	Teaching Objectives	Teaching Contents and Teaching Hours	
			Lesson Contents	Hrs
5	Contract management of construction projects	1. Learn about contract management of construction projects in general and about entering into contracts of construction projects; 2. Master the major contents of *Model Contract of Construction Projects*, executing contracts of construction projects, and taking responsibilities in non-compliance	21. An introduction to contract management of construction projects	6
			22. Entering into contracts of construction projects	
			23. The major content of *Model Contract of Construction Projects*(★)	
			24. Executing contracts of construction projects(★)	
			25. Responsibilities of non-compliance(★)	
			26. Settle contract disputes in construction projects	
6	Claim management in construction projects	1. Learn general ideas of claims, claiming skills and strategies of the contractor, counter-claims and cases of claim; 2. Master ways to calculate claim	27. An introduction to claims	4
			28. Calculation of claims(★)	
			29. Claiming skills and strategies of the contractor	
			30. Counter-claims	
			31. Cases of claim	
Total				32
Notes: Mark the important points with ★				

5. 1　Teaching Resources

Jiang Nu. *Bidding and Contract Management of Construction Project*, Dalian University of Technology Press, 2015.

Other resources:

Yin Shikui, Yu Yongqiang, Chen Haowen. *Protection Against Legal Risks in Construction contract*. China Commercial Publishing House, 2007.

Jiang Chenguang. *Compiling Approaches and Instances of Bidding and Tender Documents of Construction Projects*. Chemical Industry Press, 2011.

5. 2　Teachers'Qualification

(1) Work ethics with dedication to education, high moral standard and exemplification.

(2) Professional studying experience in construction management. Knowledge in bidding process. Understanding of construction contracts. Competence in case analysis of construction contract claims. Extensive and solid knowledge of the profession.

(3) Adoption of proper teaching methods in accordance to students' varied aptitudes.

5. 3　Teaching Facilities

Normal classrooms are sufficient for the curriculum. Multimedia is not necessary.

六、考核评价

(一)考核方式

建议采取闭卷考试的方式考核。

(二)考核说明

建议考核采取平时成绩占总成绩30%,期末成绩占总成绩70%的考核构成比例方式,并且案例分析、计算是本门课考核的重要题型,其中第四、五、六章为考核的重点章节。

6. Evaluation/Examination

6.1 Evaluation Approaches

Closed-book examinations are suggested.

6.2 Evaluation Details

In the total score, daily performance accounts for 30 percent while final examination, 70 percent. The examination will focus on Chapter 4, 5 and 6 with most questions requiring case analysis and calculation.

四川建筑职业技术学院

课程标准

课程名称：　　　　工程经济

课程代码：　　　　030188

课程学分：　　　　2

基本学时：　　　　32

适用专业：　　工程造价（中英合作办学项目）

执　笔　人：　　　　李　磊

编制单位：　　　建筑经济教研室

审　　　核：　　工程管理系（院）（签章）

批　　　准：　　　教务处（盖章）

Sichuan College of Architectural Technology

Curriculum Standard

Curriculum Name: Engineering Economy

Curriculum Code: 030188

Credits: 2

Teaching Hours: 32

Applicable Specialty: Construction Cost

(Sino-British Cooperative Program)

Compiled by: Li Lei

On Behalf of: Teaching & Research Section of Building Economy

Reviewed by: The Department of Engineering Management

Approved by: Teaching Affairs Office

《工程经济》课程标准

一、课程概述

《工程经济》是工程造价专业(中英合作办学项目)的专业基础课,在第4学期开设。

工程经济学是由技术科学、经济学、管理科学相互渗透融合而成的一门边缘学科,具有综合性、系统性、可预测性、实践性等特点。本课程研究各种工程技术方案的经济效益,研究各种技术在使用过程中如何以最小的投入获得预期产出,如何以最低的寿命周期成本实现产品、作业以及服务的必要功能。通过教学,学生应掌握工程经济的基本原理、方法和技能;能够研究、分析和评价各种技术实践活动,为决策层选择最优技术方案提供理论依据。

该课程需要一定的数学基础,通常以《经济数学》《计算机应用基础》《工程造价概论》为先修课程,后续课程有《造价控制》《BIM工程计价软件应用》等。

二、课程设计思路

本课程教学内容和课程体系的构建应与当前工程造价行业的发展和对工程造价专业技术人才的需求相联系,以相关职业活动来引导组织教学,按照以能力为本位、以职业实践为主线、以应用为中心构建课程体系。课程内容主要包括资金时间价值、工程经济分析基本要素、工程项目经济评价指标与方法、不确定性分析、财务评价、价值工程等几部分。

在教学过程中,将执业资格证书与职业能力所必需的理论知识点有机融入各教学单元,突出应用性、实践性原则。按照职业岗位的要求,在理论学习的基础上着重训练学生的岗位工作技能,强化应用理论知识解决实际问题的能力。通过完整的训练,帮助学生实现知识整合贯穿与职业能力的全面提高,给学生未来工作岗位能力的形成打下坚实的基础。

本课程建议总课时32学时,其中社会实践与实训占10%左右。

Engineering Economy Curriculum Standard

1. Curriculum Description

Engineering Economy is a professional basic course for the specialty of construction cost (Sino-British Cooperative Program), and it is delivered in the fourth semester.

Engineering economics is an interdisciplinary course which is composed of technology science, economics and management science. This course studies the economic benefits of various engineering technology solutions, and studies how to obtain the expected output with the minimum input in the use of various technologies, and how to realize the necessary functions of products, operations and services with the minimum life cycle costs. Through teaching, students should master the basic principles, methods and skills of engineering economy, and they should be able to study, analyze and evaluate various technical practice activities, and provide theoretical basis for decision makers to choose the optimal technical scheme.

This course requires a certain mathematical foundation, and usually takes *Economic Mathematics*, *Computer Application Foundation* and *Introduction to Construction Cost* as the previous delivered courses. The following courses include *Cost Control*, *Application of BIM Construction Cost Software*, etc.

2. Curriculum Design Ideas

The construction of the contents and the course system should be related to the development of the current construction cost industry and the demand for construction cost professional and technical personnel. The curriculum system should be based on ability, professional practice and application. The content mainly includes: capital time value, basic elements of engineering economic analysis, economic evaluation indexes and methods of engineering projects, uncertainty analysis, financial evaluation, value engineering and other parts.

In the process of teaching, the necessary theoretical knowledge points of practicing qualification certificate and professional ability are organically integrated into each teaching unit, and the principles of application and practice are highlighted. According to the requirements of professional positions, and the basis of theoretical learning, emphasis is placed on training students'job skills and strengthening their ability to apply theoretical knowledge to solve practical problems. Through complete training, we can help students realize the integration of knowledge and improve their professional ability in an all-round way, and lay a solid foundation for the formation of their future job ability.

There should be 32 hours for this course, of which social practice and practical training account for about 10%.

三、课程目标

通过系统的学习,学生能够理解和掌握工程经济的基本概念、基本理论,了解这一学科的基本知识框架和分析逻辑;能够运用工程经济学的基本原理去观察、分析和解释现实生活中建设项目的投资估算、方案比选、盈利能力、偿债能力、项目风险等问题;能够对成熟的技术和新技术进行经济性地分析、比较与评价,从经济的角度为技术的采用与发展提供决策依据。

(一)知识目标

(1)理解工程经济的基本概念;

(2)理解资金时间价值的概念、掌握等值计算的方法;

(3)掌握经济效果评价的内容及方法;

(4)掌握方案评选的方法;

(5)掌握不确定性分析的方法;

(6)掌握价值工程分析的方法;

(7)掌握财务评价的方法;

(二)素质目标

(1)具备一定的沟通和组织能力;

(2)具备一定的经济分析能力;

(3)能运用工程经济理论解决实际问题;

(4)培养勤学好问、诚实、严谨、细心的治学态度;

(5)培养投资理财意识。

(三)能力目标

(1)能够进行基本经济计算;

(2)能够进行项目方案比选;

(3)能够进行工程项目财务评价;

(4)能够利用专业书籍获取所需知识;

(5)能建立、完善专业知识体系;

(6)能解决工作实际问题;

(7)能不断自我提升以适应高速发展的经济社会。

3. Curriculum Objectives

Through systematic study, students can understand and master the basic concepts and theories of engineering economy, and understand the basic knowledge framework and analytical logic of this course. Students can use the basic principles of engineering economics to observe, analyze and explain the real life construction project investment estimation, program selection, profitability, solvency, project risk and other issues. They can analyze, compare and evaluate mature technology and new technology economically, and provide decision-making basis for the adoption and development of technology from the perspective of economy.

3.1 Knowledge Objectives

After learning this course, students will be able to

(1) understand the basic concepts of engineering economics;

(2) understand the concept of time value of capital and master the method of equivalent calculation;

(3) master the contents and methods of economic effect evaluation;

(4) master the method of program selection;

(5) master the method of uncertainty analysis;

(6) master the methods of value engineering analysis;

(7) master the methods of financial evaluation.

3.2 Quality Objectives

After learning this course, students will have the quality of

(1) well communication and organization skills;

(2) certain economic analysis;

(3) solving practical problems with engineering economic theory;

(4) taking studious, inquisitive, honest, rigorous and careful attitude towards learning;

(5) sensitive awareness of investment and financial management.

3.3 Ability Objectives

After learning this course, students will have the ability of

(1) performing basic economic calculations;

(2) project program comparison and selection;

(3) financial evaluation of engineering projects;

(4) using professional books to acquire required knowledge;

(5) establishing and improving professional knowledge system;

(6) solving practical problems in work;

(7) continuously improving themselves to adapt to the rapid development of economy and society.

四、课程内容与学时分配

教学内容与学时安排表

序号	教学情境/任务/项目/单元	教学目标	教学内容及学时分配			
			知识点	学时	实践项目	学时
1	总论	掌握工程经济的概念;理解工程、技术与经济之间的关系	1. 工程、技术、经济的概念	1		
			2. 经济效果与经济效益的区别和联系(★)			
2	工程经济分析基本要素	掌握工程项目建设总投资的构成;熟悉资金筹措的方式;了解投资、收入、成本费用、折旧、利润、税金等经济要素	3. 建设工程项目总投资(★)	3		
			4. 资金筹措的方式(★)			
			5. 折旧的计算方法(★)			
			6. 利润的构成及分配			
3	资金时间价值与等值计算	理解现金流量的概念;掌握现金流量图的画法;理解时间价值的概念;掌握等值计算的方法	7. 现金流量图(★)	6	按揭贷款两大还款方式比选(★)	3
			8. 资金时间价值(★)			
			9. 名义利率与实际利率(★)			
			10. 等值计算与应用(★)			

4. Curriculum Contents and Teaching Hours

Table of Teaching Contents and Teaching Hours

Item	Topic	Teaching Objectives	Teaching Contents and Teaching Hours			
			Lesson Contents	Hrs	Student Activities	Hrs
1	General	Grasp the concept of engineering economy; Understand the relationship between engineering, technology and economics	1. Concepts of engineering, tec-hnology and economy	1		
			2. Difference and connection between economic effect and eco-nomic benefit(★)			
2	Basic elements of engineering economic analysis	Master the composition of total construction investment; Familiar with financing methods; Understand investment, income, cost, depreciation, profit, tax and other economic factors	3. Total investment of constr-uction project(★)	3		
			4. Ways of financing(★)			
			5. Calculation method of dep-reciation(★)			
			6. Composition and distribu-tion of profits			
3	Time value and equivalent calculation of capital	Understand the concept of cash flow; Master the drawing method of cash flow chart; Understand the concept of time value; Master the method of equivalent calcula-tion	7. Cash flow chart(★)	6	Mortgage loan repayment of two major ways to choose(★)	3
			8. Time value of capital(★)			
			9. Nominal and real interest rates(★)			
			10. Equivalent calculation and application(★)			

序号	教学情境/任务/项目/单元	教学目标	教学内容及学时分配			
			知识点	学时	实践项目	学时
4	工程项目经济评价指标与方法	掌握静态评价指标、动态评价指标的构成及计算方法;能够进行独立方案和互斥方案的评价与决策	11. 投资项目评价指标体系的分类	6		
			12. 静态投资回收期、动态投资回收期的概念、计算方法及评价准则(★)			
			13. 净现值、净现值率、净年值、费用现值、费用年值、内部收益率、差额投资内部收益率的概念、计算方法及评价准则(★)			
			14. 互斥方案的比选(★)			
5	不确定性与风险分析	了解不确定性及风险分析的概念;掌握盈亏平衡分析、敏感性分析、概率分析在实际问题中的应用	15. 不确定及风险分析的概念	5		
			16. 盈亏平衡分析(★)			
			17. 敏感性分析(★)			
			18. 概率分析(★)			
6	财务评价	理解财务评价的概念;掌握财务评价的基本方法和步骤	19. 盈利能力评价指标	4		
			20. 偿债能力评价指标			
7	价值工程	了解价值工程的含义及经济意义;理解价值工程的基本分析方法;掌握价值工程在工程领域的应用	21. 价值、功能和全寿命周期成本的概念	4		
			22. 价值工程对象的选择			
			23. 功能评价的基本程序(★)			
			24. 价值工程在项目方案优选中的应用(★)			
合计				29		3

备注:重要知识点标注★

continued

Item	Topic	Teaching Objectives	Teaching Contents and Teaching Hours			
			Lesson Contents	Hrs	Student Activities	Hrs
4	Economic evaluation index and method of engineering project	Master the composition and calculation method of static evaluation index and dynamic evaluation index; Ability to evaluate and make decisions on independent and mutually exclusive solutions	11. Classification of evaluation index system of investment projects	6		
			12. Concepts, calculation methods and evaluation criteria of static investment payback period and dynamic investment payback period(★)			
			13. Net present value, net present value ratio, net annual value, cost present value, cost annual value, internal rate of return, internal rate of return of differential investment concept, calculation method and evaluation criteria(★)			
			14. Mutual exclusion scheme comparison(★)			
5	Uncertainty and risk analysis	Understand the concept of uncertainty and risk analysis; Master the application of breakeven analysis, sensitivity analysis and probability analysis in practical problems	15. Concept of uncertainty and risk analysis	5		
			16. Break-even analysis(★)			
			17. Sensitivity analysis(★)			
			18. Probability analysis(★)			
6	Financial evaluation	Understand the concept of financial evaluation; Master the basic methods and steps of financial evaluation	19. Profitability evaluation index	4		
			20. Indicators of solvency evaluation			
7	Value engineering	Understand the meaning and economic significance of value engineering; Understand the basic analytical methods of value engineering; Master the application of value engineering in engineering field	21. Concept of value, function and life-cycle costs	4		
			22. Selection of value engineering objects			
			23. Basic procedures for functional evaluation(★)			
			24. Application of value engineering in the optimization of project scheme(★)			
Total				29		3
Notes: Mark the important points with ★						

五、教学实施建议

(一)教学参考资料

吴全利.《建筑工程经济》,重庆大学出版社,2004。

冯为民、付晓灵.《工程经济学》,北京大学出版社,2006。

赵彬.《工程技术经济》,高等教育出版社,2003。

刘晓君.《工程经济学》(第三版),中国建筑工业出版社,2015。

(二)教师素质要求

主讲《工程经济》课程的教师不但应具有较为深厚的经济学、管理学和工程管理专业的基础理论知识,而且应具备丰富的项目实践经验。

(三)教学场地、设施要求

实训课程要求使用多媒体教室,能充分利用多媒体设备,积极调动音像、互联网等各类教学资源;运用现代教学技术加强学生对知识的掌握,培养学生解决实际问题的能力。

六、考核评价

1. 考核的形式

闭卷考试。

2. 考核评价

成绩评定采用期末考试成绩与平时成绩(均为百分制)加权确定,期末考试成绩与平时成绩分别占70%与30%,平时成绩根据出勤、作业、提问、课堂展示等综合评定。

5. Teaching Suggestions

5. 1 Teaching Resources

Wu Quanli. *Economics of Construction Engineering.* Chongqing University Press, 2004.

Feng Weimin, Fu Xiaoling. *Engineering Economics,* Peking University press, 2006.

Zhao Bin. *Engineering Technology and Economy,* Higher Education Press, 2003.

Liu Xiaojun. *Engineering Economics,* 3rd Edition. China Architeckure & Building Press, 2015.

5. 2 Teachers'Qualification

The teachers who teach *Engineering Economy* should not only have profound basic theoretical knowledge of economics, management and engineering management, but also have rich project practice experience.

5. 3 Teaching Facilities

Practical training units require the use of multimedia classroom, which can make full use of multimedia equipment, actively mobilize audio and video, Internet and other teaching resources. The use of modern teaching technology can strengthen students'mastery of knowledge and cultivate students'ability to solve practical problems.

6. Evaluation

6. 1 Assessment Method

Examination without referencing material on the spot.

6. 2 Assessment Statement

The final examination result and the daily performance (both are 100 points) are combined to determine the final examination, and the final examination accounts for 70% and the daily performance accounts for 30% . The daily performance is evaluated according to the comprehensive evaluation of attendance, homework, questions and classroom presentation.

四川建筑职业技术学院

课程标准

课程名称: 建筑电气设备安装工艺与识图及 Revit 建模基础

课程代码: 150485

课程学分: 2

基本学时: 32

适用专业: 工程造价(中英合作办学项目)

执笔人: 刘晓满

编制单位: 安装工程造价教研室

审核: 工程管理系(院)(签章)

批准: 教务处(盖章)

Sichuan College of Architectural Technology

Curriculum Standard

Curriculum Name: Building Electrical Installation, Drawing

Reading and Revit Building Model

Curriculum Code: 150485

Credits: 2

Teaching Hours: 32

Applicable Specialty: Construction Cost

(Sino-British Cooperative Program)

Compiled by: Liu Xiaoman

On Behalf of: Teaching & Research Section of Installation Cost

Reviewed by: The Department of Engineering Management

Approved by: Teaching Affairs Office

《建筑电气设备安装工艺与识图及 Revit 建模基础》课程标准

一、课程概述

《建筑电气设备安装工艺与识图及 Revit 建模基础》课程是工程造价专业实践性很强的课程。主要任务是学习运用 BIM 技术进行建筑安装工程实体模型的建立,使学生全面和系统地获得 BIM 技术和 Revit 建模软件的相关知识,培养学生的科学思想和研究方法,使学生在软件应用、逻辑思维和解决问题的能力等方面都得到基本而系统的训练,为以后工作奠定必要的基础。

《建筑电气设备安装工艺与识图及 Revit 建模基础》课程是工程造价及建筑工程管理等相关专业教学计划中必修的课程之一。它是以《建筑制图》《安装工程施工技术》《建筑 CAD》等课程为基础,通过对 Revit 软件的学习,使学生能够独立完成单位工程的模型建立,以及后续对模型的相关管理操作。

二、课程设计思路

通过"边学边练"的教学思路,教师先进行教学演示,然后学生实践操作,完成指定项目的教学方法,结合实际项目的实战情况,让学生在掌握 Revit 建模软件平台的基本操作的基础上,具备对 BIM 行业纵观全局的能力。教学内容包括讲解练习的实例项目和实际工程项目作为作业练习。另外,结合工程造价国家资源库的课程资源进行自学选学和作业布置,实现教学的多元化设计。

三、课程目标

(一)知识目标

(1)掌握建筑电气设备安装工艺与识图基础;

(2)了解 BIM 技术在建筑行业的应用前景和意义;

Building Electrical Installation, Drawing Reading and Revit Building Model Curriculum Standard

1. Curriculum Description

Building Electrical Installation, Drawing Reading and Revit Building Model is the course with strong practicality of construction cost. The main task is to learn how to use BIM technology to build the physical model of construction and installation engineering; it enables students to comprehensively and systematically acquire relevant knowledge of BIM technology and Revit modeling software, cultivate students'scientific thoughts and research methods, and enable students to get basic and systematic training in software application, logical thinking and problem-solving ability, laying a necessary foundation for future work.

Building Electrical Installation, Drawing Reading and Revit Building Model is one of the required courses in the teaching plan of construction cost and construction engineering management. It is based on courses such as *Construction Drawing*, *Installation Engineering Construction Technology*, *Construction CAD*, etc. Through the study of Revit software, students can independently complete the model establishment of unit engineering, and the subsequent management and operation of the model.

2. Curriculum Design Ideas

Through the teaching idea of "learning while practicing", the teacher first conducts the teaching demonstration, and then the student practices the operation, completes the teaching method of the designated project, combines the actual situation of the project, and enables the student to master the basic operation of Revit modeling software platform and have the ability to observe the overall situation of the BIM industry. The curriculum includes an explanation of examples of exercises and practical engineering projects as homework. In addition, combined with the course resources of the national resource library of construction cost, self-study selection and homework arrangement are carried out to realize the diversified design of teaching.

3. Curriculum Objectives

3.1　Knowledge Objectives

After learning this course, students will be able to

(1) master the installation technology of building electrical equipment and the basis of drawing recognition;

(2) understand the application prospect and significance of BIM technology in the construction industry;

（3）熟悉 Revit 软件的界面和常规操作。

（二）素质目标

（1）初步具备辩证思维的能力；

（2）具有爱岗敬业的思想,实事求是的工作作风和创新意识；

（3）熟悉建筑工程的有关政策法规；

（4）加强职业道德的意识,认识 BIM 人员的执业权限与基本要求。

（三）能力目标

（1）明确软件教学课程的地位、性质、任务和学习方法

（2）能够独立、快速地进行强电工程、弱电工程、防雷接地工程专业的模型建立；

（3）熟练掌握模型的后期深化应用操作。

四、课程内容与学时分配

教学内容与学时安排表

序号	教学情境/任务/项目/单元	教学目标	教学内容及学时分配			
			知识点	学时	实践项目	学时
1	建筑电气设备安装工艺与识图	1. 掌握电气设备图纸的构成； 2. 掌握电气识图要点和技巧	1. 电气设备施工图构成		1. 电气设备施工图构成	4
			2. 电气系统构成		2. 电气系统构成	
			3. 电气系统材质和安装工艺介绍		3. 电气系统材质和安装工艺介绍	
			4. 电气识图		4. 电气识图	
2	BIM 概述及软件介绍	1. 了解 Revit 建模软件的基本概况； 2. 软件初始设置的方法,常用操作的设置	5. 了解工程造价软件的基本概况		5. CAD 相关操作回顾	4
			6. 软件初始设置的方法,常用操作的设置		6. 打开、关闭、保存、工程设置	
			7. 操作流程和常用技巧			

(3) be familiar with the interface and routine operation of Revit software.

3. 2 Quality Objectives

After learning this course, students will have the quality of

(1) critical thinking;

(2) loving and dedication to work, practical and realistic work style and innovative consciousness;

(3) being familiar with relevant policies and regulations of construction engineering;

(4) strengthened awareness of professional ethics and understanding of the professional competence and basic requirements of BIM personnel.

3. 3 Ability Objectives

After learning this course, students will have the ability of

(1) being clear about the status, nature, task and learning method of software teaching curriculum;

(2) establishing professional models of strong power engineering, weak power engineering and lightning protection grounding engineering independently and quickly;

(3) applying the model proficiently.

4. Curriculum Contents and Teaching Hours

Table of Teaching Contents and Teaching Hours

Item	Topic	Teaching Objectives	Teaching Contents and Teaching Hours			
			Lesson Contents	Hrs	Student Activities	Hrs
1	Installation of Building Electrical and Drawing Reading	1. Master the composition of electrical equipment drawings; 2. Master the key points and skills of electrical drawing recognition	1. Construction drawing composition of electrical equipment		1. Construction drawing composition of electrical equipment	4
			2. Composition of electrical system		2. Composition of electrical system	
			3. Introduction to material and installation process of electrical system		3. Introduction to material and installation process of electrical system	
			4. Electrical drawing reading		4. Electrical drawing reading	
2	Introduction to BIM and software	1. Understand the basic overview of Revit modeling software; 2. The method of software initial setting and the setting of common operations	5. Understand the basic situation ofconstruction cost software		5. Review CAD related operations	4
			6. The method of software initial setting and the setting of common operations		6. Project settings, such as open, close and save	
			7. Operating procedures and common skills			

续表

序号	教学情境/任务/项目/单元	教学目标	教学内容及学时分配			
			知识点	学时	实践项目	学时
3	建筑和结构模块	熟练操作软件,完成模型建立	8. 轴网、标高、对象编辑		7. 绘制和编辑轴网、标高	4
			9. 建筑柱、墙、门、窗、楼梯、楼板		8. 绘制墙,布置柱、门、窗、绘制楼板	
			10. 基础、梁、结构柱、幕墙		9. 绘制幕墙等	
4	电气模块	熟练操作软件,完成电气系统模型建立	11. 桥架系统类型设置和绘制		10. 桥架配件的载入、桥架的编辑和绘制(★)	16
			12. 电气设备		11. 配电箱的布置	
			13. 照明器具的编辑和布置		12. 灯具类型的编辑和布置	
			14. 电气装置		13. 开关和插座的布置	
			15. 线管布置		14. 从面绘制线管(★)	
5	模型深化应用	掌握电气工程模型的综合应用	16. 碰撞检查、渲染、相机、漫游、出图		15. 机电全专业模型整合,碰撞检查(★)	4
					16. 模型综合应用	
合计						32
备注	1. 教学中应结合国家最新政策文件,不拘泥于教材,为完成本课程应达到的目标,合理有效地组织教学; 2. 选择满足课程教学目标要求的施工图(施工图的类型和难易程度以学生能基本掌握为界限),完成模型建立工作; 3. 可以根据相关技能考试的考核要求调整教学内容					

continued

Item	Topic	Teaching Objectives	Teaching Contents and Teaching Hours			
			Lesson Contents	Hrs	Student Activities	Hrs
3	Building and structural modules	Proficient in software operation and model building	8. Axis network, elevation and object editing		7. Drawing and editing axles and elevations	
			9. Building columns, walls, doors, Windows, stairs and floors		8. Drawing walls, arranging columns, doors and windows, and draw floors	4
			10. Foundation, beam, structural column and curtain wall		9. Drawing the curtain wall, etc	
4	Electrical module	Proficient in software operation, complete the establishment of electrical system model	11. Bridge system type setting and drawing		10. Loading, editing and drawing of bridge accessories(★)	
			12. Electrical equipment		11. Distribution box layout	
			13. Editing and arrangement of lighting fixtures		12. Editing and arrangement of lamp types	16
			14. Electrical devices		13. Layout of switches and sockets	
			15. Wiring arrangement		14. Drawing the line tube from the surface (★)	
5	Deeply application of model	Master the comprehensive application of electrical engineering models	16. Collision checking, rendering, camera, roaming, drawing		15. Integration of mechanical and electrical professional models and collision inspection(★);	4
					16. Comprehensive application of the model	
Total						32
Note	1. Teaching should be organized reasonably and effectively in combination with the latest national policy documents and should not be confined to textbooks, so as to achieve the objectives of this curriculum; 2. Select construction drawings that meet the requirements of curriculum teaching objectives (the type and difficulty of construction drawings are based on students' basic mastery) and complete the model building; 3. The teaching content can be adjusted according to the assessment requirements of relevant examination of professional skills					

五、教学实施建议

(一)教学参考资料

工业和信息化部电子行业职业技能鉴定指导中心.《BIM 建模应用技术》,中国建筑工业出版社,2018。

(二)教师素质要求

机电安装工程相关专业的高学历人才,具备 BIM 行业从业资格的相关证书,具备机电安装工程施工技术和识图的能力,具备丰富的软件教学经验。

(三)教学场地、设施要求

高配置的软件实训室。

六、考核评价

考试题目要全面,符合学院教学大纲要求,同时要做到体现重点,难度适中,题量适度,难度及题量应按照教学要求配置。学生在完成教学内容后可参加上机实际操作考试。本课程采用机考形式,平时成绩占30%,最后考核占70%。其中,平时成绩包括出勤、上课纪律、平时作业完成情况和资源库选学的情况,综合考虑得分。

5. Teaching Suggestions

5.1 Teaching Resources

Vocational skill appraisal guidance center of the electronic industry of the ministry of industry and information technology, *BIM Modeling Application Technology.* China Architecture & Building Press, 2018.

5.2 Teachers'Qualification

Highly educated talents majoring in mechanical and electrical installation engineering, with relevant certificates of BIM industry qualification, with ability in mechanical and electrical installation engineering construction technology and map reading, and with rich experience in software teaching are preferred for the teachers of this course.

5.3 Teaching Facilities

Software training room with computers equipped with required system and software.

6. Evaluation

The evaluation should cover all the contents of this course. The implement of the evaluation should be in accordance with the requirements of the college syllabus. The assessment should not be too hard or too easy and the questions for students should be appropriate. Students in the completion of teaching content can participate in the actual operation of the computer test. This course adopts the form of computer test, and the daily performance accounts for 30% , while the final examination accounts for 70%. Among them, the daily performance includes compressive factors, such as class attendance, class discipline, the usual homework completion and the learning record in the resource library.

四川建筑职业技术学院

课　程　标　准

课程名称：建筑水暖设备安装工艺与识图及 Revit 建模基础

课程代码：　　　　　　　150484

课程学分：　　　　　　　　2

基本学时：　　　　　　　32

适用专业：　　　工程造价（中英合作办学项目）

执　笔　人：　　　　　　　刘晓满

编制单位：　　　　安装工程造价教研室

审　　　核：　　　工程管理系（院）（签章）

批　　　准：　　　　教务处（盖章）

Sichuan College of Architectural Technology

Curriculum Standard

Curriculum Name: Building Plumbing Installation, Drawing Reading

and Revit Building Model

Curriculum Code: 150484

Credits: 2

Teaching Hours: 32

Applicable Specialty: Construction Cost

(Sino-British Cooperative Program)

Compiled by: Liu Xiaoman

On Behalf of: Teaching & Research Section of Installation Cost

Reviewed by: The Department of Engineering Management

Approved by: Teaching Affairs Office

《建筑水暖设备安装工艺与识图及 Revit 建模基础》课程标准

一、课程概述

《建筑水暖设备安装工艺与识图及 Revit 建模基础》课程是工程造价专业实践性很强的课程。主要任务是学习运用 BIM 技术进行建筑安装工程实体模型的建立,使学生全面和系统地获得 BIM 技术和 Revit 建模软件的相关知识,培养学生的科学思想和研究方法,使学生在软件应用、逻辑思维和解决问题的能力等方面都得到基本而系统的训练,为以后工作奠定必要的基础。

《建筑水暖设备安装工艺与识图及 Revit 建模基础》课程是工程造价及建筑工程管理等相关专业教学计划中必修的课程之一。它是以《建筑制图》《安装工程施工技术》《建筑 CAD》等课程为基础,通过对 Revit 软件的学习,使学生能够独立完成单位工程的模型建立,以及后续对模型的相关管理操作。

二、课程设计思路

通过"边学边练"的教学思路,教师先进行教学演示,然后学生实践操作,完成指定项目的教学方法,结合实际项目的实战情况,让学生在掌握 Revit 建模软件平台的基本操作的基础上,具备对 BIM 行业纵观全局的能力。教学内容包括讲解练习的实例项目和实际工程项目作为作业练习。另外,结合工程造价国家资源库的课程资源进行自学选学和作业布置,实现教学的多元化设计。

三、课程目标

(一)知识目标

(1)掌握建筑水暖设备安装工艺与识图基础;

(2)了解 BIM 技术在建筑行业的应用前景和意义;

(3)熟悉 Revit 软件的界面和常规操作。

Building Plumbing Installation, Drawing Reading and Revit Building Model Curriculum Standard

1. Curriculum Description

Building Plumbing Installation, Drawing Reading and Revit Building Modeling is the course with strong practicality of construction cost. The main task is to learn how to use BIM technology to build the physical model of construction and installation engineering; it enables students to comprehensively and systematically acquire relevant knowledge of BIM technology and Revit modeling software, cultivate students'scientific thoughts and research methods, and enable students to get basic and systematic training in software application, logical thinking and problem-solving ability, laying a necessary foundation for future work.

Building Plumbing Installation, Drawing Reading and Revit Building Modeling is one of the required courses in the teaching plan of construction cost and construction engineering management. It is based on courses such as *Construction Drawing*, *Installation Engineering Construction Technology*, *Construction CAD*, etc. Through the study of Revit software, students can independently complete the model establishment of unit engineering, and the subsequent management and operation of the model.

2. Curriculum Design Ideas

Through the teaching idea of "learning while practicing", the teacher first conducts the teaching demonstration, and then the student practices the operation, completes the teaching method of the designated project, combines the actual situation of the project, and enables the student to master the basic operation of Revit modeling software platform and have the ability to observe the overall situation of the BIM industry. The curriculum includes an explanation of examples of exercises and practical engineering projects as homework. In addition, combined with the course resources of the national resource library of construction cost, self-study selection and homework arrangement are carried out to realize the diversified design of teaching.

3. Curriculum Objectives

3.1 Knowledge Objectives

After learning this course, students will be able to

(1) master the installation technology of building plumbing equipment and the basis of drawing recognition;

(2) understand the application prospect and significance of BIM technology in the construction industry;

(3) be familiar with the interface and routine operation of Revit software.

(二)素质目标

(1)初步具备辩证思维的能力;

(2)具有爱岗敬业的思想、实事求是的工作作风和创新意识;

(3)熟悉建筑工程的有关政策法规;

(4)加强职业道德的意识,认识 BIM 人员的执业权限与基本要求。

(三)能力目标

(1)明确软件教学课程的地位、性质、任务和学习方法;

(2)能够独立、快速地进行暖通工程、建筑室内给水排水工程、消防工程专业的模型建立;

(3)熟练掌握模型的后期深化应用操作。

四、课程内容与学时分配

教学内容与学时安排表

序号	教学情境/任务/项目/单元	教学目标	教学内容及学时分配			
			知识点	学时	实践项目	学时
1	建筑水暖设备安装工艺与识图	1. 掌握水暖设备图纸的构成; 2. 掌握水暖识图要点和技巧	1. 水暖设备施工图构成 2. 水暖系统构成 3. 水暖系统材质和安装工艺介绍 4. 水暖识图		1. 水暖设备施工图构成 2. 水暖系统构成 3. 水暖系统材质和安装工艺介绍 4. 水暖识图	4
2	BIM 概述及软件介绍	1. 了解 Revit 建模软件的基本概况; 2. 软件初始设置的方法,常用操作的设置	5. 了解工程造价软件的基本概况 6. 软件初始设置的方法,常用操作的设置 7. 操作流程和常用技巧		5. CAD 相关操作回顾 6. 打开、关闭、保存、工程设置	4

3.2　Quality Objectives

After learning this course, students will have the quality of

(1) critical thinking;

(2) loving and dedication to work, practical and realistic work style and innovative consciousness;

(3) being familiar with relevant policies and regulations of construction engineering;

(4) strengthened awareness of professional ethics and understanding of the professional competence and basic requirements of BIM personnel.

3.3　Ability Objectives

After learning this course, students will have the ability of

(1) being clear of the status, nature, task and learning method of software teaching curriculum;

(2) establishing models for HVAC engineering, building indoor water supply and drainage engineering, and fire protection engineering independently and quickly;

(3) applying the model proficiently.

4. Curriculum Contents and Teaching Hours

Table of Teaching Contents and Teaching Hours

Item	Topic	Teaching Objectives	Teaching Contents and Teaching Hours			
			Lesson Contents	Hrs	Student Activities	Hrs
1	Installation of Building Plumbing and Drawing Reading	Master the composition of plumbing equipment drawings; Master the key points and skills of plumbing drawing recognition	1. Construction drawing composition of plumbing equipment		1. Construction drawing composition of plumbing equipment	4
			2. Composition of plumbing system		2. Composition of plumbing system	
			3. Introduction to material and installation process of plumbing system		3. Introduction to material and installation process of plumbing system	
			4. Plumbing drawing reading		4. Plumbing drawing reading	
2	Introduction to BIM and software	Understand the basic overview of Revit modeling software; Master the method of software initial setting and the setting of common operations	5. Understand the basic situation of construction cost software		5. Reviewing CAD related operations	4
			6. The method of software initial setting and the setting of common operations		6. Project Settings such as open, close and save	
			7. Operating procedures and common skills			

序号	教学情境/任务/项目/单元	教学目标	教学内容及学时分配			
			知识点	学时	实践项目	学时
3	建筑和结构模块	熟练操作软件,完成模型建立	8. 轴网、标高、对象编辑		7. 绘制和编辑轴网、标高	4
			9. 建筑柱、墙、门、窗、楼梯、楼板		8. 绘制墙,布置柱、门、窗,绘制楼板	
			10. 基础、梁、结构柱、幕墙		9. 绘制幕墙等	
4	暖通模块	熟练操作软件,完成风系统模型建立	11. 风道系统类型设置和绘制		10. 风管的编辑和绘制(★)	8
			12. 设备和风道末端		11. 风机和空调机组的布置(★)	
			13. 风道管件和附件		12. 风口和风阀的布置(★)	
5	建筑给水排水模块	熟练操作软件,完成水系统模型建立	14. 水管道系统类型设置和绘制		13. 家用冷水、家用热水、卫生设备管道的编辑和绘制(★)	4
			15. 卫生器具的编辑和布置		14. 卫生器具的布置(★)	
			16. 管件和管道附件		15. 管件和阀门、地漏的布置(★)	
6	消防模块	熟练操作软件,完成消防系统模型建立	17. 消防管道系统类型设置和绘制		16. 喷淋和消火栓管道的编辑和绘制	4
			18. 喷头的编辑和布置		17. 喷头的布置	
			19. 管件和管道附件		18. 管件和阀门的布置	

continued

Item	Topic	Teaching Objectives	Teaching Contents and Teaching Hours			
			Lesson Contents	Hrs	Student Activities	Hrs
3	Building and structural modules	Be proficient in software operation and model building	8. Axis network, elevation and object editing		7. Drawing and editing axles and elevations	4
			9. Building columns, walls, doors, windows, stairs and floors		8. Drawing walls, arranging columns, doors and windows, and draw floors	
			10. Foundation, beam, structural column and curtain wall		9. Drawing the curtain wall, etc.	
4	Plumbing module	Be proficient in software operation and wind system modeling	11. Wind duct system type setting and drawing		10. Editing and drawing air pipes(★)	8
			12. Equipment and duct ends		11. Arrangement of fans and air conditioning curriculum(★)	
			13. Duct pipe fittings and accessories		12. Arrangement of tuyere and air valve(★)	
5	Building water supply and drainage module	Be proficient in software operation and water system modeling	14. Setting and drawing the type of water pipeline system		13. Editing and drawing household cold water, household hot water and sanitary equipment pipes(★)	4
			15. Editing and arrangement of sanitary appliances		14. Arrangement of sanitary appliances(★)	
			16. Pipe fittings and pipe accessories		15. Arrangement of pipe fittings, valves and floor drain(★)	
6	Fire control module	Be proficient in software operation, completing the establishment of fire system model	17. Type setting and drawing of fire piping system		16. Editing and drawing sprinkler and hydrant pipes	4
			18. Editing and arranging sprinkler heads		17. Arrangement of sprinkler head	
			19. Pipe fittings and pipe accessories		18. Arrangement of pipe fittings and valves	

序号	教学情境/任务/项目/单元	教学目标	教学内容及学时分配			
			知识点	学时	实践项目	学时
7	模型深化应用	掌握电气工程模型的综合应用	20. 碰撞检查、渲染、相机、漫游、出图		19. 机电全专业模型整合,碰撞检查(★) 20. 模型综合应用	4
	合计					32

备注	1. 教学中应结合国家最新政策文件,不拘泥于教材,为完成本课程应达到的目标,合理有效地组织教学; 2. 选择满足课程教学目标要求的施工图(施工图的类型和难易程度以学生能基本掌握为界限),完成模型建立工作; 3. 可以根据相关技能考试的考核要求调整教学内容

五、教学实施建议

(一)教学参考资料

工业和信息化部电子行业职业技能鉴定指导中心.《BIM 建模应用技术》(第二版),中国建筑工业出版社,2018.

(二)教师素质要求

机电安装工程相关专业的高学历人才,具备 BIM 行业从业资格的相关证书,具备机电安装工程施工技术和识图的能力,具备丰富的软件教学经验。

(三)教学场地、设施要求

高配置的软件实训室。

六、考核评价

考试题目要全面,符合学院教学大纲要求,同时要做到体现重点,难度适中,题量适度,难度及题量应满足教学要求。学生在完成教学内容后可参加上机实际操作考试。本课程采用机考形式,平时成绩占 30%,最后考核占 70%。其中,平时成绩包括出勤、上课纪律、平时作业完成情况和资源库选学的情况,综合考虑得分。

continued

Item	Topic	Teaching Objectives	Teaching Contents and Teaching Hours			
			Lesson Contents	Hrs	Student Activities	Hrs
7	Deeply application of model	Master the comprehensive application of plumbing engineering models	20. Collision checking, rendering, camera, roaming, drawing		19. Integration of mechanical and electrical professional models and collision inspection(★); 20. Comprehensive application of the model	4
Total						32
Note	1. Teaching should be organized reasonably and effectively in combination with the latest national policy documents and should not be confined to textbooks, so as to achieve the objectives of this curriculum; 2. Select construction drawings that meet the requirements of curriculum teaching objectives(the type and difficulty of construction drawings are based on students'basic mastery) and complete the model building; 3. The teaching content can be adjusted according to the assessment requirements of relevant examination of professional skills					

5. Teaching Suggestions

5. 1　Teaching Resources

Vocational skill appraisal guidance center of the electronic industry of the ministry of industry and information technology, *BIM Modeling Application Technology.* 2nd Edition. China Architecture & Building Press, 2018.

5. 2　Teachers'Qualification

Highly educated talents majoring in mechanical and electrical installation engineering, with relevant certificates of BIM industry qualification, with ability in mechanical and electrical installation engineering construction technology and map reading, and with rich experience in software teaching are preferred for the teachers of this course.

5. 3　Teaching Facilities

Software training room with computers equipped with required system and software.

6. Evaluation

The evaluation should cover all the contents of this course. The implement of the evaluation should be in accordance with the requirements of the college syllabus. The assessment should not be too hard or too easy and the questions for students should be appropriate. Students in the completion of teaching content can participate in the actual operation of the computer test. This course adopts the form of computer test, and the daily performance accounts for 30% , while the final examination accounts for 70%. Among them, the daily performance includes compressive factors, such as class attendance, class discipline, the usual homework completion and the learning record in the resource library.

四川建筑职业技术学院

课程标准

课程名称：<u>英国工程图识读</u>

课程代码：<u>210177</u>

课程学分：<u>4</u>

基本学时：<u>60</u>

适用专业：<u>工程造价（中英合作办学项目）</u>

执　笔　人：<u>托尼·夏洛克</u>

编制单位：<u>林肯学院</u>

审　　核：<u>瑞克·隆</u>

批　　准：<u>詹姆斯·福斯特</u>

Sichuan College of Architectural Technology

Curriculum Standard

Curriculum Name: Reading & Understanding Drawings in the U. K.

Curriculum Code: 210177

Credits: 4

Teaching Hours: 60

Applicable Specialty: Construction Cost

(Sino-British Cooperative Program)

Compiled by: Tony Sherlock

On Behalf of: Lincoln College

Reviewed by: Rick Long

Approved by: James Foster

《英国工程图识读》课程标准

一、课程概述

本课程为学习者提供机会了解英国所用的工程图纸,学习建筑施工行业中用到的制图技术,也让学生有机会运用手工制图技术绘制平面和立体的图纸。本课程也可以融入其他教学模块,联合展示学生对知识的理解。

本课程为学习者提供理解制图技术的机会,也让学生获得低层民用住宅的设计实践和正确的结构形式与平面布置。学习者还能了解在有详图的情况下所用资源的类型,以及这些资源是如何满足建筑设计要求的。

二、课程设计思路

本课程经过英国伦敦城市行业协会的评估认可,其实施方案能保证课程质量的完成。

该门课程以课堂讲授为基础,开展多种形式的教学活动。这些活动都要逐一评估,确保整个教学模块的教学质量。

三、学习目标

学完本课程,学习者将能够:

(1)了解手工制图过程中的主要工具、技术、媒介;

(2)理解并解释工程图纸、详图、进度表和技术说明;

(3)了解并解释利用手工制图技术绘制的工程图纸、详图和进度表。

四、课程内容和学时安排

(一)了解手工制图过程中的主要工具、技术、媒介

工具:手工制图工具(钢笔、铅笔、尺子、橡皮、可调三角尺规、圆规、模板、曲线板、绘图板、制图胶带)。

Reading & Understanding Drawings in the U. K. Curriculum Standard

1. Curriculum Description

This course provides the learners with the opportunity to develop an understanding of the drawings used in the U. K. ; it will inform the learner of the drawing techniques used within the construction industry. It also gives the learner the opportunity to produce 2D and 3D graphical drawings, using manual drafting techniques. This can then be allied to other modules to demonstrate understanding.

This curriculum will allow learners to develop an understanding of drawing techniques. It will also allow the student to gain knowledge of the practice, design and correct format and layout for a low rise domestic property. Learners will also gain knowledge of the types of resources used when providing graphical detailing, and how these compliment the design of a building.

2. Curriculum Design Ideas

The quality of the curriculum is assessed by City & Guilds and accredited for meeting its quality assurance practices.

The delivery is classroom based and will take the format of various activities, which will be individually assessed, leading to a module qualification.

3. Curriculum Objectives

(1) know and understand the main equipment, media and techniques used in the production of manual graphical information;

(2) understand and be able to interpret graphical drawings, details, schedules and specifications;

(3) understand and be able to produce graphical drawings, details, and schedules by using manual drafting techniques.

4. Curriculum Contents and Teaching Hours

4.1 Know and understand the main equipment, media and techniques used in the production of manual graphical information

Equipment: Hand drafting equipment (pens, pencils, scale rules, erasers, adjustable set squares, compasses, templates and flexible curves, drawing boards and drafting tape).

媒介：铅笔的分级（HB、H、2H）；钢笔粗细（0.2毫米、0.25毫米、0.4毫米、0.5毫米）；纸张大小（A1、A2、A3、A4）。

绘图技术：绘制直线和形状；用尺规作图；标注与尺寸；图示特点；符号的使用；英国标准；使用统一的项目信息。

（二）理解并解释工程图纸、详图、工期进度表和技术说明

工程图纸和详图：平面和立体图纸的工程信息和尺度信息、规划图纸及勘察图纸、初步草图、设计图、施工图、结构图、平面布置图、手绘图纸。

工期进度表和技术说明：技术说明信息、门窗施工进度表。

（三）理解并解释利用手工制图技术绘制的工程图纸、详图和工期进度表

工程图纸和详图：平面图、立面图、剖面图、详图、平面及立体投影、草图、视角、图例、示意图；

工期进度表和技术说明：技术说明信息、门窗施工进度表。

教学过程中将开展针对每个学习目标和具体内容的各种实践活动，并对每项活动进行评估。每位老师可根据学生的学习进度、英文水平、对内容的掌握程度自行规划、使用教学资源。学习目标决定了具体的教学活动实施情况。

五、教学实施建议

（一）教学资源

直尺、绘图纸、各类铅笔和绘画纸。

（二）教师素质

建筑专业资格证书、教师资格证书。

简历中适当素质。

（三）教学场地及设备要求

教学活动将在四川建筑职业技术学院的指定教室内展开。

（四）教学条件

PPT、投影仪、黑板、画板、文具。

六、考核评价

教学活动的评估将根据教师对学生的观察、日常问答情况、课程作业评分来确定。

Media: Grades of pencil(HB, H, 2H); Pens thick(0.2mm, 0.25mm, 0.4mm, 0.5mm); paper sizes(A1, A2, A3, A4).

Techniques: drawing lines & shapes: drawing to scale; lettering and dimensioning; graphic conventions; using symbols; eg British Standards; using co-ordinated project information.

4.2   Understand and be able to interpret graphical drawings, details, schedules and specifications

Graphical drawings and details: Constructional and dimensional data for 2D and 3D; planning and surveying drawings; preliminary sketch drawings; design drawings; production drawings; structural drawings; layout drawings and free hand sketches.

Schedules and specifications: specification information; door and window schedules.

4.3   Understand and be able to produce graphical drawings, details, and schedules by using manual drafting techniques

Graphical drawings and details: Plans; elevations; sections; details; 2D and 3D projections; sketches; perspectives; presentational charts; schematic diagrams.

Schedules and specifications: specification information; door and window schedules.

Various activities will be delivered and assessed as per the Curriculum Objectives and specification. Resources will be planned and delivered by individual teacher, depending on the pace of the students, understanding of English and grasp of the content. Therefore, specific class activities will be directed by the Learning Outcomes.

5. Teaching Suggestions

5.1   Teaching Resources

Scale rules, graph paper, various pencils and drawing paper.

5.2   Qualifications of the Teacher

Professional Chartered(Construction)qualifications and Teaching Quals.
See CV's.

5.3   Teaching Places

Teaching undertake at Sichuan College of Architectural Technology in classrooms allocated.

5.4   Teaching Facilities

PowerPoint/OHP/Chalk board/drawing boards/writing materials.

6. Evaluation/Examination

Assessment of activities undertaken by observation/question & answer/assessment of course work.

四 川 建 筑 职 业 技 术 学 院

课 程 标 准

课程名称： 建筑 CAD

课程代码： 010481

课程学分： 2

基本学时： 32

适用专业： 工程造价（中英合作办学项目）

执 笔 人： 刘 觅

编制单位： 制图教研室

审 核： 土木工程系（院）（签章）

批 准： 教务处（盖章）

Sichuan College of Architectural Technology

Curriculum Standard

Curriculum Name: _____Construction CAD_____

Curriculum Code: _____010481_____

Credits: _____2_____

Teaching Hours: _____32_____

Applicable Specialty: _____Construction Cost_____

_____(Sino-British Cooperative Program)_____

Compiled by: _____Liu Mi_____

On Behalf of: _____Teaching & Research Section of Drawing_____

Reviewed by: _____The Department of Civil Engineering_____

Approved by: _____Teaching Affairs Office_____

《建筑 CAD》课程标准

一、课程概述

计算机辅助设计(CAD)技术在许多工程领域的应用日益广泛,是提高设计质量、加速产品更新换代、增强市场竞争力的重要手段。学生掌握计算机绘图及计算机辅助设计技术对于适应专业课的学习和今后从事专业技术领域的工作具有重要的意义。《建筑 CAD》课程是工程造价(中英合作办学项目)专业的一门重要的专业基础课,适宜在第 5 学期开设。本课程以目前广泛应用的 AutoCAD 软件作为典型代表,将课堂理论教学与实践教学结合,详细介绍计算机绘图软件的使用操作原理和方法,通过教师用多媒体讲授基础理论及学生进行计算机操作,使学生熟练掌握计算机辅助设计软件——AutoCAD 绘制工程图样的方法。本课程的先修课程有《计算机基础》《画法几何》和《建筑制图》。

二、课程设计思路

通过本课程的学习,学生能够进一步开发形象思维能力,获得运用计算机绘制工程图样的能力,使学生的综合图形表达能力和设计能力得到进一步提高,为将来从事工程设计、CAD 应用和开发打下基础。

本课程贯穿项目化教学理念和教学方法,课程有三个训练项目:导引项目为简单平面图,课内师生共同学习 CAD 入门知识;主体项目 A 和主体项目 B 图形结构相似,为典型结构的居民楼建筑施工图,包括平面图、立面图、详图等,难度系数相同,课内师生共同绘制项目 A,系统学习 CAD 绘图指令的对话过程和指令应用技巧等,项目 B 学生课外独立训练,进度与项目 A 平行推进;最后是拓展项目,不同专业方向的学生,自主绘制相应专业方向施工图,如装饰专业绘制室内设计施工图、市政专业绘制道路施工图等。

Construction CAD **Curriculum Standard**

1. Curriculum Description

Computer aided design(CAD)technology is widely used in many engineering fields. It is an important means to improve design quality, accelerate product updating and enhance market competitiveness. Computer graphics and computer-aided design skills are of great significance to students for adapting to the study of professional courses and future work in the field of professional technology. *Construction CAD* is an important basic professional course for the specialty of construction cost (Sino-British Cooperative Program). It is suitable to be delivered in the fifth semester. This course takes AutoCAD software which is widely used at present as a typical representative, combines theory and practice in teaching, and introduces the principles and methods of using computer graphics software in detail. Multimedia teaching and computer operation enable students to master the methods of engineering drawings by AutoCAD. The delivered units are *Computer Foundation*, *Descriptive Geometry* and *Architectural Drawing*.

2. Curriculum Design Ideas

Through this course, students'image thinking ability can be further developed, so that they can acquire the ability of making engineering drawings by computer, and the students'comprehensive graphics expression ability and design ability can be further improved, which lays a foundation for future engineering design, CAD application and development.

This course runs through the concept and method of project-based teaching. There are three training items in the course: the guiding project is simple plan, which helps teachers and students further the basic knowledge of CAD; The main project A and the main project B have similar graphic structure, and their construction drawings are residential buildings drawings with typical structures, including plans, elevations, details and so on, which have the same difficulty coefficients. The teachers and students draw project A together in class to systematically study the dialogue process and application skills of CAD drawing instructions, while students in charge of project B independently carry out training after class and ensure that its progress is in accordance with A. Finally, as for expansion projects, students of different professional directions, independently draw corresponding professional construction drawings, such as interior design construction drawings for decoration majors, road construction drawings for municipal majors, and so on.

三、课程目标

学生能够掌握 AutoCAD 常规绘图命令、编辑指令、文字和尺寸标注指令的对话步骤,综合运用 CAD 指令绘制建筑平面图、立面图、详图等。在绘图过程中,强化理解"制图规范、正投影原理、建筑施工图"等基础知识,提升建筑专业识图能力,并养成遵守标准规范的质量意识和一丝不苟的严谨的工作态度等。

(一)知识目标

(1)了解 AutoCAD 的主要功能;

(2)熟练掌握 AutoCAD 常用的绘图命令;

(3)掌握常用的编辑和修改命令,快速准确地进行绘图和标注。

(二)素质目标

(1)培养辩证思维的能力;

(2)具有严谨的工作作风和敬业爱岗的工作态度;

(3)培养自觉遵守职业道德和行业规范标准。

(三)能力目标

(1)能够独立运用 AutoCAD 绘制一般的建筑平面图、立面图、剖面图和详图;

(2)能够独立运用 AutoCAD 编辑修改一般的建筑工程图样,能进行工程图样的标注。

四、课程内容与学时分配

教学内容与学时安排表

序号	教学情境/任务/项目/单元	教学目标	教学内容及学时分配			
			知识点	学时	实践项目	学时
1	AutoCAD 基础知识	掌握 AutoCAD 的界面及命令基本操作,掌握坐标系与应用	1. AutoCAD 概述及软件界面 2. 图形文件管理 3. 命令的基本操作(★) 4. 坐标系与坐标输入方式(★)	2	1. 基本几何图形的绘制练习	2

3. Curriculum Objectives

Students can master the operation steps of AutoCAD routine drawing commands, editing instructions, text and dimensioning instructions, and use CAD commands to draw building plans, elevations, details and so on. In the process of drawing, we should strengthen the understanding of the basic knowledge of drawing norms, or thographic projection principles and construction drawings, enhance the drawing reading ability of architecture major, and cultivate the quality consciousness of abiding by the standard norms and a meticulous working attitude.

3. 1 Knowledge Objectives

After learning this course, students will be able to

(1) understand the main functions of AutoCAD;

(2) use skillfully the drawing commands commonly used in AutoCAD;

(3) grasp the common editing and modifying commands, draw and mark quickly and accurately.

3. 2 Quality Objectives

After learning this course, students will have the quality of

(1) critical thinking;

(2) forming rigorous working style and dedicated attitude towards work;

(3) self-awareness in abiding professional ethics and standards of professional norms.

3. 3 Ability Objectives

After learning this course, students will have the ability of

(1) using AutoCAD to draw general building plans, elevations, sections and details;

(2) using AutoCAD to edit and modify general architectural drawings, and mark engineering drawings.

4. Curriculum Contents and Teaching Hours

Table of Teaching Contents and Teaching Hours

Item	Topic	Teaching Objectives	Teaching Contents and Teaching Hours			
			Lesson Contents	Hrs	Student Activities	Hrs
1	Basic knowledge of AutoCAD	Master the AutoCAD's interface and basic command of its operation; Master coordinate system and application	1. Overview of AutoCAD and software interface	2	1. Drawing practice of basic geometric figures	2
			2. Management of drawing			
			3. Basic operations of commands (★)			
			4. Coordinate system and coordinate input mode (★)			

序号	教学情境/任务/项目/单元	教学目标	教学内容及学时分配			
			知识点	学时	实践项目	学时
2	绘图环境及图层管理	掌握图层管理器的应用,掌握绘图环境的设置	5. 设置绘图环境 6. 图层的管理(★) 7. 图形显示控制 8. 图形样板文件	2	2. 绘制标准图框 3. 建筑工程图样板文件	2
3	绘制平面图形	掌握平面图形的绘制方法	9. 绘制基本图形 10. 精确绘制图形(★)	2	4. 绘制户型图练习	2
4	编辑平面图形	掌握平面图形的常用编辑方法	11. 选择对象 12. 基本编辑命令(★) 13. 高级编辑命令	2	5. 户型图编辑练习	2
5	辅助绘图命令与工具	掌握文本、填充、图块和查询的方法及运用	14. 文本标注与编辑 15. 图案填充和编辑 16. 图块操作(★) 17. 设计中心 18. 数据查询	2	6. 项目综合训练	2
6	尺寸标注	掌握不同类型尺寸标注的方法和标注样式的设置	19. 标注样式设置 20. 尺寸标注方法(★) 21. 编辑尺寸标注	2	7. 户型图标注练习	2
7	图纸布局与打印输出	掌握图纸布局和输出的设置	22. 从模型空间输出图形 23. 从图纸空间输出图形(★)	2	8. 建筑工程图样出图练习	2
8	建筑工程图绘图	掌握用ACAD进行建筑制图的方法和步骤	24. 建筑工程图一般绘图步骤与方法(★)	2	9. 建筑工程图项目综合训练	2
合计				16		16
备注:重要知识点标注★						

continued

Item	Topic	Teaching Objectives	Teaching Contents and Teaching Hours			
			Lesson Contents	Hrs	Student Activities	Hrs
2	Drawing Environment and Layer Management	Master the application of Layer; Manager and the setting of drawing environment	5. Setting up the drawing environment	2	2. Draw standard drawing frame	2
			6. Layer management(★)			
			7. Graphic display control		3. Template file of architectural engineering	
			8. Graphic template file			
3	Drawing Plans	Master the Drawing method of plans	9. Drawing basic plans	2	4. Practice of drawing house floor plan	2
			10. Drawing plans accurately(★)			
4	Editing Plans	Master common Editing Methods of Graphic figures	11. Choose the object	2	5. Editing exercises of house floor plan	2
			12. Basic editing command(★)			
			13. Advanced editing command			
5	Aided Drawing Commands and Tools	Master the methods and application of text, filling, drawing block and query	14. Text annotation and editing	2	6. Project comprehensive training	2
			15. Patterns filling and editing			
			16. Drawing block operation(★)			
			17. Design center			
			18. Data query			
6	Dimensioning	Master the method of different types of dimensioning and the setting of dimensioning style	19. Dimensioning style settings	2	7. Exercise of house floor plan annotation	2
			20. Dimensioning method(★)			
			21. Editing dimensioning			
7	Drawing Layout and Print Output	Master the layout and output settings of drawings	22. Output figures from model space	2	8. Drawing exercises of architectural engineering drawings	2
			23. Output figures from drawing space(★)			
8	Architectural engineering drawings	Master the methods and steps of building drawing with CAD	24. General drawing procedures and methods of architectural engineering drawings(★)	2	9. Comprehensive training of architectural engineering drawing project	2
Total				16		16
Note	Mark important points with ★					

五、教学实施建议

(一)教学参考资料

唐英敏、吴志刚、李翔.《AutoCAD 建筑绘图教程》(第二版),北京大学出版社,2014。

董祥国.《建筑 CAD 技能实训》,中国建筑工业出版社,2016。

郭慧.《AutoCAD 建筑制图教程》(第三版),北京大学出版社,2018。

(二)教师素质要求

本课程授课教师应具有本课程或相关课程的教师资格,具有工程类专业背景,有良好的课堂管理组织和建筑工程图识图、绘图能力。

(三)教学场地、设施要求

计算机教室。

六、考核评价

本课程宜以实践操作的方式进行考核,成绩构成比例按《四川建筑职业技术学院学生学业考核办法》执行,各教学项目考核成绩比例如下表:

序号	考核内容	成绩构成比例/%
1	AutoCAD 基础知识	5
2	绘图环境及图层管理	10
3	绘制平面图形	20
4	编辑平面图形	20
5	辅助绘图命令与工具	10
6	尺寸标注	20
7	图纸布局与打印输出	5
8	建筑工程图绘图	10
	合计	100

5. Teaching Suggestions

5.1 Teaching Resources

Tang Yingmin, Wu Zhigang, Li Xiang. *AutoCAD Architectural Drawing Course.* 2nd Edition. Peking University Press, 2014.

Dong Xiangguo. *Practical Training for Construction CAD.* China Architecture & Building Press, 2016.

Guo Hui. *AutoCAD Architectural Drawing Course.* 3rd Edition. Peking University Press, 2018.

5.2 Teachers'Qualification

Teachers should have the relative qualifications for this course or related course; they should have the education background of engineering, and have good abilities of organizing classes and reading and drawing architectural engineering drawings.

5.3 Teaching Facilities

Computer laboratory classroom.

6. Evaluation

This course should be assessed in a practical way. The score of the course is based on the *Regulation of Academic Evaluation for Students in SCAT.* The evaluation contents are as follows:

Item	Evaluation Contents	Proportion/%
1	Basic knowledge of AutoCAD	5
2	Drawing Environment and Layer Management	10
3	Drawing plans	20
4	Editing plans	20
5	Aided Drawing Commands and Tools	10
6	Dimensioning	20
7	Drawing Layout and Print Output	5
8	Architectural engineering drawings	10
Total		100

四 川 建 筑 职 业 技 术 学 院

课 程 标 准

课程名称： __BIM 技术基础__

课程代码： __900077__

课程学分： __1__

基本学时： __18__

适用专业： __学院所有工程类专业__

执 笔 人： __彭笑川__

编制单位： __建筑信息管理教研室__

审　　核： __工程管理系（院）__

批　　准： __（签章）__

Sichuan College of Architectural Technology

Curriculum Standard

Curriculum Name: Fundamentals of BIM Technology

Curriculum Code: 900077

Credits: 1

Teaching Hours: 18

Applicable Specialty: All the specialties related with engineering

Compiled by: Peng Xiaochuan

On Behalf of: Teaching & Research Section of

Architecture Information Management

Reviewed by: The Department of Engineering Management

Approved by: Teaching Affairs Office

《BIM 技术基础》课程标准

一、课程概述

本课程是为工程类专业开设的一门专业必修课。主要介绍 BIM 技术概念及软件 Revit 软件的使用,是学院为普及 BIM 技术在建筑生产管理中的应用而开设的公共基础课。通过本课程的学习,学生将了解 BIM 技术的发展及应用概况,以及 Revit 建模软件的基本操作方法和应用优点。通过实际工程案例,理解和掌握精确构建 BIM 项目模型并提取模型信息的流程和方法。通过信息化模型的应用,为将来从事建筑设计、建筑施工、建筑造价咨询等工作打下基础。

在本课程之前要进行专业基础课程《建筑 CAD》《建筑识图》等的学习。以便为后续专业课程的学习打下基础。

二、课程设计思路

本课程根据常规建筑面积(3 000 m² 以下)建筑施工图纸、结构施工图纸,通过 Revit 软件分别完成建筑信息模型、结构信息模型的建立。首先,通过案例讲解软件的基本操作及属性控制。其次,介绍信息化标准,对所有构件进行统一规范的命名。再次,通过项目流程管理讲解构件建立的过程及操作流程。最后,讲解模型信息管理、修改及提取、从而全面地掌握 Revit 软件的基本应用。

三、课程目标

在具有相应的专业基础知识后,熟练掌握 Revit 软件的操作及建筑、结构、模型的建立及运用。

(一)知识目标

(1)掌握基本建筑、结构、构件的定义、编辑、修改操作方法;
(2)掌握建筑、结构、模型建立的基本流程;
(3)掌握特殊构件(异形构件)的建立方法;

Fundamentals of BIM Technology Curriculum Standard

1. Curriculum Description

This is a required course for the specialty of construction cost. It is to introduce the concept of BIM and the usage of the software Revit. In order to popularize the application of the technology of BIM in construction engineering and construction management, this course is delivered in our college as public basic course. Through this course, students can understand the origin, development and application of BIM technology and the basic operation methods and application advantage of the software of BIM. Through the practical cases, students can understand and master the process and method for accurately building a BIM project model and extracting model information. Through the application of the information model, it will lay the foundation for their future work in architectural design, construction, construction cost consulting and so on.

The delivered courses are the professional basic courses, such as *Construction CAD*, *Construction Drawing Reading*, etc. This course lays a foundation for the study of the following professional courses.

2. Curriculum Design Ideas

Based on the construction drawings and structure construction drawings of common buildings (construction areas are less than 3,000 m²), this course can teach us the way to construct the building information model and structure information model with Revit software. Firstly, the course introduces the basic operation methods and attributes control of the software through practical cases. Secondly, it introduces the information standard with uniform specification of all components. Thirdly, it explains the process of building components and operational through project process management. Finally, it explains the information management, modification and extraction of different models. Thus, students can master the basic application of Revit software.

3. Curriculum Objectives

After learning the basic professional knowledge, students can skillfully apply the Revit software and can construct and apply the building and structure models.

3.1 Knowledge Objectives

After learning this course, students will be able to

(1) master the concept, editing and modification of the basic construction, structure and component;

(2) master the basic process of setting building and structure models;

(3) master the setting methods for special components (special-shaped);

（4）掌握外部族文件的加载方法；

（5）掌握模型数据在软件之间的传递方法。

（二）素质目标

（1）培养做事细致的作风；

（2）培养求真、求实的务实精神；

（3）培养高效做事的素质；

（4）建立团队协作，互相沟通的职业素质。

（三）能力目标

（1）具有按照建筑施工图，结构施工图进行建模的能力；

（2）具有依据 BIM 模型修改、提取、管理信息的能力；

（3）具有检查建筑、结构模型定位问题的能力；

（4）具有检查建筑、结构碰撞问题的能力；

（5）具有通过互联网查阅资料解决问题的能力。

四、课程内容与学时分配

教学内容与学时安排表

序号	教学情境/任务/项目/单元	教学目标	教学内容及学时分配			
			知识点	学时	实践项目	学时
1	BIM 技术概念及发展、BIM 技术的国内外应用概况	了解 BIM 技术的概念和发展历程、国内外的应用情况	1. BIM 技术的定义	2		
			2. BIM 技术发展缘由			
			3. BIM 技术的发展阶段等			
2	BIM 技术平台的概况、BIM 技术应用标准及要求	主要了解我国 BIM 技术发展的水平及应用平台、标准及规范	4. BIM 技术平台主要有 Autodesk 公司旗下品牌、广联达公司、斯维尔公司等企业	2		
			5. 我国 BIM 技术应用规范及标准，以及建模精度要求，应用点、应用面要求（★）			

(4) master the loading method of external family files;

(5) master the transfer methods of template date among software.

3.2 Quality Objectives

After learning this course, students will have the quality of

(1) developing meticulous work style;

(2) pragmatic spirit of seeking truth and being realistic;

(3) working efficiently;

(4) cooperation and communication.

3.3 Ability Objectives

After learning this course, students will have the ability of

(1) setting model according to the building construction drawings and structure construction drawings;

(2) modifying, extracting and managing information based on BIM model;

(3) checking the positioning of building and structure model;

(4) solving the problem of building and structure model collision;

(5) assessing data through the Internet to solve problems.

4. Curriculum Contents and Teaching Hours

Table of Teaching Contents and Teaching Hours

Item	Topic	Teaching Objectives	Teaching Contents and Teaching Hours			
			Lesson Contents	Hrs	Student Activities	Hrs
1	Concept and development of BIM technology, BIM technology application at home and abroad	Understand the concept and development history of BIM technology, BIM technology application at home and abroad	1. The concept of BIM technology	2		
			2. The origin and the reasons for the development of BIM technology			
			3. The development stages of the BIM technology			
2	Overview of BIM technology platform, application standards and requirements of BIM technology	Generally understand the development levels and application platform, req-uirements and codes of BIM technology	4. The BIM technology platform mainly includes brands such as Autodesk, Guanglianda and Swell	2		
			5. Application requirements and codes of BIM technology in China, requirements for modeling accuracy, application points and application areas (★)			

序号	教学情境/任务/项目/单元	教学目标	教学内容及学时分配			
			知识点	学时	实践项目	学时
3	Revit 建筑模型构建	了解并掌握 Revit 建筑构件的定义、编辑、布置等	6. 轴网、标高	4	1. 依据建筑案例工程完成全部建筑构件的建立	
			7. 建筑构件：墙、柱、门、窗、楼梯、楼地面、屋顶、台阶等建筑构件的使用(★)		2. 提取并保存部分信息成果	
4	Revit 结构模型构建	掌握 Revit 结构构件的定义、编辑、布置等	8. 轴网、标高(单独)	4	3. 依据结构案例工程完成全部结构构件的建立	
			9. 结构构件：剪力墙、结构柱、结构梁、结构楼梯、结构板、屋顶结构、雨棚结构等结构构件的使用(★)		4. 提取并保存部分信息成果	
5	Revit 特殊构件构建	掌握 Revit 异形构件的定义、编辑、布置等；主要讲解体量、族、特殊造型及幕墙构件控制	10. 特殊构件：建筑的立面线条样式；结构异形梁、板、柱结构	4	5. 依据 BIM 应用中的建筑、结构特殊构件图纸，完成特殊构件的定义及编辑	
			11. 不能通过常规建模实现的其他造型等(★)		6. 提取并保存构件信息	

continued

Item	Topic	Teaching Objectives	Teaching Contents and Teaching Hours			
			Lesson Contents	Hrs	Student Activities	Hrs
3	Revit building model construction	Understand and master the definition, editing and layout of different construction components in Revit software	6. Axis net and elevation	4	1. Construct the building model of different components based on practical construction project case	
			7. Revit building model construction of different components: walls, columns, doors, windows, staircases, floors, ceilings and steps(★)		2. Extract and save some information of the model	
4	Revit Structure Model construction	Master the definition, editing and layout of different structure components with Revit	8. Axis net and elevation (structure model)	4	3. Construct the structure model of different components based on practical structure project case	
			9. Revit structure model construction of different components: shear wall, structural column, structural beam, structural staircase, structural panel, roof structure, and canopy structure(★)		4. Extract and save some information of the model	
5	Revit Model construction of special components	Master the definition, editing and layout of special components with Revit; Master the construction of special components with Revit: volumes, family files, special models and curtain walls	10. Revit model construction of special components: facade line style of the building and structural beam, structural panel and structural column	4	5. Complete the definition and editing of special components based on the drawings of special components in building and structure in BIM	
			11. Other models that can be constructed with conventional modeling(★)		6. Extract and save some information of the model	

序号	教学情境/任务/项目/单元	教学目标	教学内容及学时分配			
			知识点	学时	实践项目	学时
6	Revit 模型综合应用	独立练习,巩固相关知识,熟练软件操作能力		2	7. 根据指定图纸,独立完成建设项目的建筑、结构全部模型	
					8. 按要求提取模型信息	
合计				18		
备注:重要知识点标注 ★						

五、教学实施建议

(一)教学参考资料

李恒、孔娟.《Revit 2015 中文版基础教程》,清华大学出版社 2015。

廖小烽、王君峰.《Revit 2013/2014 建筑设计火星课堂》,人民邮电出版社,2013。

王全杰.《办公大厦建筑工程图》,重庆大学出版社,2012。

(二)教师素质要求

任课教师不仅需要具有建筑工程识图及软件应用能力,还需要拥有由后期模型应用要求提取、修改及更新建筑、结构信息的能力,并对项目信息进行有效应用。

六、考核评价

本课程期末成绩采用实际操作考核的方式进行,总评成绩由期末成绩和平时成绩构成。成绩构成比例原则按《四川建筑职业技术学院学生学业考核办法》执行,最后按照百分制评定。

continued

Item	Topic	Teaching Objectives	Teaching Contents and Teaching Hours			
			Lesson Contents	Hrs	Student Activities	Hrs
6	The comprehensive application of Revit model	Practice independently, consolidate relevant knowledge, and be proficient in software operation		2	7. Independently construct the building and structure model of construction project based on its drawings	
					8. Extract some required information of the mode	
Total				18		
Notes: Mark the important points with ★						

5. Teaching Suggestions

5.1 Teaching Resources

Li Heng, Kong Juan. *Basic Course of Revit* 2015 *in Chinese*. Tsinghua University Press, 2015.

Liao Xiaofeng, Wang Junfeng. *Huoxing Classroom of Revit* 2013/2014 *Architectural Design*. People's Post and Telecommunications Publishing House, 2013.

Wang Quanjie. *Office Building Construction Drawing*. Chongqing University Press, 2012.

5.2 Teachers'Qualification

Teachers should not only have the qualification of reading construction drawings and software application, but also should have the ability to extract, modify, and update building and structure information inpost application of models and effectively apply project information.

6. Evaluation

The final examination of this unit is based on the practical operation of the Revit software. The total score consists of the daily performance and the final examination. According to the *Regulation of Academic Evaluation for Students in SCAT*. The final score is graded into a hundred-mark system.

四 川 建 筑 职 业 技 术 学 院

课 程 标 准

课程名称：　　　　　企业经营管理

课程代码：　　　　　900078

课程学分：　　　　　1

基本学时：　　　　　18

适用专业：　工程造价（中英合作办学项目）

执 笔 人：　　　　　韩　杰

编制单位：　　建筑经济管理教研室

审　　　核：　　经济管理系(院)(签章)

批　　　准：　　教务处(盖章)

Sichuan College of Architectural Technology

Curriculum Standard

Curriculum Name: Construction Enterprise Management

Curriculum Code: 900078

Credits: 1

Teaching Hours: 18

Applicable Specialty: Construction Cost

(Sino-British Cooperative Program)

Compiled by: Han Jie

On Behalf of: Teaching & Research Section of

Construction Economy Management

Reviewed by: The Department of Economic Management

Approved by: Teaching Affairs Office

《企业经营管理》课程标准

一、课程概述

《企业经营与管理》是工程造价专业的一门专业选修课。建议本课程的先修课程是经济学基础《管理心理学》《施工企业会计》等专业基础课,后续课程是《建筑企业统计》《材料成本核算》《工程经济》。

本课程适宜在第1学期开设。

二、课程设计思路

本课程以职业岗位为导向;以适应职业岗位对企业管理人才的需求为出发点;以提高教学质量、增强教学效果为直接目的;以建设结构合理、素质过硬的教师队伍为长远支持;以教学改革、科研创新为动力;以特色铸造高度,从高质量、高水平中提炼出精品。

以工学结合为切入点,做好课程建设规划;根据职业岗位的需求情况,搭建职业行动能力框架;修订人才培养方案,建立科学合理的课程标准;实施任务导向,开发学习领域,进行教学情景设计;以建构主义为基础,以必需、够用为度,开发设计以学生为本的"情境式"教学包;采用引导启发式教学和提升情景案例教学,力求理论联系实际,学以致用,加强学生对将来社会工作岗位的适应能力;配备结构合理的教学团队,不断完善网络课堂;积极运用现代教育技术提高整体教学水平,扩大授课信息量。

三、课程目标

企业经营与管理课程教学所要培养的能力总目标是基层管理岗位的综合管理技能与素质。通过本课程教学与训练培养学生的六大关键能力,即企业环境分析能力、目标管理与计划能力、决策与战略管理能力、组织设计与协调能力、领导与沟通的能力、控制与信息处理能力。

通过教学过程强化学生能力培养,学生能系统掌握企业基础知识、管理学的基础理论,全面理解企业经营与管理实践中面临的基本问题,并能够熟练应用管理学的基本原理进行分析和解决,提高管理技能。

Construction Enterprise Management Curriculum Standard

1. Curriculum Description

Construction Enterprise Management is the professional elective course for the specialty of construction cost. The courses delivered are *The Basis of Economics*, *Management Psychology*, *Construction Enterprise Accounting*, etc. and the followed courses are *Construction Enterprise Statistics*, *Material Cost Accounting*, *Engineering Economy* and some other courses.

This course is delivered in the first semester.

2. Curriculum Design Ideas

This course is oriented by professional positions, starting with the need for enterprise management talents in vocational posts. The direct objective of the course is to improve teaching quality and enhancing teaching effectiveness and the long-term basis is to build a well-structured and well-qualified teachers'team. Teaching reform and scientific research and innovation are the driving force, and we tried to refine this course with characteristics and high quality.

Taking the combination of learning and practicing as the starting point, the course structure is well designed and planed. The core vocational competence is based on the needs of vocational posts. The professional training program is revised to make more reasonable curriculum standard. The design of this course is oriented by task, and it aims at expanding knowledge and imitating practical vocational situations. Based on the theory of structuralism, the principle of selecting necessary material, we develop and design the student-oriented "contextual" teaching package. We strive to link the theory with practice and learn with methods of guided heuristic teaching and the promotion of case teaching to strengthen students'adaptability to future posts. This course is delivered by qualified teaching group. The teachers make good use of information technology to improve the overall teaching level and expand the amount of teaching information.

3. Curriculum Objectives

The overall objective of this course is to develop students'comprehensive management skills and capabilities in the basic management posts. Through this course, students will develop their six core capabilities: enterprise environmental analysis capabilities, goal management and planning capabilities, decision-making and strategic management capabilities, organizational design and coordination capabilities, leadership and communication capabilities, information control and processing capabilities.

In the process of learning, students'capabilities will be strengthened and they will systematically master the basic knowledge of enterprises, management theory basis, comprehensively understand the basic problems faced in the practice of enterprise management and skillfully analyze and solve problems with the basic theory of enterprise management.

(一)知识目标

(1)了解企业的概念及其特征,了解企业组织形式按不同标准划分的各种类型,理解现代企业制度的概念、特征以及内容,理解企业文化的含义、构成和特征,了解企业环境的分类、内容和环境分析的内容。

(2)理解管理的定义和性质,了解企业管理的基本任务,了解管理者的不同分类和工作任务,理解管理者应具备的技能。

(3)理解目标设置的基本原则,理解目标管理的概念、特点和基本过程,了解计划工作的概念、内容、制定的方法和程序,理解计划工作的有效性权变因素。

(4)掌握决策的概念和程序,区分确定型决策、风险型决策和不确定型决策,了解企业战略的概念、特征和构成要素,了解战略管理的概念、分类和过程。

(5)理解组织结构的基本形式和设计原则,掌握管理幅度和管理层次的关系,了解职权的类型与职权关系,理解授权的定义和过程。

(6)理解领导的实质和领导权力的构成,了解各种领导理论,理解激励的基本原理和各种激励理论应用原则,理解沟通的基本模式和有效沟通的原则。

(7)理解管理控制的目标和内容,了解管理控制的各种分类方法和类型,了解管理控制过程,了解各种控制技术与方法。

(二)素质目标

(1)有崇高的敬业精神和良好的职业道德。

(2)有扎实的专业技能和高度的行动力与执行力。

(3)有高度的整合能力和良好的规范化能力。

(4)有积极的开拓精神和良好的团队合作精神。

(5)有良好的创新意识和高度的意志力与承受能力。

(6)有明确的职业目标和职业发展规划。

(三)能力目标

(1)能进行企业登记的网上预评和网上申报,能识别企业所面临环境的不确定性程度,能针对企业的内外部环境进行 SWOT 分析。

3.1   Knowledge Objectives

After learning this course, students will be able to

(1) understand the concept of enterprises and their characteristics, understand the various types of enterprise organization forms divided by different standards, understand the concept, characteristics and content of modern enterprise system, understand the meaning, composition and characteristics of corporate culture and understand the classification, content and analysis of the enterprise environment.

(2) understand the definition and nature of management, understand the basic tasks of enterprise management, understand the different classification and job tasks of managers and understand the basic skills managers should have.

(3) understand the basic principles of goal setting, understand the concepts, characteristics and processes of goal management, understand the concepts, content, methods and procedures of planning work, and understand the factors that govern the effectiveness of planning.

(4) master the concepts and procedures of decision-making, distinguish between deterministic, risk-based and uncertain decision-making, understand the concepts, characteristics and constituent elements of enterprise strategy, and understand the concept, classification and process of strategic management.

(5) understand the basic form and design principles of organizational structure, master the relationship between the level of management and management level, understand the type of authority and the relationship of authority and understand the definition and process of authorization.

(6) understand the essence of leadership and the composition of leadership power, understand the various leadership theories, understand the basic principles of motivation and the application principles of various motivation theories and understand the basic mode of communication and the principle of effective communication.

(7) understand the objectives and content of management control, understand the various classification methods and types of management control, understand the management control process and understand the various control techniques and methods.

3.2   Quality Objectives

After learning this course, students will have the quality of

(1) professionalism and work ethic.

(2) mastering solid professional knowledge and quick response in action and execution.

(3) complete integration and standardization.

(4) developingpioneering spirit and good team spirit.

(5) innovation and a high degree of willpower and tolerance.

(6) setting clear career objectives and career development plans.

3.3   Ability Objectives

After learning this course, students will have the ability of

(1) making online reservation and declaration for enterprise registration, identifying the degree of uncertainty faced by the enterprise, and carrying out SWOT analysis based on the internal and external environment of the enterprise.

（2）能划分企业管理的基本职能,能承担管理者的人际关系的任务、信息方面任务、决策的任务,能基本具备履行管理职能所需的相应技能。

（3）能针对管理中的不同的人性假设,运用相应的管理方式和要求,能从不同的管理理论得到启发,并能应用于实际。

（4）能正确制定企业目标,并进行目标管理,能分析一份企业计划书,理解计划的分类和制定的方法。

（5）能进行确定性决策、风险型决策和不确定型决策,能把企业未来的生存和发展问题作为制定战略的出发点和归宿。

（6）能分析某一企业的组织结构形式并提出修改意见,能找到增加管理幅度、减少管理层次、使组织结构扁平的好方法,能进行力场分析,降低组织变革的阻力。

（7）能运用各种领导理论、解决实际管理问题,能灵活运用各种激励方法,能消除管理沟通障碍、进行有效沟通。

（8）能确定关键控制点、强调例外控制,能正确对待工作中出现的偏差,并进行有效控制,能使正式组织控制、非正式组织控制和自我控制趋于一致。

四、课程内容与学时分配

教学内容与学时安排表

序号	教学情境/任务/项目/单元	教学目标	知识点	学时	实践项目	学时
1	企业管理概述	了解现代企业的相关概念及特征、现代企业制度等基础知识	1. 企业概述 2. 现代企业制度 3. 企业组织结构(★) 4. 企业管理的基本原理(★)	2		
2	企业战略管理	了解企业战略的概念、特征、作用和分类等	5. 企业战略管理概述 6. 企业战略管理环境分析 7. 企业总体战略与竞争战略(★) 8. 企业战略管理的步骤	2		

（2）dividing the basic functions of enterprise management, assuming the task of optimizing interpersonal relationships, controlling information and making decisions as managers and mastering the basic skills needed to perform the management function.

（3）applying some management methods and requirements to different people with different personalities, and mastering different management theories and applying them.

（4）setting the objectives of the enterprise and managing the objectives, analyzing an enterprise plan, understanding the classification of the plan and the method of development.

（5）making certain decisions, risk-based decisions and uncertain decisions, and taking the future survival and development of enterprises as the starting point and destination of strategy.

（6）analyzing the organizational structure of a certain enterprise and proposing suggestions for revising, finding a good way to increase the management range, reducing the level of management and making the organizational structure flat and carrying out force field analysis and reducing the resistance of organizational change.

（7）using various leadership theories, solving practical management problems, flexibly using various incentive methods, eliminating the barriers to management communication and making effective communication.

（8）identifying key control points, emphasizing exception control, correctly treating deviations in the work, and making formal organizational control, informal organization control and consistent self-control tendency.

4. Curriculum Contents and Teaching Hours

Table of Teaching Contents and Teaching Hours

Item	Topic	Teaching Objectives	Teaching Contents and Teaching Hours			
			Lesson Contents	Hrs	Student Activities	Hrs
1	Introduction to Construction Enterprise Management	Understand the basics of modern enterprise concepts and characteristics, modern enterprise system, etc.	1. Introduction to enterprises	2		
			2. Introduction to modern enterprise system			
			3. Enterprise organization structure(★)			
			4. Basic theories of enterprise management(★)			
2	Enterprise strategic management	Understand the concept, characteristics, role and classification of enterprise strategy	5. Introduction to enterprise strategy	2		
			6. Analysis of enterprise strategic management environment			
			7. The overall strategy and the competitive strategy of the enterprise(★)			
			8. Steps to enterprise strategic management			

序号	教学情境/任务/项目/单元	教学目标	教学内容及学时分配			
			知识点	学时	实践项目	学时
3	企业经营决策与计划	掌握企业管理决策和计划的概念、内容、程序和方法,熟悉计划的编制程序等	9. 企业经营决策(★)	2		
			10. 企业经营计划(★)			
4	企业质量管理	掌握企业质量管理的相关概念以及发展史,了解企业质量管理的七种统计分析方法	11. 质量管理概述	2		
			12. 质量管理体系与认证			
			13. 质量管理中统计分析方法			
5	企业物流管理	了解和掌握物流管理的概念及发展阶段,掌握采购的策略,熟悉各种运输方式的特点,能够科学合理地组织运输和库存管理	14. 企业物流管理概述	2		
			15. 商品采购管理			
			16. 商品运输管理			
			17. 商品库存管理			
6	企业人力资源管理	掌握人力资源管理的概念、主要内容及特点,掌握工作分析的概念,了解招聘、人力资源培训类别及反馈和绩效管理的基本内容	18. 人力资源管理概述	3		
			19. 工作分析与人力资源规划(★)			
			20. 人力资源的招聘与培训(★)			
			21. 薪酬管理与绩效考评(★)			

continued

Item	Topic	Teaching Objectives	Teaching Contents and Teaching Hours			
			Lesson Contents	Hrs	Student Activities	Hrs
3	Enterprise management decision and planning	Master the concepts, contents, procedures and methods of enterprise management decisions and plans, and be familiar with the process of making plans	9. Enterprise management decision(★)	2		
			10. Enterprise management planning(★)			
4	Enterprise quality management	Master the relevant concepts and development history of enterprise quality management and understand the seven statistical analysis methods of enterprise quality management	11. Introduction to quality management	2		
			12. Quality management system and certification			
			13. Statistical analysis method in quality management			
5	Enterprise logistics management	Understand and master the concept and development stage of logistics management; master procurement strategies, be familiar with the characteristics of various modes of transportation and organize the transportation and inventory management scientifically and reasonably	14. Introduction to enterprise logistics management	2		
			15. Commodity procurement management			
			16. Commodity transport management			
			17. Commodity inventory management			
6	Enterprise human resources management	Master the concept, main content and characteristics of human resources management; master the concept of job analysis, understand the categories of recruitment and human resources training and understand the basic content of feedback and performance management	18. Introduction to human resources management	3		
			19. Job analysis and human resources planning(★)			
			20. The recruitment and training of human resources(★)			
			21. Compensation Management and Performance Evaluation(★)			

序号	教学情境/任务/项目/单元	教学目标	教学内容及学时分配			
			知识点	学时	实践项目	学时
7	企业财务管理	理解财务管理和企业筹资的概念,掌握筹资的各种方式和优缺点,了解成本和利润的管理及财务分析方法,掌握财务指标的分析和运用。	22. 企业财务管理概述	3		
			23. 企业筹资管理(★)			
			24. 企业成本与利润管理(★)			
			25. 企业财务分析(★)			
8	其他企业管理理论	了解市场营销的管理过程各环节的基本知识,能够正确运用各种市场营销策略,掌握生产管理、信息管理、企业文化的基本概念和特点,了解创新和管理创新的模式。	26. 营销管理	2		
			27. 生产管理			
			28. 信息管理			
			29. 企业文化			
			30. 管理创新(★)			
合计				18		
备注:重要知识点标注★						

五、教学实施建议

(一)教学参考资料

李渠建.《企业管理基础》,高等教育出版社,2013。

(二)教师素质要求

具有教师资格,以及经济管理类专业背景,或从事过经济管理方面的教学或社会实践工作,非常熟悉企业经营管理过程当中的要求,同时具备教师的职业素养。

(三)教学场地、设施要求

教学场地可以是多媒体教室,也可以是普通教室。

continued

Item	Topic	Teaching Objectives	Teaching Contents and Teaching Hours			
			Lesson Contents	Hrs	Student Activities	Hrs
7	Enterprise financial management	Understand the concept of financial management and corporate financing, master the various ways and disadvantages of financing, understand the management of costs and profits and financial analysis methods, and master the analysis and application of financial indicators	22. Introduction to enterprise financial management	3		
			23. Corporate financing management(★)			
			24. Enterprise Cost and Profit Management(★)			
			25. Corporate financial analysis (★)			
8	Other theories about enterprise management	Understand the basic knowledge of marketing management process, be able to correctly use various marketing strategies, master the basic concepts and characteristics of production management, information management, industry culture and understand the model of innovation and management innovation	26. Marketing management	2		
			27. Production management			
			28. Information management			
			29. Corporate culture			
			30. Management Innovation(★)			
		Total		18		
Notes: Mark the important knowledge points with ★						

5. Teaching Suggestions

5.1　Teaching Resources

Li Qujian. *Enterprise Management Basis*. Higher Education Press, 2013.

5.2　Teachers'Qualification

Teachers for this course should have the teachers'qualification certificate; they should have an education background in economy management, have the experience of teaching courses about economy management or certain social practice, be familiar with the requirements in the process of enterprise management and they should have the professional qualities.

5.3　Teaching Facilities

This course can be delivered in multimedia classrooms or common classrooms.

六、考核评价

(一) 考核方式

建议本门课程考核采用笔试闭卷或开卷。

(二) 考核说明

理论课平时成绩(课堂表现、出勤、作业)占 30%,期末卷面成绩占 70%。考核重点放在"企业经营决策与计划""企业人力资源管理"和"企业财务管理"三章,其他章节次之。但是试题覆盖面应包括各个章节。

6. Evaluation

6.1 Evaluation Method

Examination without reference materials or examination with certain materials.

6.2 Evaluation Instruction

The daily performance accounts for 30% of the total score and the final examination accounts for 70%. The focus of the evaluation is "enterprise management decision and planning" "enterprise human resources management" and "enterprise financial management". The evaluation should cover all the contents of this course.

四川建筑职业技术学院

课 程 标 准

课程名称: 外教课堂学习法

课程代码: 210038

课程学分: 1

基本学时: 16

适用专业: 工程造价(中英合作办学项目)

执 笔 人: 伍慧卿

编制单位: 中丹项目教学部

审 核: 国际技术教育学院

批 准: 教务处(盖章)

Sichuan College of Architectural Technology

Curriculum Standard

Curriculum Name : Learning Strategy in Foreign Teachers'Classrooms

Curriculum Code :   210038

Credits :   1

Teaching Hours :   16

Applicable Specialty :   Construction Cost

                    (Sino-British Cooperative Program)

Compiled by :   Wu Huiqing

On Behalf of : Teaching & Research Section of Sino-Denmark Program

Reviewed by :   The Department of ISTE

Approved by :   Teaching Affairs Office

《外教课堂学习法》课程标准

一、课程概述

本课程是针对中外合作办学项目专业开设的一门学习指导课。主要目的是为解决合作办学项目专业学生即将进入外教专业课的学习,对于外教惯用的教学方法、考核方法不熟悉以及沟通技能欠缺,从而给专业学习带来困难和障碍的问题,专门开设的特色课程。

本课程适宜在第1、2学期开设。

二、课程设计思路

本课程以培养学生的国际化视野,提升跨文化交际能力为导向,促进和帮助学生树立人文交流意识、鼓励跨文化交流的师生关系建立,采用理论教学和项目教学法,指导学生掌握外教的课堂教学及考评方法,通过模拟实践提升外教课堂学习技能。

三、课程目标

通过本课程教学与训练培养学生的四个关键能力,即正确看待文化冲突的能力、跨文化交流的能力、自我学习的能力、与外教沟通的能力。

通过教学过程强化学生能力培养,使学生了解跨文化交流的基础理论,理解跨文化冲突的基本表现,熟悉外教的课堂教学基本方法,提高自身的学习能力和与外教的沟通交流能力。

(一)知识目标

(1)了解西方教育体系和现代教育趋势,对中西方的教育发展、教育观念及教育资源现状加深理解。

(2)理解跨文化交际的概念,理解跨文化冲突的根源和表现,理解外事交往的基本原则。

(3)理解外教教学的常用教学方法、考评方法。

Learning Strategy in Foreign Teachers' *Classrooms* **Curriculum Standard**

1. Curriculum Description

This course is a compulsory and featured tutorial one for construction cost specialty students (Sino-British Cooperative Program). It aims to help Chinese students get prepared and grasp proper learning strategies before entering the foreign teachers'classroom for the acquisition of specialized courses, tackling problems in cross-cultural communication, unfamiliar teaching style and methods, and different, assessment criteria from Chinese teachers. It is delivered for two semesters of one year since the freshmen admission.

2. Curriculum Design Ideas

This course is designed on ideas of helping students broaden their international horizons, improve cross-cultural communication ability, promote students awareness of cultural exchange, encourage the establishment of friendship between foreign teachers and Chinese students. This course encompasses theoretical teaching and task-based learning, guiding students to acquire the classroom teaching and evaluation methods of foreign teachers, and enhances learning skills in foreign teachers'classrooms through lots of simulation activities.

3. Curriculum Objectives

This course aims to cultivate students'four key abilities, namely, the ability to view cultural conflicts correctly, the ability of cross-cultural exchange, the ability of autonomous learning, and the ability to communicate with foreign teachers.

The teaching process aims to strengthen the students'ability training, so that students can understand the basic theory of cross-cultural communication, the common phenomena of cross-cultural conflicts, and be familiar with the basic methods of foreign teaching, improving their ability of learning and communicating with foreign teachers.

3.1 Knowledge Objectives

After learning this course, students will be able to

(1) understand the system of Western education and the trend of modern education; deepen students'understanding of the educational development, ideas and resources of China and the West.

(2) understand ideas of cross-cultural exchange, the reasons and symptoms of cross-cultural conflicts and the basic principles of diplomacy.

(3) acquire the common teaching and assessment methods used by foreign teachers.

（4）掌握相应的学习策略,掌握学习风格的测试、学习计划的制定、时间的有效管理、主动学习和自我反思与评价的方法。

（5）掌握团队建设的定义和方法。

（6）掌握与外教沟通的方法和技能。

（二）素质目标

（1）树立跨文化交流的意识,培养跨文化交流的能力。

（2）理解文化冲突的缘由并能克服跨文化冲突的障碍。

（3）提高学习的自我管理意识和团队合作。

（4）树立与外教积极沟通的意识。

（三）能力目标

（1）能正确看待中外教育、文化的差异,理解跨文化交际的意义并主动消除文化冲突的能力;

（2）学习自我管理和评价的能力;

（3）团队建设的能力;

（4）与外教沟通的能力。

四、课程内容与学时分配

教学内容与学时安排表

序号	教学情境/任务/项目/单元	教学目标	教学内容及学时分配			
			知识点	学时	实践项目	学时
1	世界教育概述	了解中西方教育的状况,对比教育差异	1. 中国教育框架	2		
			2. 西方教育思想			
			3. 对比中西方教育的异同			
2	跨文化交际与课堂教学	了解跨文化交际、冲突的定义、原则和能力培养方法,了解外事纪律	4. 跨文化交际的定义	2		
			5. 跨文化冲突的定义			
			6. 解决跨文化冲突的方法			

(4) grasp proper learning strategies, the learning styles test index, the planning of learning schedules, effective time management, autonomous learning and methods of reflection and self-evaluation.

(5) understand the definition and the method of team building.

(6) master skills and methods of communicating with foreign teachers.

3.2 Quality Objectives

After learning this course, students will have the quality of

(1) cross-cultural communication;

(2) understanding the reasons of cultural conflicts and learning to overcome the obstacles;

(3) improving the awareness of self-management in study and the teamwork ability;

(4) fostering the awareness of active communication with foreign teachers.

3.3 Ability Objectives

After learning this course, students will have the ability of

(1) considering the educational and cultural differences of China and other countries correctly, understanding the meaning of cross-cultural interaction and cultivating the ability to eliminate the cultural conflict actively;

(2) mastering the ability for self-management and self-assessment;

(3) developing the ability for team building;

(4) developing the ability for communicating with foreign teachers.

4. Curriculum Contents and Teaching Hours

Table of Teaching Contents and Teaching Hours

Item	Topic	Teaching Objectives	Teaching Contents and Teaching Hours			
			Lesson Contents	Hrs	Student Activities	Hrs
1	Introduction to the world education	Understand the educational situations in China and the West; compare the differences in education of home and abroad	1. China's educational framework	2		
			2. Western educational thoughts			
			3. Compare the educational differences between home and abroad			
2	Crosscultural communication and classroom teaching and learning	Understand the definition of cultural exchanges and conflicts as well as the communication principle and ability cultivation methods; understand the diplomatic disciplines	4. Definition of cultural exchanges	2		
			5. Definition of cultural conflicts			
			6. Solutions of cultural conflicts			

序号	教学情境/任务/项目/单元	教学目标	教学内容及学时分配			
			知识点	学时	实践项目	学时
3	外教课堂教学法及考评方法	了解外教课堂常用教学及考核方法	7. 外教课堂常用教学方法 8. 外教课堂常用考评方法	2		
4	学习策略	了解自身的学习风格、树立目标、制订计划、时间管理和自我反思与评价、团队建设	9. 学习风格测试 10. 学习目标建立 11. 学习计划制订 12. 时间管理 13. 主动学习与自我评价 14. 团队建设	4		
5	外教沟通技巧	掌握与外教沟通的技巧	15. 书面沟通 16. 电话沟通 17. 面对面沟通	2		
6	实践操作	主题演讲、外教沟通	18. 分组主题英文演讲 19. 分组与外教话题沟通	4		
合计				16		
备注	鉴于学时的限制和本专业的教学层次要求,教师在讲授过程中应注意不能只停留于表面,而是需要对重要知识点进行拓展、加深,挖掘表格背后的支撑知识,练习的灵活性加大,难度加深					

五、教学实施建议

(一) 教学参考资料

伍慧卿.《外教课堂学习策略》,自编讲义。

(二) 教师素质要求

具有教师资格及海外留学背景,从事管理类专业学科的老师,同时具备教师的职业素养。

(三) 教学场地、设施要求

教学场地可以是多媒体教室,也可以是普通教室。

continued

Item	Topic	Teaching Objectives	Teaching Contents and Teaching Hours			
			Lesson Contents	Hrs	Student Activities	Hrs
3	Teaching and assessment methods in foreign teachers' classrooms	Understand those commonly used teaching and assessment methods in foreign teachers'classrooms	7. Teaching methods	2		
			8. Assessment methods			
4	Learning strategies	Understand one's learning styles; targets setting; making plans; time management; reflection and self evaluation; team building	9. Learning style test	4		
			10. Setting up learning goals			
			11. Making learning plans			
			12. Time management			
			13. Autonomous learning and self evaluation			
			14. Team building			
5	Communicative skills with foreign teachers	Grasp communicative skills with foreign teachers	15. Communication by writing	2		
			16. Communication by phone			
			17. Face to face communication			
6	Simulation practice	Keynote speech; communication practice with foreign teachers	18. Keynote speeches of groups	4		
			19. Communication practice in groups with foreign teachers			
Total				16		
Note	In view of the limitations of teaching hours and the students'academic level requirements of the major, teachers should not focus on the shallow teaching of knowledge, but need to expand and deepen those key points, excavate the background knowledge of this table, increase the flexibility and difficulty of simulation practice					

5. Teaching Suggestions

5.1 Teaching Resources

Wu Huiqing. *Learning Strategy in Foreign Teachers'Classroom*. Self-tailored Handbook.

5.2 Teachers'Qualification

Teachers must have teacher qualification certificate, with overseas studying background, engaged in management disciplines, and the professional ethics.

5.3 Teaching Facilities

This course can be delivered in either multimedia classrooms or common classrooms.

六、考核评价

(一)考核方式

建议本门课程考核采用小论文(案例分析、影评、对比研究等)和实操。

(二)考核说明

理论课平时成绩(课堂表现、出勤、作业)占20%,小论文占20%,实操占60%(各30%)。

6. Evaluation

6.1 The Method of Assessment

It is suggested that small essays (like case analysis, film review, comparative research, etc.) be submitted in the course of learning for this unit assessment plus simulation practice performance.

6.2 Assessment Requirement

The final grade score is the total of the learning process assessment (including the usual attendance, in-class performance and homework, accounting for 20%), small essays scores (accounting for 20%) and the two simulations practice grades (accounting for 60% , each of 30%).

四 川 建 筑 职 业 技 术 学 院

课 程 标 准

课程名称：　　　　装配式建筑概论

课程代码：　　　　010761

课程学分：　　　　1

基本学时：　　　　18

适用专业：　　工程造价(中英合作办学项目)

执 笔 人：　　　　吴俊峰

编制单位：　　　　施工教研室

审　　核：　　土木工程系(院)(签章)

批　　准：　　　教务处(盖章)

Sichuan College of Architectural Technology

Curriculum Standard

Curriculum Name: Overview of Prefabricated Construction

Curriculum Code: 010761

Credits: 1

Teaching Hours: 18

Applicable Specialty: Construction Cost

(Sino-British Cooperative Program)

Compiled by: Wu Junfeng

On Behalf of: Teaching & Research Section of Construction

Reviewed by: The Department of Civil Engineering

Approved by: Teaching Affairs Office

《装配式建筑概论》课程标准

一、课程概述

随着建筑行业的转型升级,建筑工业化(装配式建筑)是一个重要的发展趋势。为了加强建筑相关专业的学生对装配式建筑的认识,适应行业发展的需要,《装配式建筑概论》成为建筑相关专业所开设的一门新型主要专业基础课程。它的作用是通过学习装配式施工概述、吊装机具设备工具、装配式施工工艺、设备吊装技术、装配式建筑施工安全管理等知识,培养学生对装配式建筑有全面的了解和认知。

由于装配式建筑概论课程开设的专业是建筑相关专业,并且本课程学习涉及其他课程知识,因此装配式建筑概论课程均开设于第4学期。

装配式建筑概论课程的先修课程涉及《建筑工程制图》《建筑工程测量》《建筑材料》《建筑力学》《房屋建筑学》《建筑施工组织》《建筑工程预算》;装配式建筑概论课程的后续课程主要涉及《装配式施工与施工机械》《工程吊装技术》等课程。

二、课程设计思路

装配式建筑是建筑产业现代化发展的必然途径,对其相关基本的概念性知识有必要进行普及。为加强建筑相关专业的学生对建筑行业转型升级、建筑产业现代化的认识,必须强化装配式建筑的概念性知识学习。为适应建筑转型升级和建筑产业现代化发展的需要,本课程通过对装配式施工概述、吊装机具设备工具、装配式施工工艺、设备吊装技术、装配式建筑施工安全管理的学习,培养学生对装配式建筑全面的了解和认知能力。

由于装配式建筑概论课程理论性强、实践性强、综合性大、社会性广,工程施工中许多技术问题的解决,均要涉及相关学科知识的综合运用。因此,要求学生拓宽知识专业面、扩大知识面,要有牢固的专业基础理论和知识,并自觉地进行运用到本课程中。

Overview of Prefabricated Construction Curriculum Standard

1. Curriculum Description

With the transformation and upgrading of the construction industry, prefabricated construction became an important trend. In order to enhance students'understanding of prefabricated building and meet the needs of industry development, *Overview of Prefabricated Construction* is delivered. It is a new type of professional basic course for all architecture-related specialties. It aims at cultivating students'comprehensive understanding of prefabricated building by learning the knowledge of overview of prefabricated construction, hoisting equipment and tools, prefabricated technology, equipment hoisting technology, and construction safety management of prefabricated building.

This course is related to architecture and involves knowledge of other courses, so it is delivered in the fourth semester.

The delivered courses are *building engineering drawing*, *building engineering surveying*, *building materials*, *building mechanics*, *building architecture*, *construction organization*, *building engineering budget*, etc. The courses followed are mainly *prefabricated construction and construction machinery*, *hoisting technique* and so on.

2. Curriculum Design Ideas

Prefabricated construction is an inevitable way for the modernization of construction industry. It is necessary to popularize its basic conceptual knowledge. In order to enhance the students'understanding of the transformation and upgrading of the construction industry and the modernization of the construction industry, it is necessary to strengthen the conceptual knowledge learning of prefabricated building. In order to meet the needs of building transformation and upgrading and the modernization of construction industry, in this course, students will learn the overview of prefabricated construction, hoisting equipment and tools, prefabricated technology equipment hoisting technology, and construction safety management of prefabricated building to cultivate students'comprehensive understanding and cognitive ability of prefabricated building.

The course is quite theoretical, practical, comprehensive and social. The solutions of many technical problems in engineering construction need to use relevant knowledge. Therefore, students are required to broaden their knowledge and professional areas, have solid professional basic theory and knowledge, and consciously apply them in this course.

教学中必须实现本课程与装配式建筑施工岗位要求的衔接。由于本课程是纯理论课程,因此可在教学中运用现代信息技术教学,调动学生的积极性,通过视频、动画、照片等展现本课程的教学内容,有利于课堂教学与装配式建筑施工实际相结合,让学生更容易理解和掌握。

三、课程目标

通过对本门课程的装配式施工概述、吊装机具设备工具、装配式施工工艺、设备吊装技术、装配式建筑施工安全管理的学习,培养学生对装配式建筑具有全面的了解和认知能力。

(一)知识目标

(1)了解装配式混凝土结构的发展意义,懂得国内外装配式建筑发展现状,知道装配式建筑发展前景;

(2)熟悉吊装机具设备工具的使用与注意事项,熟悉设备吊装技术与吊装技术方案的编制;

(3)掌握装配式混凝土结构构件的生产、运输与堆放、现场吊装、连接等,了解水、电、暖等预留与预埋;

(4)掌握装配式混凝土结构质量验收与控制,了解常见质量问题处理,以及安全措施;

(5)掌握装配式施工的相关施工工艺,了解装配式建筑的管理特征;

(6)熟悉装配式建筑的施工安全管理。

(二)素质目标

(1)培养学生具有从事装配式建筑所必需的专业基础知识,能适应建筑工业化对学生毕业后从事职业岗位的新要求;

(2)让学生拓宽知识面,增强自学能力;养成严谨务实的工作作风;

(3)培养学生开阔、敏捷的思路,具有创新精神、自觉学习的态度;

(4)培养学生养成安全管理的意识;

(5)培养综合运用所学知识分析问题和解决问题的能力。

(三)能力目标

(1)通过对本课程的学习,学生能充分认识装配式建筑的内涵,对装配式建筑的相关基本概念知识十分清楚。

In the teaching process, we must realize the connection between this course and the post requirements of prefabricated construction. Because this course is purely theoretical, we can use modern information technology in the teaching to arouse the enthusiasm of student. We can use video, animation, photographs and other ways to present teaching content, which facilitates the combination of classroom teaching and prefabricated construction and helps students easily understand and master the knowledge.

3. Curriculum Objectives

Students will learn the overview of prefabricated construction, hoisting equipment and tools, prefabricated technology equipment hoisting technology, and construction safety management of prefabricated building to cultivate students'comprehensive understanding and cognitive ability of prefabricated building.

3.1　Knowledge Objectives

After learning this course, students will be able to

(1) understand the significance of the prefabricated concrete structure, understand the development status of the prefabricated building at home and abroad, and know the development prospects of the prefabricated building;

(2) be familiar with the use of hoisting equipment and tools and points for attention, and master equipment hoisting technology and hoisting technology program preparation;

(3) master the production, transportation and stacking of prefabricated concrete structural components, site hoisting and connection, and understand the reservation and embedding of water, electricity, heating, etc. ;

(4) master the quality acceptance and control of prefabricated concrete structure, common quality problems, and safety measures;

(5) master the relevant construction technology of prefabricated construction and understand the management characteristics of prefabricated construction;

(6) be familiar with construction safety management of prefabricated building.

3.2　Quality Objectives

After learning this course, students will have the quality of

(1) engaging in prefabricated building with professional basic knowledge, and adapting to the new requirements of building industrialization to engage in professional posts after graduation;

(2) broadening knowledge and enhancing their self-study ability, and developing a rigorous and pragmatic style of work;

(3) being open-minded, agile, innovative and self-conscious in learning attitude;

(4) having awareness of safety management;

(5) analyzing and solving problems by using the knowledge comprehensively.

3.3　Ability Objectives

After learning this course, students will have the ability of

(1) understanding the connotation of prefabricated building, and having a clear understanding of the basic conceptual knowledge of prefabricated building.

（2）能够深刻体会装配式混凝土建筑是以构件预制化生产、装配式施工为生产方式，以设计标准化、构件部品化、施工机械化为特点，能够整合设计、生产、施工等整个产业链，实现建筑产品节能、环保、全寿命周期价值最大化的可持续发展的新型建筑生产方式。

（3）能够运用装配式混凝土结构构件的生产、运输与堆放、现场吊装、连接、质量验收与问题处理等基本知识，为今后从事装配式建筑工作打下扎实的基础。

四、课程内容与学时分配

教学内容与学时安排表

序号	教学情境/任务/项目/单元	教学目标	教学内容及学时分配			
			知识点	学时	实践项目	学时
1	装配式施工概述	1. 了解装配式混凝土结构的发展意义； 2. 懂得国内外装配式建筑发展现状； 3. 知道装配式建筑发展前景	1. 装配式混凝土结构的发展意义	1		
			2. 国内外装配式建筑发展现状			
			3. 装配式建筑发展前景			
2	吊装机具设备工具	1. 了解钢丝绳的种类，掌握钢丝绳的技术性能和容许用拉力计算，掌握绳卡等附件正确使用； 2. 了解索具与吊具的概念、吊钩的分类，掌握吊钩的技术规格和吊钩的操作方法； 3. 了解滑车的类型和构造，掌握使用滑车时的注意事项； 4. 了解电动卷扬机的类型、选择和布置，掌握卷扬机的安装和使用卷扬机时的注意事项	4. 钢丝绳及其附件（★）	2		
			5. 索具、吊具的常用端部配件（★）			
			6. 滑轮组	2		
			7. 平衡梁			
			8. 常用机具			

（2）understanding that the prefabricated concrete building is a new type of sustainable building production mode, which takes prefabricated production and assembled construction as its production mode, standardizes the design, fractionalizes the components and mechanizes the construction, integrates the whole industrial chain of design, production and construction, and realizes the sustainable development of energy-saving, environmental protection and life cycle value maximization of building products.

（3）using the basic knowledge of production, transportation and stacking of assembled concrete structural components, site hoisting, connection, quality acceptance and problem handling to lay a solid foundation for future work in prefabricated construction.

4. Curriculum Contents and Teaching Hours

Table of Teaching Contents and Teaching Hours

Item	Topic	Teaching Objectives	Teaching Contents and Teaching Hours			
			Lesson Contents	Hrs	Student Activities	Hrs
1	Introduction to prefabricated construction	1. Understand the development significance of prefabricated concr-ete structure; 2. Understand the development status of prefabricated building at home and abroad; 3. Know the development pros-pects of prefabricated building	1. Development significance of assembly concrete structures	1		
			2. Development status of prefabricated building at home and abroad			
			3. Development prospects of prefabricated buildings			
2	Lifting machinery and equipment	1. Understand the types of wire rope, master the technical performance of wire rope and allowable tension calculation, and master the correct way of using rope clips and other accessories; 2. Understand the concept of rigging and spreader, the classification of hook, master the technical specifications of hook and its operation method; 3. Understand the type and structure of the pulley and master its points for attention. 4. Understand the types, selection and arrangement of electric hoisters, and master the installation and use of hoisters	4. Wire rope and its accessories（★）	2		
			5. Common end fittings of rigging and spreader（★）	2		
			6. Pulley block			
			7. Balanced Beam			
			8. Common Machineries and Tools			

续表

序号	教学情境/任务/项目/单元	教学目标	教学内容及学时分配			
			知识点	学时	实践项目	学时
3	施工工艺	1. 了解装配式混凝土结构构件； 2. 熟悉预制混凝土构件工厂生产工艺； 3. 掌握预制构件的运输与堆放； 4. 掌握预制混凝土构件的连接和装配式混凝土结构施工工艺； 5. 了解水、电、暖等预留与预埋； 6. 掌握装配式混凝土结构质量验收与控制，了解常见质量问题处理，以及安全措施	9. 装配式混凝土结构构件	3		
			10. 预制混凝土构件工厂生产（★）			
			11. 预制构件运输（★）	3		
			12. 施工现场生产（★）			
4	设备吊装技术	1. 了解液压同步提升系统，懂得液压提升系统设计基本方法，掌握液压提升设备的安装与液压提升系统试运行调试； 2. 了解自行式起重机基本知识，懂得自行式起重机的技术使用与安全管理，掌握自行式起重机常用吊装工艺； 3. 了解塔式起重机的分类、特点、基本结构和参数，知道塔式起重机的选择与计算，掌握塔式起重机安装； 4. 知道桅杆式起重机设计，掌握桅杆式起重机的吊装工艺； 5. 熟悉吊装技术方案的编制	13. 整体液压同步提升技术简介	3		
			14. 自行式起重机			
			15. 塔式起重机（★）			
			16. 桅杆式起重机设计及吊装工艺	3		
			17. 吊装技术方案的编制（★）			

continued

Item	Topic	Teaching Objectives	Teaching Contents and Teaching Hours			
			Lesson Contents	Hrs	Student Activities	Hrs
3	Construction Technology	1. Understand the components of assembled concrete structures; 2. Be familiar with the production process of precast concrete components in the factory; 3. Master the transportation and stacking of prefabricated components; 4. Master the connection of precast concrete components and the construction technology of assembled concrete structures; 5. Understand the reservation and embedding of water, electricity and heating; 6. Master quality acceptance and control of assembled concrete structure. understand common quality problems and safety measures	9. Assembled concrete structural components	3		
			10. Precast concrete components in the factory(★)			
			11. Transportation of prefabricated components(★)	3		
			12. Production in construction site(★)			
4	Hoisting Technology	1. Understand the hydraulic synchronous lifting system and the basic design methods of hydraulic lifting system, and master the installation of hydraulic lifting equipment and the commissioning of hydraulic lifting system; 2. Understand the basic know-ledge of self-propelled crane, the technical use and safety management of self-propelled crane, and master the common hoisting technology of selfpropelled crane; 3. Understand the classification, characteristics, basic structure and parameters of tower crane, master the selection and calculation of tower crane, and its installation; 4. Understand the design of mast crane and the lifting technology of mast crane; 5. Be familiar with the preparation of hoisting technical scheme	13. Brief Introduction of Integrated Hydraulic Synchronized Lifting Technology	3		
			14. Self-propelled Crane			
			15. Tower crane(★)			
			16. Design and hoisting technology of mast crane	3		
			17. Compilation of hoisting technical scheme(★)			

序号	教学情境/任务/项目/单元	教学目标	教学内容及学时分配			
			知识点	学时	实践项目	学时
5	装配式建筑施工安全	1. 熟悉生产安全管理； 2. 掌握预制构件运输、起吊及支撑防护施工安全措施； 3. 懂得环境安全	18. 生产安全管理	1		
			19. 预制构件运输、起吊及支撑防护(★)			
合计				18		
备注	重要知识点标注★					

五、教学实施建议

(一)教学参考资料

推荐教材：

陈文元、戴安全.《装配式建筑与设备吊装技术》,中国建筑工业出版社,2018。

参考书目：

陈群、蔡彬清、林平.《装配式建筑概论》,中国建筑工业出版社,2017。

郭学明.《装配式混凝土结构建筑的设计、制作与施工》,机械工业出版社,2017。

(二)教师素质要求

进行本门课程教学的教师应具有土木工程专业背景,有丰富的理论与工程实践教学能力,具有一定的建筑、结构与施工专业知识,对施工机械有一定的了解。

(三)教学场地、设施要求

本课程涉及 PC 构件厂、预制构件场内外运输、装配式构件现场施工等实践教学环节,应通过大量的图片、动画、录像等来呈现,以缩短学生的认知进程。因此,进行教学的教室应为多媒体教室。

六、考核评价

成绩构成比例原则按《四川建筑职业技术学院学生学业考核办法》执行。

(1)本课程总评成绩由平时成绩和期末成绩综合确定。其中平时成绩占30%,期末成绩占70%。

continued

Item	Topic	Teaching Objectives	Teaching Contents and Teaching Hours			
			Lesson Contents	Hrs	Student Activities	Hrs
5	Construction Safety of Assembled Buildings	1. Be familiar with production safety management; 2. Master transportation, lifting and supporting protection of prefabricated components; 3. Understand environmental safety.	18. Production Safety Management	1		
			19. Transportation, lifting and support protection of prefabricated components(★)			
Total				18		
Note: Mark important points with ★						

5. Teaching Suggestions

5.1 Teaching Resources

Recommended textbooks:

Chen Wenyuan, Dai Anquan. *The Technology of Prefabricated Building and Equipment Lifting.* China Architecture & Building Press, 2018.

References:

Chen Qun, Cai Binqing, Lin Ping. *Overview of Prefabricated Construction.* China Architecture & Building Press, 2017.

Guo Xueming, *Design, Production and Construction of Assembled Concrete Structural Buildings.* China Machine Press, 2017.

5.2 Teachers'Qualification

Teachers of this course should have a professional background in civil engineering, rich professional theory and engineering practice, have certain professional knowledge in architecture, structure and construction, and know some construction machineries.

5.3 Teaching Facilities

This course involves practical teaching such as PC component factory, transportation of prefabricated components inside and outside the field, construction of prefabricated components on site, etc. In order to shorten the cognitive process of students, we can use a large number of pictures, animations, videos, and other media in the teaching. Therefore, the classroom for teaching should be multimedia classroom.

6. Evaluation

The score of the course is based on the *Regulation of Academic Evaluation for Students in SCAT*.

(1) The total score is consisted of the daily performance and the final examination. The daily performance accounts for 30% of the total score and the final examination accounts for 70%.

（2）考核方式为考查，采用笔试考试。

（3）根据关于《建筑工程技术专业教学资源库》使用办法（试行）的规定，对于符合关于"建筑工程技术专业教学资源库"使用办法（试行）的情况，本课程的考核成绩也可由卷面成绩、平时成绩和学习资源库成绩综合确定。其中卷面成绩占60%，平时成绩占30%，学习资源库成绩占10%。

(2) The final examination is carried out with test papers.

(3) According to the regulations of the *Teaching Resource Bank for Construction Engineering Technology Specialty* (trial implementation), the total score of this course can also be determined by the final examination, the daily performance and the score of the learning resource bank. Among them, the final examination accounts for 60% of the total score, the daily performance accounts for 30% and the score of the learning resource bank accounts for 10%.

四川建筑职业技术学院

课程标准

课程名称：　　　　Revit 建模基础

课程代码：　　　　150389

课程学分：　　　　　2

基本学时：　　　　　32

适用专业：　　工程造价（中英合作办学项目）

执　笔　人：　　　　李剑心

编制单位：　　　建筑工程造价教研室

审　　　核：　　工程管理系（院）（签章）

批　　　准：　　　教务处（盖章）

Sichuan College of Architectural Technology

Curriculum Standard

Curriculum Name: Revit Modeling Basis

Curriculum Code: 150389

Credits: 2

Teaching Hours: 32

Applicable Specialty: Construction Cost

     (Sino-British Cooperative Program)

Compiled by: Li Jianxin

On Behalf of: Teaching & Research Section of Construction Cost

Reviewed by: The Department of Engineering Management

Approved by: Teaching Affairs Office

《Revit 建模基础》课程标准

一、课程概述

本课程是工程造价(中英合作办学项目)专业的一门专业技术课程,是在完成了《建筑识图》《建筑结构基础与识图》等课程之后的一门专业软件应用课程。它主要从 Revit 软件应用出发,讲授使用软件完成工程建模与具体应用实务的课程,具有复习巩固学生工程造价和 BIM 相关知识,以及结合应用提高工作质量的作用,为激烈的市场竞争打下坚实的基础。该课程为专业课,在第 4 学期开设。本课程的先修课程为《建筑识图造》和《建筑结构基础与识图》,后续课程为《BIM 技术基础》。

二、课程设计思路

该课程根据国家"十三五"规划、行指委的《工程造价教学基本要求》、四川建筑职业技术学院《工程造价人才培养方案》,以及社会需求进行设置,通过学习 Revit 软件的实际操作,能够独立依照图纸完成房屋建筑模型的建立。该课程全部为实际操作课程,理论:实践为0:32,上课地点应为安装了 Revit 软件的计算机房。

三、课程目标

能够独立操作 Revit 软件完成建(翻)模,并且能够处理软件使用过程中遇到的基本问题。

(一)知识目标

(1)了解 BIM 技术在项目中的应用;

(2)掌握 Revit 建模技术在 BIM 技术中的应用。

(二)素质目标

(1)培养学生团结协作的精神;

(2)树立爱岗敬业的作风;

(3)养成学生初步的职业操守和素养。

Revit Modeling Basis **Curriculum Standard**

1. Curriculum Description

This is a professional technical course for the specialty of construction cost(Sino-British Cooperative Program). It is a professional software application course delivered after the learning of *Construction Drawing Reading*, *Basic Building Structure and Drawing Reading* and so on. It mainly introduces the application of Revit software and focuses on the construction of engineering modeling and its practical application with software. This course can review and strengthen the relative knowledge of construction cost and BIM, improve the quality of its work with application and lay a solid foundation for fierce market competition. This is a professional course delivered in the fourth semester. The delivered courses are *Construction Drawing Reading* and *Basic Building Structure and Drawing Reading*. The unit followed is *Fundamentals of BIM Technology*.

2. Curriculum Design Ideas

The design of this course is based on *The Basic Requirements for the Specialty of Construction Cost* issued by vocation guiding committee in the 13th Five-year plan, professional training program for the specialty of construction cost (Sino-British cooperative program) and the needs of society. After the practical operation of the Revit software, students can construct the model of buildings independently according to the drawings. The contents of the lessons are about the practical operation of the software and the student's activity accounts for all 32 teaching hours. The classrooms for the learning and teaching of this unit should be computer labs with Revit software.

3. Curriculum Objectives

Through this course, students can complete modeling with Revit software and can cope with the problems in the operation of the software.

3.1 Knowledge Objectives

After learning this course, students will be able to

(1)understand the application of BIM technology in practical project;

(2)master the modeling by Revit in BIM.

3.2 Quality Objectives

After this course, students will have the quality of

(1)being united and being cooperative;

(2)having devoted and responsible work style;

(3)taking initial professional ethics and quality.

(三)能力目标

(1)能独立运用 Revit 软件进行建(翻)模;

(2)能独立处理 Revit 软件建模过程中的基本问题。

四、课程内容与学时分配

教学内容与学时安排表

序号	教学情境/任务/项目/单元	教学目标	教学内容及学时分配			
			知识点	学时	实践项目	学时
1	BIM 概述	了解 BIM 技术;了解基于 Autodesk Revit 平台的建模流程;熟悉 Revit 软件及其常用命令	1. 基于 Autodesk Revit 平台的建模流程简介	2		2
			2. Revit 软件及其常用命令介绍		1. Revit 软件及其常用命令介绍	
2	创建模型	熟悉项目准备工作;掌握项目标高和轴网的创建;掌握结构模型创建;掌握建筑模型创建;掌握其他构件创建	3. 创建项目标高和轴网	12	2. 创建项目标高和轴网(★)	12
			4. 基础创建		3. 结构模型创建(★)	
			5. 柱创建			
			6. 梁创建			
			7. 板创建			
			8. 楼梯创建			
			9. 其他结构构件创建			
			10. 墙创建		4. 建筑模型创建(★)	
			11. 门			
			12. 窗			

3.3 Ability Objectives

After learning this course, students will have the ability of

(1) using Revit software independently to build(turn) models;

(2) handling the basic problems in the Revit software modeling process independently.

4. Curriculum Contents and Teaching Hours

Table of Teaching Contents and Teaching Hours

| Item | Topic | Teaching Objectives | Teaching Contents and Teaching Hours | | | |
			Lesson Contents	Hrs	Student Activities	Hrs
1	Overview of BIM	Understand BIM technology; Understand the modeling process by Autodesk Revit platform; be familiar with Revit software and its common operation commands	1. Brief introduction to the modeling process by Autodesk software	2		2
			2. Introduction to Revit software and its common operation commands		1. Introduction to Revit software and its common operation commands	
2	Creating the model	Be familiar with the preparatory work of project; master the creating methods for elevation and axis net; master the method in creating structure model; master the method in creating building model; master the method in creating the models of other components	3. Create the elevation and axis net of the project	12	2. create the elevation and axis net of the project(★)	12
			4. Create the model of the foundation		3. Create the structure model(★)	
			5. Create the model of column			
			6. Create the model of beam			
			7. Create the model of pane			
			8. Create the model of staircase			
			9. Create the model of other structure components			
			10. Create the model of walls		4. Create the building model(★)	
			11. Create the model of doors			
			12. Create the model of windows			

序号	教学情境/任务/项目/单元	教学目标	教学内容及学时分配			
			知识点	学时	实践项目	学时
2	创建模型	熟悉项目准备工作;掌握项目标高和轴网的创建;掌握结构模型创建;掌握建筑模型创建;掌握其他构件创建	13. 台阶创建 14. 散水创建 15. 其他建筑构件创建 16. 其他构件创建	12	4. 建筑模型创建(★) 5. 其他构件创建	12
3	其他常用功能应用	掌握房间标注和标准层绘制;熟悉渲染与漫游操作;熟悉明细表统计操作	17. 房间标注和标准层绘制 18. 渲染与漫游 19. 明细表统计	2	6. 房间标注和标准层绘制 7. 渲染与漫游 8. 明细表统计	2
合计				16		16
备注:重要知识点标注★						

五、教学实施建议

(一)教学参考资料

王婷.《全国 BIM 技能培训教程 Revit 初级》,中国电力出版社,2015。

胡煜超.《Revit 建筑建模与室内设计基础》,机械工业出版社,2017。

(二)教师素质要求

工程造价专业的教师,具有图学会 BIM 建模师者优先。

(三)教学场地、设施要求

需要安装了 Revit 软件并且能够流畅运行的计算机房。

continued

Item	Topic	Teaching Objectives	Teaching Contents and Teaching Hours			
			Lesson Contents	Hrs	Student Activities	Hrs
2	Creating the model	Be familiar with the preparatory work of project; master the creating methods for elevation and axis net; master the method in creating structure model; master the method in creating building model; master the method in creating the models of other components	13. Create the model of steps		4. Create the building model(★)	
			14. Create the model of apron			
			15. Create the model of other construction components			
			16. Create the model of other components		5. Create the model of other components	
3	Other commonly used functions	Master the methods in room annotation and standard layer drawing; be familiar with the operation of rendering and roaming; be familiar with the operation of statistics of the part list	17. Room annotation and standard layer drawing	2	6. Room annotation and standard layer drawing	2
			18. Rendering and roaming		7. Rendering and roaming	
			19. Statistics of the part list		8. Statistics of the part list	
Total				16		16
Notes: Mark the important knowledge points with ★						

5. Teaching Suggestions

5.1 Teaching Resources

Wang Ting. *National BIM Skills Training Course Revit Elementary*, China Electric Power Press, 2015.

Hu Yuchao. *The Basis of Revit Architecture Modeling and Interior Design*, China Machine Press, 2017.

5.2 Teachers'Qualification

Teachers should be the majors of construction cost, and those with the certificate of BIM modeler issued by China Graphics Society are preferred.

5.3 Teaching Facilities

This course should be delivered in a computer lab with Revit software and computers that can run smoothly with the software.

六、考核评价

(一) 模型建立

(1) 根据给定图纸创建结构模型;

(2) 根据给定图纸创建建筑模型;

(3) 根据给定图纸创建其他构件。

(二) 模型应用

(1) 对创建好的模型进行房间标注和标准层绘制;

(2) 对创建好的模型进行明细表统计。

考核方式:上机实际操作

成绩构成:成绩构成比例原则按《四川建筑职业技术学院学生学业考核办法》执行。

6. Evaluation

The course contents covered in the assessment are creating the model and application of the model.

6. 1 In creating the model

(1) create the structure model based on certain drawing;

(2) create the building model based on certain drawing;

(3) create the model of other components.

6. 2 In the application of the model

(1) make room annotation and draw standard layer in created models;

(2) make statistics of the part list in the created model.

Evaluation method: practical operation of the software on computer

The score of the course is based on the *Regulation of Academic Evaluation for Students in SCAT.*

四 川 建 筑 职 业 技 术 学 院

课 程 标 准

课程名称：   **BIM 工程计价软件应用**

课程代码：   150397

课程学分：   1

基本学时：   18

适用专业：   工程造价（中英合作办学）

执 笔 人：   侯　兰

编制单位：   建筑工程造价教研室

审　　核：   工程管理系(院)(签章)

批　　准：   教务处(盖章)

Sichuan College of Architectural Technology

Curriculum Standard

Curriculum Name: Application of BIM Construction Cost Software

Curriculum Code: 150397

Credits: 1

Teaching Hours: 18

Applicable Specialty: Construction Cost

(Sino-British Cooperative Program)

Compiled by: Hou Lan

On Behalf of: Teaching & Research Section of Construction Cost

Reviewed by: The Department of Engineering Management

Approved by: Teaching Affairs Office

《BIM 工程计价软件应用》课程标准

一、课程概述

本课程适用于造价(土建普通班、实验班、中英造价)专业,《BIM 工程计价软件应用》课程属于考查课程,是工程造价专业的一门操作应用型专业技术课程,主要研究如何利用 BIM 软件计算工程造价(只包含计价部分)。

课程任务包括总结工程造价计价原理、BIM 计价软件的界面操作介绍、计价软件编制工程量清单、计价软件编制招标控制价、计价软件编制投标报价、软件编制工程结算价款等几个部分的内容。课程在讲解过程中,需要学生具备工程清单计价、定额计价等基础,课程实践性强,需要结合生产实际讲解。

先修课程包括《工程识图与构造》《道路工程施工技术》《桥梁工程施工技术》等。

二、课程设计思路

本课程根据工程造价专业的教育理念及建筑行业企业岗位职业技能、综合素质要求等进行设计。设置依据是工程造价专业人才培养大纲和实施计划;课程框架、教学目标、教学内容的设计和选取根据工程造价专业人员素质、技能要求、造价工程师考试大纲等进行。

三、课程目标

学习完本课程,学生能掌握工程造价计价原理,能熟练利用相关软件编制工程量清单、招标控制价、投标报价、工程结算,并能分析三个阶段工程造价编制方法的异同。同时,形成独立使用 BIM 计价软件编制各种造价文件的能力。

(一)知识目标

(1)熟悉工程造价计价原理;

(2)掌握工程量清单的组成;

(3)熟悉招标控制价的编制依据和程序;

Application of BIM Construction Cost Software **Curriculum Standard**

1. Curriculum Description

This course is applicable for the specialty of construction cost (direction of civil engineering, experimental classes and Sino-British cooperative program). It is a course of practical application and professional technology. This course mainly gives knowledge about how to calculate construction cost (the pricing part) based on BIM software.

The main contents of this course are the theory of construction cost pricing, the operation interface of BIM pricing software, the compiling of bills of quantities by the pricing software, the compiling of bid control price by the pricing software, the compiling of the bidding price by the pricing software and the compiling of the settlement price by the pricing software. In learning this course, students should have the basic knowledge of the pricing of bills of quantities, quota principle and so on. This course is quite practical and should be analyzed with practices.

The delivered courses are *Basic Building Structure and Drawing Reading*, *Road Engineering Construction Technology*, *Bridge Construction Technology*, etc.

2. Curriculum Design Ideas

The design of this course is based on the basic education theory, the professional core competence for positions in construction industry and the comprehensive ability development for the specialty of construction cost. The curriculum standard is based on the outline of professional training program and implementation plan of it. The design and selection of curriculum framework, teaching objectives and teaching contents are based onthe core competence and skill requirements for the specialty of construction cost and test syllabus for cost engineer, etc.

3. Curriculum Objectives

Through this course, students can strengthen the basic theory of construction cost pricing and they can be proficient in compiling bills of quantities, bid control price, the bidding price and the settlement price. At the same time, they can analyze the similarities and differences in the compiling methods for the three stages and compile different pricing files by the BIM pricing software independently.

3.1 Knowledge Objectives

After learning this course, students will be able to

(1) be familiar with the basic theory of construction cost pricing;

(2) master the components of bills of quantities;

(3) be familiar with the methods and process in compiling bid control price;

（4）掌握投标报价的编制依据和程序；

（5）掌握工程结算的编制依据和程序；

（6）掌握招标控制价、投标报价、工程结算的区别。

（二）素质目标

（1）通过专业介绍培养学生对专业的热爱、对工程造价岗位的热爱，为培养高尚的职业道德打下基础；

（2）培养学生的组织、协调能力；

（3）培养学生分析问题、解决问题的能力；

（4）培养学生的学习能力。

（三）能力目标

（1）能在没有老师直接指导下利用计价软件完成软件工程量清单的编制工作；

（2）能在没有老师直接指导下利用计价软件完成软件招标控制价的编制工作；

（3）能在老师指导下完成投标报价和工程结算价的编制工作。

四、课程内容与学时分配

教学内容与学时安排表

序号	教学情境/任务/项目/单元	教学目标	教学内容及学时分配			
			知识点	学时	实践项目	学时
1	工程造价计价原理	回忆、掌握工程造价编制原理	1. 工程造价组成	2		
			2. 分部分项工程费计算			
			3. 措施项目费计算			
			4. 其他项目费计算			
			5. 规费、税金计算			
2	BIM 计价软件界面介绍	熟悉 BIM 计价软件界面	6. 软件种类介绍	2		
			7. 界面介绍			
			8. 常规应用介绍			

(4) be familiar with the methods and process in bidding price;

(5) be familiar with the methods and process in compiling settlement price;

(6) master the differences between bid control price, bidding price and settlement price.

3.2 Quality Objectives

After learning this course, students will have the quality of

(1) continuous passion for this specialty and the love for the posts of this specialty and a noble professional ethics;

(2) organization and cooperation;

(3) analyzing and solving problems;

(4) developing learning ability.

3.3 Ability Objectives

After learning this course, students will have the ability of

(1) compiling the bills of quantities with pricing software by themselves;

(2) compiling the bid control price by pricing software by themselves;

(3) compiling the bidding price and settlement price with the guidance of teacher.

4. Curriculum Contents and Teaching Hours

Table of Teaching Contents and Teaching Hours

Item	Topic	Teaching Objectives	Teaching Contents and Teaching Hours			
			Lesson Contents	Hrs	Student Activities	Hrs
1	Basic theory of construction cost pricing	Recall the basic theories and master the methods of compilingconstruction cost files	1. The components of construction cost	2		
			2. The calculation of construction cost in different parts or different items of the construction project			
			3. The calculation of step item cost			
			4. The calculation of other item cost			
			5. The calculation of stipulated fees and taxes			
2	Introduction to the operation interface of BIM pricing software	Be familiar with the interface operation of BIM pricing software	6. Introduction to the software	2		
			7. Introduction to its operation interface			
			8. Introduction to its common application			

序号	教学情境/任务/项目/单元	教学目标	教学内容及学时分配			
			知识点	学时	实践项目	学时
3	工程量清单编制、招标控制价编制	掌握 BIM 计价软件编制工程量清单、招标控制价的方法	9. 分部分项工程费计算（★）	6	工程造价计算案例分析	4
			10. 措施项目费计算（★）			
			11. 其他项目费计算（★）			
			12. 规费、税金计算（★）			
			13. 材料价差调整、人工费调整、相关信息填写（★）			
			14. 利用软件快速生成清单（★）			
4	投标报价、工程结算价编制	掌握 BIM 计价软件编制投标报价、工程结算的方法	15. 投标报价编制	4		
			16. 工程结算编制			
			17. 招标控制价、投标报价、工程结算价编制异同分析（★）			
合计				14		4
备注:重要知识点标注★						

五、教学实施建议

(一)教学参考资料

四川省建设工程造价管理总站.《四川省建设工程工程量清单计价定额》,中国计划出版社,2015。

中华人民共和国住房和城乡建设部.《建设工程工程量清单计价规范》(GB 50500—2013),中国计划出版社,2013。

(二)教师素质要求

工程造价专业课老师,应具备教师资格证,应为工程造价专业或相关专业人才具有相应的执业资格和一定的社会实践和专业实践能力,有一定的专业教学经验。

continued

Item	Topic	Teaching Objectives	Teaching Contents and Teaching Hours			
			Lesson Contents	Hrs	Student Activities	Hrs
3	The compiling of bills of quantities and bid control price	Master the methods of compiling of bills of quantities and bid control price by BIM pricing software	9. The calculation of construction cost in different parts or different items of the construction project(★)	6	The analysis of practical construction cost pricing project	4
			10. The calculation of step item cost(★)			
			11. The calculation of other item cost(★)			
			12. The calculation of stipulated fees and taxes(★)			
			13. The adjustment of material prices, labor cost and the filling of other files(★)			
			14. Quickly build a the bills of quantities with pricing software (★)			
4	The compiling of bidding price and settlement price	Master the methods of compiling of bidding price and settlement price by BIM pricing software	15. Compile the bidding price	4		
			16. Compile the settlement price			
			17. Analyze the similarities and differences of bid control price, bidding price and settlement price (★)			
Total				14		4
Notes: Mark the important points with ★						

5. Teaching Suggestions

5.1 Teaching Resources

Construction Cost Management Station in Sichuan Province. *Pricing for Bills of Quantities Quota in Construction Engineering of Sichuan Province*. China Planning Press, 2015.

Ministry of Housing and Urban-Rural Development of the People's Republic of China. *Pricing Specification for Construction Projects* (GB 50500—2013). China Planning Press, 2013.

5.2 Teachers'Qualification

Teachers for this course should have the teachers'qualification certificate; they should have the education background of construction cost or other relative majors, have the relative qualification certificates and certain social or professional practice and they should have some professional teaching experience.

(三) 教学场地、设施要求

由于需要正版专业软件,故该课程必须在工程造价实训室进行。

六、考核评价

(一) 考核方式

上机操作、闭卷。

(二) 考核说明

总评成绩的构成比例原则按《四川建筑职业技术学院学生学业考核办法》执行:平时:期末 = 3:7。

考核重点应覆盖的范围:工程造价计价原理、BIM 计价软件的界面操作介绍、计价软件编制工程量清单、计价软件编制招标控制价、计价软件编制投标报价、软件编制工程结算价款等几个部分的内容。

考核的原则要求:以是否掌握 BIM 计价软件常规操作,是否能用 BIM 计价软件熟练编制工程量清单、招标控制价、投标报价、工程结算价为标准。

5. 3　Teaching Facilities

This course should be delivered in construction cost training room with BIM software.

6. Evaluation

6. 1　Evaluation Method

Practical operation on computer without reference materials.

6. 2　Evaluation Instruction

The score of the course is based on the *Regulation of Academic Evaluation for Students in SCAT*. The daily performance accounts for 30% of the total score and the final examination accounts for 70%.

The focus of the evaluation is the basic knowledge of construction cost pricing, the operation interface of BIM pricing software, the compiling of bills of quantities by the pricing software, the compiling of bid control price by the pricing software, the compiling of the bidding price by the pricing software and the compiling of the settlement price by the pricing software.

The principle for the evaluation: the main intention is to test whether the students master the basic operation of BIM and whether they can compile the bills of quantities, bid control price, bidding price and settlement price by BIM pricing software.

四川建筑职业技术学院

课 程 标 准

课程名称： 建筑工程预算

课程代码： 150169

课程学分： 4

基本学时： 64

适用专业： 工程造价（中英合作办学项目）

执 笔 人： 高红艳

编制单位： 建筑工程造价教研室

审　　核： 工程管理系（院）（签章）

批　　准： 教务处（盖章）

Sichuan College of Architectural Technology

Curriculum Standard

Curriculum Name: Building Construction Budget

Curriculum Code: 150169

Credits: 4

Teaching Hours: 64

Applicable Specialty: Construction Cost

(Sino-British Cooperative Program)

Compiled by: Gao Hongyan

On Behalf of: Teaching & Research Section of Construction Cost

Reviewed by: The Department of Engineering Management

Approved by: Teaching Affairs Office

《建筑工程预算》课程标准

一、课程概述

本课程是工程造价(中英合作办学项目)专业的专业课程。本课程着重研究建筑产品的生产成果与生产消耗之间的定量关系,是研究如何合理地确定建筑工程造价的一门综合性、实践性较强的应用型课程。本课程介绍建筑工程定额及预算的组成;分析建筑工程造价的构成;讨论建筑安装工程造价的计算方法;讲解工程竣工结算的编制方法;训练施工图预算的编制技能。先修课程及其顺序是《建筑材料》《建筑识图与构造》《建筑施工工艺》。

二、课程设计思路

本课程根据工程造价专业的教育理念及建筑行业企业岗位职业技能、综合素质要求等进行设计。设置依据是工程造价专业人才培养大纲和实施计划;课程框架、教学目标、教学内容的设计和选取根据工程造价专业人员素质、技能要求、造价工程师考试大纲等进行;教学资源主要利用工程造价专业资源库、四川工程造价信息等。

三、课程目标

(一)知识目标

(1)掌握建筑工程预算的编制原理及基本方法;

(2)了解建筑工程预算在工程管理工作中的作用;

(3)了解建筑工程定额的分类、作用;

(4)掌握预算定额的内容组成及应用方法;

(5)熟悉建筑工程费用的组成及其内容,会正确进行发承包价格的计算;

(6)掌握施工图预算的编制技能。

(二)素质目标

(1)树立爱岗敬业的思想,自觉遵守职业道德及行业规范;

Building Construction Budget Curriculum Standard

1. Curriculum Description

This course is a professional one in construction cost(Sino-British Cooperative Program). This course focuses on the quantitative relationship between the production results of construction products and production consumption, and is a comprehensive and application-oriented course on how to reasonably determine the cost of construction projects. This course introduces the composition of construction project quota and budget, analyzes the composition of construction cost, discusses the calculation method of construction cost, explains the compilation method of project completion settlement, and trains the preparation skills of construction drawing budget. The prerequisite courses were *Building Materials*, *Building Structure and Drawing Reading*, and *Construction Engineering Technology*.

2. Curriculum Design Ideas

This course is designed according to the educational concept of construction cost specialty, vocational skills and comprehensive abilities required by construction industry and other enterprises. The setting is based on the course description and teaching plan of construction cost specialty. The design and selection of curriculum framework, teaching objectives and teaching contents are in accordance with the qualification and skills requirements of construction cost professionals, the test syllabus of cost engineers and so on. The teaching and learning resources include the resource library for construction cost in SCAT, the construction cost information in Sichuan province, etc.

3. Curriculum Objectives

3.1　Knowledge Objectives

After learning this course, students will be able to

(1) master the preparation principle and basic method of building construction budget;

(2) understand the role of building construction budget in engineering management;

(3) understand the classification and function of construction engineering quota;

(4) master the content composition and application method of the budget quota;

(5) be familiar with the composition of construction costs and their contents, and be able to calculate the contract price correctly;

(6) master the preparation skills of the building drawing budget.

3.2　Quality Objectives

After learning this course, students will have the quality of

(1) dedication and commitment, and consciously abiding professional ethics and industry norms;

（2）执行工程造价工作的各项法规、政策；

（3）明确工程造价人员的执业权限与基本要求。

（三）能力目标

（1）能正确使用预算定额；

（2）能正确换算预算定额；

（3）能正确使用费用定额；

（4）能编制材料单价和人工单价；

（5）能熟练地编制施工图预算；

（6）能编制工程结算和处理相关业务。

四、课程内容与学时分配

教学内容与学时安排表

序号	教学情境/任务/项目/单元	教学目标	教学内容及学时分配			
			知识点	学时	实践项目	学时
1	建筑工程预算概述	了解建筑工程计价理论；工程造价原理；了解建筑工程基础知识	1. 建筑工程计价理论	2		
			2. 工程造价原理			
			3. 建筑工程基础知识			
2	施工图预算编制原理	熟悉施工图预算编制的依据、程序	4. 施工图预算编制的依据	2		
			5. 施工图预算编制的程序			
3	列项	掌握并记忆常用的建筑工程项目，并按图会划分项目（分项工程项目）	6. 施工图预算列项基本方法	4	讲练结合，完成一套 500 m² 左右的建筑工程列项	2
			7. 列项实例			
4	建筑面积计算	掌握建筑面积计算规则及其计算方法，正确理解计算规则的含义，能依据计算规则正确进行建筑面积计算	8. 建筑面积计算规则识读	2	讲练结合，完成习题集建筑面积计算习题	2
			9. 建筑面积计算实例			

(2) implementing the regulations and policies of the construction cost work;

(3) clarifying the practice authority and basic requirements of the construction cost personnel.

3.3 Ability Objectives

(1) using budget quotas correctly;

(2) converting the budget quota correctly;

(3) using the cost quotas correctly;

(4) preparing material unit price and labor unit price;

(5) preparing the building drawing budget skillfully;

(6) preparing engineering settlement and handle relevant business.

4. Curriculum Contents and Teaching Hours

Table of Teaching Contents and Teaching Hours

Item	Topic	Teaching Objectives	Teaching Contents and Teaching Hours			
			Lesson Contents	Hrs	Student Activities	Hrs
1	Building Construction budget	Understand the theory of construction project pricing, the principle of construction cost, and the basic knowledge of construction engineering	1. The theory of construction project pricing	2		
			2. The principle of construction cost			
			3. The basic knowledge of construction engineering			
2	Principle of budget preparation for construction drawings	Be Familiar with the basis and procedure of building drawing budget preparation	4. The basis of building drawing budget preparation	2		
			5. The Procedure			
3	Project lists	Master & memorize commonly used construction projects, and divide projects according to the drawings (sub-projects)	6. Basic methods of building drawings budget list	4	Lecture & practice combined; complete a project list of a 500-square-meter building	2
			7. Cases of project lists			
4	Calculation of Building Area	Master the calculation rules of building area and its calculation method, correctly understand the meaning of calculation rules, and be able to calculate the building area according to the calculation rules	8. Read the calculation rules of building area	2	Lecture & practice combined; complete the relevant exercise assignment	2
			9. Examples of calculation			

序号	教学情境/任务/项目/单元	教学目标	教学内容及学时分配			
			知识点	学时	实践项目	学时
5	工程量	掌握并记忆常用项目的建筑工程量计算规则及其计算方法,正确理解计算规则的含义,能依据定额计算规则正确进行工程量的计算	10. 建筑工程量计算方法、步骤	18	讲练结合,完成一套500 m² 左右室内建筑工程工程量计算	18
			11. 建筑工程量计算实例			
6	直接费计算及工料分析	熟悉直接费计算及工料分析的方法	12. 直接费组成及工料分析的方法	2	讲练结合,完成一套500 m² 左右室内建筑工程直接费计算及工料分析	2
			13. 实例直接费计算及工料分析			
7	建筑安装工程费用	了解建筑工程费用的组成、内容;掌握建筑工程费用标准的确定方法;能依据费用定额及所给条件计算建筑工程费用	14. 建筑工程费用的组成、内容	4	给定条件确定装饰工程费用标准	2
			15. 建筑工程费用标准的确定方法		给定条件计算建筑工程造价	
			16. 建筑工程费用计算			
8	工程结算	了解建筑工程工程结算	17. 建筑工程工程结算的原理及方法	4		
合计				38		26
备注						

continued

Item	Topic	Teaching Objectives	Teaching Contents and Teaching Hours			
			Lesson Contents	Hrs	Student Activities	Hrs
5	Quantity of engineering	Master and memorize the calculation rules of quantity of engineering and calculation method of common projects, correctly understand the meaning of calculation rules, and be able to calculate the quantity of engineering according to the calculation rules of quota.	10. The calculation methods and steps of quantity of engineering	18	Lecture & practice combined; complete a set of engineering quantity calculation about a 500-square-meter indoor construction	18
			11. Calculation examples of engineering quantity			
6	cost calculation and quantity analysis	Be familiar with the methods of cost calculation and quantity analysis	12. The composition of cost and the quantity analysis method	2	Lecture & practice combined; complete a set of cost calculation and quantity analysis about a 500-square-meter indoor construction	2
			13. Examples of cost calculation and quan-tity analysis			
7	Construction Installation construction cost	Understand the composition and content of construction costs; Master the method of determining the standard of construction cost; Be able to calculate the construction cost according to the cost quota and the conditions given	14. The composition and content of construction costs	4	Determine the decorative construction cost standard on the given conditions	2
			15. Determine the method of construction cost standard		Calculate the construction cost on the given conditions	
			16. Calculate the construction cost			
8	Project settlement	Understand the construction project settlement	17. The principle and methods of construction project settlement	4		
Total				38		26
Note						

五、教学实施建议

(一)教学参考资料

袁建新.《建筑工程预算》第五版,中国建筑工业出版社,2015。

《四川省建设工程工程量清单计价定额》建筑与装饰工程、附录。

工程造价教学资源库。

《建筑工程预算》国家级精品课程。

(二)教师素质要求

任课教师应具备教师资格证,应为工程造价专业或相关专业人才具有相应执业资格和一定的社会实践和专业实践能力,有一定的专业教学经验。

(三)教学场地、设施要求

实施该课程教学,无须特殊环境,一般教室或多媒体教室均可。

六、考核评价

(一)考核方式

闭卷(可以带定额)。

(二)考核说明

总评成绩的构成比例原则按《四川建筑职业技术学院学生学业考核办法》执行:平时:期末 = 3:7。

考核重点应覆盖的范围:建筑工程预算定额应用、建筑工程费用、建筑工程计量、建筑工程造价文件编制。

考核的原则要求:以是否掌握建筑工程计量与计价编制方法为考核重点,突出动手能力的培养。

5. Teaching Suggestions

5.1　Teaching Resources

Yuan Jianxin. *Building Construction Budget.* 5th Edition. China Architecture & Building Press, 2015.

Pricing Quota of Construction Engineering Quantity List in Sichuan Province, the chapters of *Construction and Decorative Engineering & Appendix.*

The teaching resources library for *Construction Engineering Cost.*

The national excellent online course of *Construction Project Budget.*

5.2　Teachers'Qualification

Teachers should have the professional ethics and the teacher qualification certificate; they should have the education background of construction cost and have the relative qualification certificates and certain social or professional practices; they should have some professional teaching experience.

5.3　Teaching Facilities

This course can be delivered in either common classrooms or multimedia classrooms.

6. Evaluation

6.1　The Method of Assessment

Closed-book examination(only the *Quotas* booklet is allowed to carry and for reference in testing).

6.2　Assessment Requirement

The evaluation of this course is in accordance with the *Regulation of Academic Evaluation for Students in SCAT*; the total score is consisted of the daily performance(accounts for 30%) and the final examination(accounts for 70%).

The highlights in examination include: the application of construction project budget quota, construction engineering cost, construction engineering measurement, and construction project cost document preparation.

The principle of examination is to check whether students master the compiling methods of measurement and valuation for construction engineering, and to highlight the cultivation of hands-on ability.

四 川 建 筑 职 业 技 术 学 院

课 程 标 准

课程名称： 建筑工程预算实训

课程代码： 500160

课程学分： 3

基本学时： 2 周

适用专业： 工程造价(中英合作办学项目)

执 笔 人： 高红艳

编制单位： 建筑工程造价教研室

审 核： 工程管理系(院)(签章)

批 准： 教务处(盖章)

Sichuan College of Architectural Technology

Curriculum Standard

Curriculum Name: Practical Training for Building Construction Budget

Curriculum Code: 500160

Credits: 3

Teaching Hours: 2 Weeks

Applicable Specialty: Construction Cost

(Sino-British Cooperative Program)

Compiled by: Gao Hongyan

On Behalf of: Teaching & Research Section of Construction Cost

Reviewed by: The Department of Engineering Management

Approved by: Teaching Affairs Office

《建筑工程预算实训》课程标准

一、课程概述

建造工程预算实训,是建筑工程预算课程的重要实践教学环节之一,是工程造价(中英合作办学项目)专业的专业课。通过课程实训,学生可以系统掌握建筑工程预算的内容、要求、格式、编制步骤和编制方法,巩固在建筑工程预算课程中所学的理论知识,培养学生分析解决建筑工程预算实际问题的能力。本课程的先修课程为《建筑工程预算》。

二、课程设计思路

本课程根据工程造价专业的教育理念及建筑行业企业岗位职业技能、综合素质要求等进行设计。设置依据是工程造价专业人才培养大纲和实施计划;课程框架、教学目标、教学内容的设计和选取根据工程造价专业人员素质、技能要求、造价工程师考试大纲等进行设计;教学资源主要利用工程造价专业资源库、四川工程造价信息等。

三、课程目标

通过建造工程预算实训,学生应当掌握建筑工程预算定额的构成及使用方法,能熟练使用定额;掌握建造工程预算方法,能确定建筑工程造价。

(一)知识目标

(1)掌握建筑工程定额的组成及使用方法,掌握建筑工程费用的组成、内容及取费标准的确定;

(2)掌握建筑工程计量与计价的方法。

(二)素质目标

(1)树立爱岗敬业的思想,自觉遵守职业道德行业规范;

(2)通过建筑工程预算实训,培养学生专业兴趣和工作热情。

Practical Training for Building Construction Budget Curriculum Standard

1. Curriculum Description

The building construction budget training is one of the important practical teaching links of the construction engineering budget course, and it is the professional course for construction cost specialty (Sino-British Cooperative Program). Through this course, the student can master the contents, requirements, format, compiling steps and methods of the building construction budget, consolidate the theoretical knowledge learned in the building construction budget unit, and get students' ability to analyze and solve the practical problems of the building construction budget trained. The prerequisite course was *Building Construction Budget*.

2. Curriculum Design Ideas

This course is designed according to the educational concept of construction cost specialty, vocational skills and comprehensive abilities required by construction industry and other enterprises. The setting is based on the course description and teaching plan of construction cost specialty. The design and selection of curriculum framework, teaching objectives and teaching contents are in accordance with the qualification and skills requirements of construction cost professionals, the test syllabus of cost engineers and so on. The teaching and learning resources include the resource library for construction cost in SCAT, the construction cost information in Sichuan province, etc.

3. Curriculum Objectives

Through this course, students should grasp the composition and application method of the building construction budget quota, be able to use the quota skillfully, master the building construction budget method, and can determine the building construction cost.

3.1 Knowledge Objectives

After learning this course, students will be able to

(1) master the composition and application method of the building construction quota, master the composition, content and the determination of the charging standard of the building construction cost;

(2) master the method of measurement and valuation of construction engineering.

3.2 Quality Objectives

After learning this course, students will have the quality of

(1) dedication and commitment, and consciously abiding professional ethics and industry norms;

(2) continuous professional interest and enthusiasm for work.

(三)能力目标

(1)能熟练应用建筑工程预算定额；

(2)能计算建筑工程量；

(3)能确定建筑工程造价。

四、课程内容与学时分配

教学内容与学时安排表

序号	教学情境/任务/项目/单元	教学目标	教学内容及学时分配		
			知识点	实践项目	学时/d
1	实训准备	完成实训准备工作	1. 设计文件	1. 搜集、准备、整理相关实训资料	1
			2. 计价依据	2. 熟悉图纸	
2	列项计算工程量	完成给定项目工程量计算	3. 如何列项	3. 给定项目工程量的计算	9
			4. 工程量方法、步骤		
3	计算工程造价	完成给定项目工程造价计算	5. 工程造价的组成	4. 计算工程造价	3
			6. 工程造价的确定		
4	形成工程造价文件	形成工程造价文件	7. 工程造价文件的内容，装订顺序	5. 完成建筑工程计量与计价实训成果	1
合计					14
备注					

五、教学实施建议

(一)教学参考资料

《四川省建设工程工程量清单计价定额》建筑与装饰工程、附录；

一套 1 000 m² 以内的公共建筑设计图纸；

工程造价教学资源库；

《建造工程预算》省级精品课程。

3.3 Ability Objectives

After learning this course, students will have the ability of

(1) using building construction budget quotas skillfully;

(2) calculating the building construction quantity;

(3) determining the building construction cost.

4. Curriculum Contents and Teaching Hours

Table of Teaching Contents and Teaching Hours

Item	Topic	Teaching Objectives	Teaching Contents and Teaching Hours		
			Lesson Contents	Student Activities	Training Day/d
1	Preparation for practical training	Prepare well for the practical training	1. Files and regulations for practical training	1. Collect, prepare and arrange the relative materials for practical training	1
			2. The basis for calculating cost	2. Be familiar with the construction drawing	
2	List and calculate the engineering quantity	Complete the given list calculation of engineering quantity	3. Work out a list	3. Calculate the engineering quantity of a given list	9
			4. The method and steps of calculation		
3	Calculate the construction cost	Complete the given list calculation of construction cost	5. The composition of construction cost	4. Calculate the construction cost	3
			6. Determine the construction cost		
4	Compile the document of construction cost	Compile the document of construction cost	The content of construction cost document and the compiling sequence	5. Complete the achievement of the practical training for the measurement and valuation of construction engineering	1
Total					14
Note					

5. Teaching Suggestions

5.1 Teaching Resources

Pricing Quota of Construction Engineering Quantity List in Sichuan Province, the chapters of *Construction and Decorative Engineering & Appendix*.

A set of public building design drawings within 1,000 square meters; the teaching resources library for *Construction Engineering Cost*.

The national excellent online course of *Construction Project Budget*.

(二)教师素质要求

任课教师应具备教师资格证,应为工程造价专业或相关专业人才具有相应执业资格和一定的社会实践和专业实践能力,有一定的专业教学经验。

(三)教学场地、设施要求

实施该课程教学,无须特殊环境,一般教室即可。

六、考核评价

(1)建造工程预算实训成绩主要按学生上交的实训成果:一套完整的建筑工程工程造价文件、指导教师检查的情况及出勤情况,综合评定。

(2)课程设计成绩按五级计算(优、良、中、及格、不及格),凡不及格者,按学校有关规定执行。

5.2 Teachers'Qualification

Teachers should have the professional ethics and the teacher qualification certificate; they should have the education background of construction cost and have the relative qualification certificates and certain social or professional practices; they should have some professional teaching experience.

5.3 Teaching Facilities

This course can be delivered in common classrooms.

6. Evaluation

(1) The evaluation score of this course relies mainly on the results of practical training presented by students: a complete set of building construction cost documents, the attendance and the daily performance checked by the tutor, comprehensive assessment, etc.

(2) The score is graded into five-level-system of excellent, good, moderate, poor and failing. Anyone who fails shall be subject to the relevant provisions of the college.

四 川 建 筑 职 业 技 术 学 院

课 程 标 准

课程名称：　　　　　　钢筋工程量计算

课程代码：　　　　　　150186

课程学分：　　　　　　2.5

基本学时：　　　　　　40

适用专业：　　　工程造价(中英合作办学项目)

执 笔 人：　　　　　　潘桂生

编制单位：　　　　建筑工程造价教研室

审　　核：　　　工程管理系(院)(签章)

批　　准：　　　　　教务处(盖章)

Sichuan College of Architectural Technology

Curriculum Standard

Curriculum Name: Calculation of Reinforcement Quantity

Curriculum Code: 150186

Credits: 2. 5

Teaching Hours: 40

Applicable Specialty: Construction Cost

(Sino-British Cooperative Program)

Compiled by: Pan Guisheng

On Behalf of: Teaching & Research Section of Construction Cost

Reviewed by: The Department of Engineering Management

Approved by: Teaching Affairs Office

《钢筋工程量计算》课程标准

一、课程概述

本课程是技术性、综合性较强的课程,是工程造价专业(中英合作办学项目)学生的专业课。该课程的先修课程有《建筑材料》《结构力学》《建筑施工技术》《建筑构造和识图》等,后续课程有《建筑工程预算》《建筑工程工程量清单计价》《竣工结算》等,通过本课程的学习,学生能够根据建筑工程施工图纸及相关标准图集,准确计算建筑工程各类构件的钢筋工程量,为后续课程的学习过程中钢筋算量打下良好基础。工程造价中英合作办学专业该课程在第 4 学期开设。

二、课程设计思路

本课程根据工程造价专业(中英合作办学项目)的教育理念及建筑行业、企业岗位职业技能、综合素质要求等进行设计。设置依据是工程造价(中英合作办学项目)专业人才培养大纲和实施计划;依据该课程所包含的知识点和建筑物的主要结构构成进行课程框架的搭建和教学内容的设计;教学目标的确定和选取根据工程造价专业人员素质、技能要求、造价工程师考试大纲等进行设计;教学资源主要为工程造价专业资源库、四川工程造价信息等。

三、课程目标

通过该课程的学习,学生应该学会使用平法图集并具备一定的自学能力,能够根据相关标准图集读懂建筑工程施工图纸并能准确计算出建筑工程各类构件的钢筋工程量。

(一)知识目标

(1)了解钢筋工程量计算的基本知识、钢筋工程量计算依据和钢筋工程量计算的基本方法;

Calculation of Reinforcement Quantity **Curriculum Standard**

1. Curriculum Description

This is a technical and comprehensive course and it is the professional course for the major of construction cost (Sino-British Cooperative Program). The delivered courses were *Building Materials*, *Structural Mechanics*, *Building Construction Technology*, *Basic Building Structure and Drawing Reading*, and so on. The courses followed are *Building Construction Budget*, *Pricing for Bills of Quantities of Building Construction*, *Project Final Account* and so on. Through the study of this course, students can accurately calculate the quantity of reinforcement in different components of building construction based on different construction drawings and relative standard collective drawings. This course is delivered in the fourth semester for the specialty of construction cost(Sino-British Cooperative Program).

2. Curriculum Design Ideas

The design of this course is based on the educational concept of the specialty of construction cost (Sino-British Cooperative Program), vocational skills and comprehensive abilities required by construction industry and other enterprises; the course is designed in accordance with the professional training program and the teaching plan of the specialty of construction cost; the construction of the course structure and the design of teaching plan are based on the necessary contents for the course and the main structure of buildings; the determination and selection of the teaching objectives are based on the requirements for the specialty of construction cost, the test syllabus of cost engineers and so on; the teaching and learning resources are the resource library for construction cost in SCAT, the construction cost information in Sichuan province and so on.

3. Curriculum Objectives

Through this course, students should learn to read drawings by ichnographic representing method and have certain self-learning ability, should have the ability to read construction engineering drawings and accurately calculate the quantity of reinforcement in different components of building construction based on relative standard collective drawings.

3. 1 Knowledge Objectives

After learning this course, students will be able to

(1) understand the basic knowledge, the calculation basis and the basic calculation methods for calculating the quantity of reinforcement;

(2)熟悉住房和城乡建设部颁发的《建设工程工程量清单计价规范》与计量规范中钢筋工程量的计算规则和钢筋工程项目划分的基本规定；

(3)掌握建筑工程的平法识图和钢筋工程量计算的基本方法及技巧。

(二)素质目标

(1)培养学生的专业兴趣及对本专业的热爱；

(2)培养学生踏实肯干、不畏辛苦的工作作风；

(3)加强学生的职业道德教育。

(三)能力目标

(1)能熟练使用16G101系列平法图集；

(2)能在没有教师的直接指导下自行使用16G101图集读懂实际工程中各构件的钢筋设置情况；

(3)能在没有教师的直接指导下参考16G101图集独立完成实际工程的钢筋准确算量。

四、课程内容与学时分配

序号	教学情境/任务/项目/单元	教学目标	教学内容及学时分配			
			知识点	学时	实践项目	学时
1	概述	了解钢筋工程量计算的依据；了解钢筋的分类；掌握钢筋工程量计算基础知识及钢筋工程量计算基本方法	1. 工程量计算的依据 2. 钢筋分类 3. 钢筋工程量计算基础知识(★) 4. 钢筋工程量计算方法	4		

(2) be familiar with the reinforcement calculation codes and the regulations for the division of reinforcement project in pricing for bills of quantities and measurement specification authorized by Ministry of Housing and Urban-Rural Development of the People's Republic of China;

(3) master the basic methods and skills in reading drawings by ichnographic representing method and calculation of reinforcement quantity.

3.2　Quality Objectives

After learning this course, students will have the quality of

(1) continuous interest and love for this specialty;

(2) devoted and responsible work style;

(3) good professional ethics.

3.3　Ability Objectives

After learning this course, students will have the ability of

(1) mastering the competence of reading drawings by ichnographic representing method of 16G101 series skillfully;

(2) reading and understanding the quantity of reinforcement in different components of building construction based on the drawings by ichnographic representing method of 16G101 by themselves;

(3) calculating the quantity of reinforcement in different components of building construction based on drawings by ichnographic representing method 16G101 by themselves.

4. Curriculum Contents and Teaching Hours

Table of Teaching Contents and Teaching Hours

Item	Topic	Teaching Objectives	Teaching Contents and Teaching Hours			
			Lesson Contents	Hrs	Student Activities	Hrs
1	Introduction	Understand the basis for calculating reinforcement quantity; understand the classification of different reinforcement; master the basic knowledge of calculating reinforcement and the basic methods of calculating reinforcement quantity	1. The basis for calculating construction quantity	4		
			2. The classification of reinforcement			
			3. The basic knowledge of calculating reinforcement quantity(★)			
			4. The basic methods of calculating reinforcement quantity			

序号	教学情境/任务/项目/单元	教学目标	教学内容及学时分配			
			知识点	学时	实践项目	学时
2	钢筋工程量计算	掌握钢筋工程量计算规则;学会平法图集的使用;能够熟练计算出各构件的钢筋工程量	5. 基础钢筋平法识读	24	1. 基础钢筋平法识读及钢筋工程量计算	12
			6. 基础钢筋工程量计算(★)			
			7. 柱钢筋平法识读		2. 柱钢筋平法识读及钢筋工程量计算	
			8. 柱钢筋工程量计算(★)			
			9. 梁钢筋平法识读		3. 梁钢筋平法识读及钢筋工程量计算	
			10. 梁钢筋工程量计算(★)			
			11. 板钢筋平法识读		4. 板钢筋平法识读及钢筋工程量计算	
			12. 板钢筋工程量计算(★)			
	合计			28		12

备注:重要知识点标注★

五、教学实施建议

(一)教学参考资料

教材:王武齐.《钢筋工程量计算》(第二版),中国建筑工业出版社,2018。

必备辅助资料:16G101 图集。

其他辅助资料:《房屋建筑与装饰工程工程量计算规范》(GB 50854—2013)、《建设工程工程量清单计价规范》(GB 50500—2013)。

continued

Item	Topic	Teaching Objectives	Teaching Contents and Teaching Hours			
			Lesson Contents	Hrs	Student Activities	Hrs
2	Calculation of Reinforcement Quantity	Master the rules in calculating reinforcement quantity; master the usage of drawings by ichnographic representing method; accurately calculate the quantity of reinforcement in different components of building construction	5. Reading drawings of foundation reinforcement by ichnographic representing method	24	1. Reading drawings of foundation reinforcement by ichnographic representing method and calculate the quantity of foundation reinforcement	12
			6. Calculating the quantity of foundation reinforcement (★)			
			7. Reading drawings of bar reinforcement by ichnographic representing method		2. Reading drawings of bar reinforcement by ichnographic representing method and Calculate the quantity of bar reinforcement	
			8. Calculating the quantity of bar reinforcement(★)			
			9. Reading drawings of beam reinforcement by ichnographic representing method		3. Reading drawings of beam reinforcement by ichnographic representing method and calculate the quantity of beam reinforcement	
			10. Calculating the quantity of beam reinforcement(★)			
			11. Reading drawings of plate reinforcement by ichnographic representing method		Reading drawingsof plate reinforcement by ichnographic representing method and calculate the quantity of plate reinforcement	
			12. Calculating the quantity of plate reinforcement(★)			
Total				28		12
Note: Mark the important points with ★						

5. Teaching Suggestions

5. 1　Teaching Resources

Textbook: Wang Wuqi. *Calculation of Reinforcement Quantity.* 2nd Edition. China Architecture & Building Press, 2018.

Required auxiliary materials: Reading drawings by ichnographic representing method, 16G101 series.

Recommended auxiliary materials: *Specification for Calculation of Construction and Decoration Works* (GB 50854—2013) and *Specification for the Bill of Quantities of Construction Project* (GB 50500—2013).

(二)教师素质要求

任课教师应具备一定的职业道德素质,取得教师资格证,应为工程造价专业或相关专业人才并具有相应执业资格和一定的社会实践和专业实践能力,有一定的专业教学经验。

(三)教学场地、设施要求

实施该课程教学,一般教室即可,也可适当参观一下工程实际。

六、考核评价

本课程考核方式为考试,考核的重点是建筑工程各构件钢筋工程量的计算能力。考核范围应覆盖平法图集的应用、平法识图及各构件钢筋工程量计算方法等内容。

该课程成绩构成比例原则是按《四川建筑职业技术学院学生学业考核办法》执行,总评成绩具体构成比例为:平时:期末 =3∶7。

5. 2　Teachers'Qualification

Teachers should have the professional ethics and the teachers'qualification certificate; they should have the education background of construction cost and have the relative qualification certificates and certain social or professional practices; they should have some professional teaching experience.

5. 3　Teaching Facilities

This course can be delivered in common classrooms, and if necessary, we can visit the practical construction sites.

6. Evaluation

The focus of the evaluation is the ability to calculate the quantity of reinforcement in different components of building construction. The evaluation should cover the contents of the usage of drawings by ichnographic representing method, the reading of the drawings by ichnographic representing method and the calculation of the quantity of reinforcement in different components.

The score of the course is based on the *Regulation of Academic Evaluation for Students in SCAT*. The daily performance accounts for 30% of the total score and the final examination accounts for 70%.

四川建筑职业技术学院

课 程 标 准

课程名称：　　　钢筋工程量计算实训

课程代码：　　　500105

课程学分：　　　1

基本学时：　　　30

适用专业：　　工程造价（中英合作办学项目）

执　笔　人：　　　潘桂生

编制单位：　　建筑工程造价教研室

审　　　核：　　工程管理系（院）（签章）

批　　　准：　　　教务处（盖章）

Sichuan College of Architectural Technology

Curriculum Standard

Curriculum Name: Practical Training for Calculation of

Reinforcement Quantity

Curriculum Code: 500105

Credits: 1

Teaching Hours: 30

Applicable Specialty: Construction Cost

(Sino-British Cooperative Program)

Compiled by: Pan Guisheng

On Behalf of: Teaching & Research Section of Construction Cost

Reviewed by: The Department of Engineering Management

Approved by: Teaching Affairs Office

《钢筋工程量计算实训》课程标准

一、课程概述

本课程是技术性、综合性较强的课程,是工程造价专业(包括安装、市政方向工程造价班、中英工程造价班)学生的专业课。该课程的先修课程有《建筑材料》《结构力学》《建筑施工技术》《建筑构造和识图》等,后续课程有《建筑工程预算》《建筑工程工程量清单计价》《竣工结算》等,通过本课程的学习,学生能够根据建筑工程施工图纸及相关标准图集,准确计算建筑工程各类构件的钢筋工程量,为后续课程的学习过程中钢筋算量打下良好基础。造价安装专业该课程在第3学期开设,工程造价(中英合作办学项目)专业该课程在第4学期开设,造价市政专业该课程在第5学期开设。

二、课程设计思路

本课程根据工程造价专业(中英合作办学项目)的教育理念及建筑行业、企业岗位职业技能、综合素质要求等进行设计。设置依据是工程造价(安装、市政方向)、工程造价(中英合作办学项目)专业人才培养大纲和实施计划;依据该课程所包含的知识点和建筑物的主要结构构成进行课程框架的搭建和教学内容的设计;教学目标的确定和选取根据工程造价专业人员素质、技能要求、造价工程师考试大纲等进行设计;教学资源主要为工程造价专业资源库、四川工程造价信息等。

三、课程目标

通过该课程的学习,学生应该学会使用平法图集并具备一定的自学能力,能够根据相关标准图集读懂建筑工程施工图纸并能准确计算出建筑工程各类构件的钢筋工程量。

(一)知识目标

(1)了解钢筋工程量计算的基本知识、钢筋工程量计算依据和钢筋工程量计算的基本方法;

Practical Training for Calculation of Reinforcement Quantity Curriculum Standard

1. Curriculum Description

This is a comprehensive and practical course, and it is an important practical training for the course of calculation of reinforcement quantity. It is the professional course for the specialty of construction cost (Sino-British Cooperative Program). Through the practical training of calculation of reinforcement quantity, students can master the methods in calculating reinforcement quantity, and can strengthen the capacity and confidence in calculating reinforcement quantity by themselves. The delivered courses were *Structural Mechanics*, *Construction Engineering Technology*, *Building Structure and Drawing Reading*, *Calculation of Reinforcement Quantity* and so on. The courses followed are *Building Construction Budget*, *Pricing for Bills of Quantities*, *Engineering Settlement* and so on. This course is delivered in the fourth semester for the specialty of construction cost (Sino-British Cooperative Program).

2. Curriculum Design Ideas

The design of this course is based on the educational concept of the specialty of construction cost (Sino-British Cooperative Program), vocational skills and comprehensive abilities required by construction industry and other enterprises; the course is designed in accordance with the professional training program and teaching plan of the specialty of construction cost, the construction of the course structure and the design of teaching plan are based on the necessary contents for the course and the main structure of buildings; the determination and selection of the teaching objectives are based on the requirements for the specialty of construction cost, the test syllabus of cost engineers and so on; the teaching and learning resources are the resource library for construction cost in SCAT, the construction cost information in Sichuan province and so on.

3. Curriculum Objectives

Through this course, students should learn to read drawings by ichnographic representing method and should have the ability to read construction engineering drawings and accurately calculate the quantity of reinforcement in different components of building construction based on drawings by ichnographic representing method.

3.1 Knowledge Objectives

After learning this course, students will be able to

(1) understand the basic knowledge, the calculation basis and the basic calculation methods for calculating the quantity of reinforcement;

（2）熟悉住房和城乡建设部颁发的《建设工程工程量清单计价规范》与计量规范中钢筋工程量的计算规则和钢筋工程项目划分的基本规定；

（3）掌握建筑工程的平法识图和钢筋工程量计算的基本方法及技巧。

(二) 素质目标

（1）培养学生的专业兴趣及对本专业的热爱；

（2）培养学生踏实肯干、不畏辛苦的工作作风；

（3）加强学生的职业道德教育。

(三) 能力目标

（1）能熟练使用16G101系列平法图集；

（2）能在没有教师的直接指导下自行使用16G101图集读懂实际工程中各构件的钢筋设置情况；

（3）能在没有教师的直接指导下独立参考(16G101图集)完成实际工程的钢筋准确算量。

四、课程内容与学时分配

教学内容与学时安排表

序号	教学情境/任务/项目/单元	教学目标	教学内容及学时分配			
			知识点	学时	实践项目	学时
1	实训准备	完成实训准备工作，包括领取实训资料及其他所需资料的搜集	1. 设计文件		1. 搜集、准备、整理相关实训资料	4
			2. 钢筋工程量计算依据		2. 熟悉图纸	
2	钢筋工程量计算	掌握钢筋工程量计算规则；学会平法图集的使用；能够熟练计算出各构件的钢筋工程量	3. 基础钢筋工程量计算（★）		3. 基础钢筋平法识读及钢筋工程量计算	22

(2) be familiar with the reinforcement calculation codes and the regulations for the division of reinforcement project in pricing for bills of quantities and measurement specification authorized by Ministry of Housing and Urban-Rural Development of the People's Republic of China;

(3) master the basic methods and skills in reading drawings by ichnographic representing method and calculation of reinforcement quantity.

3.2 Quality Objectives

After learning this course, students will have the quality of

(1) continuous interest and love for this specialty;

(2) devoted and responsible work style;

(3) good professional ethics.

3.3 Ability Objectives

After learning this course, students will have the ability of

(1) mastering the competence of reading drawings by ichnographic representing method of 16G101 series skillfully;

(2) reading and understanding the quantity of reinforcement in different components of building construction based on the drawings by ichnographic representing method of 16G101 by themselves;

(3) calculating the quantity of reinforcement in different components of building construction based on drawings by ichnographic representing method 16G101 by themselves.

4. Curriculum Contents and Teaching Hours

Table of Teaching Contents and Teaching Hours

Item	Topic	Teaching Objectives	Teaching Contents and Teaching Hours		
			Lesson Contents	Student Activities	Hrs
1	Preparation for practical training	Prepare well for the practical training, including the receiving of training materials and the collecting of other materials	1. Files and regulations for the practical training	1. Collecting, preparing and arranging the relative materials for practical training	4
			2. The basis for calculating reinforcement quantity	2. Being familiar with the construction drawing	
2	Calculation of Reinforcement Quantity	Master the rules in calculating reinforcement quantity; learn the usage of drawings by ichnographic representing method; skillfully calculate the reinforcement quantity of different components	3. Calculation of the quantity of foundation reinforcement(★)	3. Reading drawings of foundation reinforcement by ichnographic representing method and calculating the quantity of foundation reinforcement	22

序号	教学情境/任务/项目/单元	教学目标	教学内容及学时分配			
			知识点	学时	实践项目	学时
2	钢筋工程量计算	掌握钢筋工程量计算规则;学会平法图集的使用;能够熟练计算出各构件的钢筋工程量	4. 柱钢筋工程量计算（★）		4. 柱钢筋平法识读及钢筋工程量计算	22
			5. 梁钢筋工程量计算（★）		5. 梁钢筋平法识读及钢筋工程量计算	
			6. 板钢筋工程量计算（★）		6. 板钢筋平法识读及钢筋工程量计算	
			7. 总结不同钢筋计算方法		7. 总结不同钢筋计算方法	
3	钢筋工程量汇总并形成钢筋工程量计算成果文件	汇总钢筋总消耗量及单位钢筋耗用量	8. 汇总出工程中每种钢筋的消耗量		8. 钢筋工程量汇总	4
			9. 汇总出该工程钢筋总消耗量			
			10. 计算出单位钢筋消耗量			
			11. 装订成册,形成钢筋工程量计算成果文件		9. 完成实训内容装订成册	
合计						30
备注:重要知识点标注★						

continued

Item	Topic	Teaching Objectives	Teaching Contents and Teaching Hours		
			Lesson Contents	Student Activities	Hrs
2	Calculation of Reinforcement Quantity	Master the rules in calculating reinforcement quantity; learn the usage of drawings by ichnographic representing method; skillfully calculate the reinforcement quantity of different components	4. Calculating the quantity of bar reinforcement (★)	4. Reading drawings of bar reinforcement by ichnographic representing method and calculating the quantity of bar reinforcement	22
			5. Calculating the quantity of beam reinforce-ment (★)	5. Reading drawings of beam reinforcement by ichnographic representing method and calculating the quantity of beam reinforcement	
			6. Calculating the quantity of plate reinforce-ment (★)	6. Reading drawings of plate reinforcement by ichnographic representing method and calculating the quantity of plate reinforcement	
			7. Summarizing the total consumption quantity of different reinforcement	7. Summarizing the calculation of different reinforcement quantity	
3	Summary of the calculation process and compiling the file of reinforcement calculation	Summarize the total consumption of reinforcement quantity and the unit consumption reinforcement quantity	8. Summarizing the total consumption of reinforcement quantity for the whole project		4
			9. Summarizing the total consumption of reinforcement quantity for the whole project		
			10. Calculating the unit consumption of reinforcement quantity	8. Compiling the materials of practical training	
			11. Compiling the files and presenting the result of calculation of reinforcement quantity	9. Compiling all the documents in practical training	
Total					30
Note: Mark the important knowledge points with ★					

五、教学实施建议

(一)教学参考资料

教材:王武齐.《钢筋工程量计算》(第二版),中国建筑工业出版社,2018。

必备辅助资料:16G101 图集。

其他辅助资料:《房屋建筑与装饰工程工程量计算规范》(GB 50854—2013)、《建设工程工程量清单计价规范》(GB 50500—2013)。

(二)教师素质要求

任课教师应具备一定的职业道德素质,取得教师资格证,应为工程造价专业或相关专业人才并具有相应执业资格和一定的社会实践和专业实践能力,有一定的专业教学经验。

(三)教学场地、设施要求

实施该课程教学,一般教室即可。

六、考核评价

本课程考核方式为考查,采取书面检查和答辩方式进行考核。根据《四川建筑职业技术学院学生学业考核办法》,《钢筋工程量计算实训》成绩按五级制打分,即优、良、中、及格、不及格,总评成绩具体构成比例为:实训纪律考核: 实训成果: 答辩成绩 = 3: 3: 4。

5. Teaching Suggestions

5.1   Teaching Resources

Textbook: Wang Wuqi. *Calculation of Reinforcement Quantity*. 2nd Edition. China Architecture & Building Press, 2018.

Required auxiliary materials: Reading drawings by ichnographic representing method, 16G101 series.

Recommended auxiliary materials: *Specification for Calculation of Construction and Decoration Works* (GB 50854—2013) and *Specification for the Bill of Quantities of Construction Project* (GB 50500—2013)。

5.2   Teachers'Qualification

Teachers should have the professional ethics and the teachers'qualification certificate; they should have an education background in construction cost and have the relative qualification certificates and certain social or professional practices; they should have some professional teaching experience.

5.3   Teaching Facilities

This course can be delivered in common classrooms.

6. Evaluation

The evaluation of this course includes the complied files and the oral presentation. According to the *Regulation of Academic Evaluation for Students in SCAT*, the total score is consisted of the attendance in practical training(30%), the compiled files of practical training(30%) and the oral presentation of the results(40%). The score is graded into five- level-system: excellent, good, moderate, poor and failing.

四川建筑职业技术学院

课程标准

课程名称： 工程量清单计价

课程代码： 150141

课程学分： 3

基本学时： 48

适用专业： 工程造价（中英合作办学项目）

执 笔 人： 袁　鹰

编制单位： 建筑工程造价教研室

审　　核： 工程管理系（院）（签章）

批　　准： 教务处（盖章）

Sichuan College of Architectural Technology

Curriculum Standard

Curriculum Name: _____ Pricing for Bill of Quantities _____

Curriculum Code: _____ 150141 _____

Credits: _____ 3 _____

Teaching Hours: _____ 48 _____

Applicable Specialty: _____ Construction Cost _____

_____ (Sino-British Cooperative Program) _____

Compiled by: _____ Yuan Ying _____

On Behalf of: Teaching & Research Section of Construction Cost

Reviewed by: _____ The Department of Engineering Managment _____

Approved by: _____ Teaching Affairs Office _____

《工程量清单计价》课程标准

一、课程概述

工程量清单计价是工程造价专业(中英合作办学项目)的一门专业必修课程。课程3学分,48学时,在第4学期开设,其先修专业课为《建筑工程预算》,后续专业课为《BIM管理》《工程造价软件应用》。本课程着重研究如何按《建设工程工程量清单计价规范》及《房屋建筑与装饰工程工程量计算规范》(GB 50854—2013)的要求,编制工程量清单、编制招标控制价和清单投标报价。

二、课程设计思路

设置依据:《建设工程工程量清单计价规范》(GB 50500—2013)《房屋建筑与装饰工程工程量计算规范》(GB 50854—2013)、《四川建筑职业技术学院人才培养方案(2017版)》。

课程框架的设计:工程造价概述、计价规范及计算规范、工程量清单的编制、招标控制价编制、投标报价编制。

教学目标的确定:掌握《建设工程工程量清单计价规范》(GB 50500—2013)、《房屋建筑与装饰工程工程量计算规范》(GB 50854—2013)的基本要求、计量计价方法;能依据施工图纸及相应条件正确地编制工程量清单;能根据工程量清单、规范、定额、工料机价格编制招标控制价和投标报价。

教学内容的选取:工程造价计价原理及编制方法、工程量清单与工程量清单计价的定义、工程量清单要素构成及其内容、工程量清单计价与定额计价的区别、工程量清单计价规范的主要内容、工程量清单计价格式、计算规范的主要内容、附录的内容和格式、计算规范的计算规则应用、工程量计算方法、工程量清单编制方法、选择定额、分析清单项目的主项和附项、计算计价工程量、确定工料机单价、编制综合单价、计算招标控制价、投标报价的编制、招标控制价与投标报价的相同点、招标控制价与投标报价的不同点。

三、课程目标

通过本课程,学生能具备基本的职业素质、具有高级职业人才所必需的思想基础、思维方式及职业道德,能掌握工程量清单计价的基本知识和基本技能,基本具备在工程造价工作岗位及相关岗位上解决实际问题的能力。

(一)知识目标

(1)掌握工程量清单计价的原理及基本知识;

(2)了解工程量清单招标投标在建设工作中的作用;

(3)掌握工、料、机单价的确定方法;

(4)熟悉《建设工程工程量清单计价规范》(GB 50500—2013);

Pricing for Bill of Quantities Curriculum Standard

1. Curriculum Description

This is a comprehensive and practical course, and a professional compulsory course for construction cost specialty(Sino-British Cooperative Program). It consists of 3 credits, 48 teaching hours, and is delivered in the fourth semester. The prerequisite course was *Building Construction Budget*, and the courses followed are *BIM Management* and *Construction Cost Software Application*. This course focuses on how to prepare bill of quantities, compile tender sum limit and offer bidding quotations upon bill of quantities according to the *Pricing Specification of Bill of Quantities in Construction Engineering* (GB 50500—2013) and the *Calculation Specification of Bill of Quantities for Building and Decoration Construction*(GB 50854—2013).

2. Curriculum Design Ideas

This course is designed according to the educational concept of construction cost specialty, vocational skills and comprehensive abilities required by construction industry and other enterprises. The setting is based on the professional training program and teaching plan of construction cost specialty. The design and selection of curriculum framework, teaching objectives and teaching contents are in accordance with the qualification and skills requirements of construction cost professionals, the test syllabus of cost engineers, etc. The teaching and learning resources include the resource library for construction cost in SCAT, the construction cost information in Sichuan province, etc.

3. Curriculum Objectives

Through this course, students should learn to acquire basic professional qualities, the ideological foundation, thinking style and professional ethics necessary for senior professional talents, to master the basic knowledge and basic skills of pricing for bill of engineering quantity, and to cultivate the ability to solve practical problems in construction cost occupations and other relevant jobs.

3.1 Knowledge Objectives

After learning this course, students will be able to

(1) master the principle and basic knowledge of pricing for bill of quantities;

(2) understand the role of using bill of quantities in tendering and bidding of construction engineering work;

(3) master the methods of deciding the unit price of labor, materials and machinery;

(4) be familiar with *Pricing Specification of Bill of Quantities in Construction Engineering*(GB 50500—2013);

（5）熟悉《房屋建筑与装饰工程工程量计算规范》（GB 50854—2013）。

（二）素质目标

（1）具备举一反三、处理综合问题的思维能力；

（2）具有爱岗敬业的思想，实事求是的工作作风和创新意识；

（3）熟悉工程造价工作的有关政策法规；

（4）加强职业道德的意识，认识工程造价人员的执业权限与基本要求。

（三）能力目标

（1）能正确使用《建设工程工程量清单计价规范》（GB 50500—2013）及《房屋建筑与装饰工程工程量清单计算规范》（GB 50500—2013）；

（2）能正确应用定额；

（3）会计算清单工程量和计价工程量；

（4）会计算或确定工料机单价；

（5）能熟练地编制综合单价；

（6）能编制工程量清单招标控制价和工程量清单投标报价。

四、课程内容与学时分配

教学内容与学时安排表

序号	教学情境/任务/项目/单元	教学目标	教学内容及学时分配			
			知识点	学时	实践项目	学时
1	概论	1. 理解工程量清单的概念； 2. 理解工程量清单计价的概念； 3. 了解工程量清单计价的内容构成； 4. 明了工程量清单计价与定额计价的不同点	1. 工程造价计价原理及编制方法、工程量清单与工程量清单计价	2	1. 工程造价基本原理习题	1
			2. 工程量清单、招标控制价及工程量清单报价编制内容（★）	2		
			3. 工程量清单计价与定额计价的区别	1		

(5) be familiar with *Calculation Specification of Bill of Quantities for Building and Decoration Construction* (GB 50854—2013).

3.2 Quality Objectives

After learning this course, students will have the quality of

(1) dealing with comprehensive issues and be able to draw inferences about other cases from one instance;

(2) devoted, responsible and practical working style and innovative spirit;

(3) being familiar with relevant policies and regulations of construction cost work;

(4) strengthened awareness of professional ethics and recognize the practice authority and basic requirements of construction cost professionals.

3.3 Ability Objectives

After learning this course, students will have the ability of

(1) using properly the *Pricing Specification of Bill of Quantities in Construction Engineering* (GB 50500—2013) and *Calculation Specification of Bill of Quantities for Building and Decoration Construction* (GB 50854—2013);

(2) using the quota correctly;

(3) calculating the bill of quantities and its pricing;

(4) calculating or determine the unit price of labor, materials and machinery;

(5) preparing proficiently comprehensive unit prices;

(6) compiling the tendering control price of bill of quantities and the bidding offer.

4. Curriculum Contents and Teaching Hours

Table of Teaching Contents and Teaching Hours

Item	Topic	Teaching Objectives	Teaching Contents and Teaching Hours			
			Lesson Contents	Hrs	Student Activities	Hrs
1	Introduction	1. Understand the concept of bill of quantities; 2. Understand the concept of pricing for bill of quantities; 3. Understand the content and composition of pricing for bill of quantities; 4. Understand the difference between the pricing for bill of quantities and the pricing for quotas.	1. The principle and compiling method of pricing for construction cost, the bill of quantities and the pricing for it	2	1. Complete the Exercise of the principal of construction cost	1
			2. The compiling content of bill of quantities, tendering control price and the bidding offer(★)	2		
			3. The difference between the pricing for bill of quantities and the pricing for quotas	1		

序号	教学情境/任务/项目/单元	教学目标	教学内容及学时分配			
			知识点	学时	实践项目	学时
2	工程量清单计价规范概述	1. 理解工程量清单计价规范的主要内容； 2. 理解工程量清单计价的本质特性； 3. 能正确解读工程量清单计价规范的术语； 4. 了解工程量清单计价格式各种表格之间的关系	4. 工程量清单计价规范的主要内容（★）	2	2. 计价规范习题	1
			5. 工程量清单计价格式	1		
3	房屋建筑与装饰工程工程量清单计算规范	1. 理解计算规范的内容； 2. 能正确解读附录中常用项目的计算规则、项目特征及其工程内容； 3. 能使用计算规范，并与定额计算规则相区别	6. 计算规范的主要内容	2	3. 计算规范习题	1
			7. 规范附录的常用项目（★）	2		
4	工程量清单编制	1. 会编制分部分项工程量清单的五要素； 2. 会应用工程量计算规则计算清单工程量； 3. 能完整编制一个单位工程的工程量清单	8. 计算规范附录中计算规则应用（★）	2	4. 计算规则应用习题	2
			9. 工程量计算（★）	2	5. 实例清单工程量计算	2
			10. 工程量清单编制（★）	3	6. 实例清单编制	1

continued

Item	Topic	Teaching Objectives	Teaching Contents and Teaching Hours			
			Lesson Contents	Hrs	Student Activities	Hrs
2	Overview of specification of pricing for bill of quantities	1. Understand the main content of specification of pricing for bill of quantities; 2. Understand the essential features of pricing for bill of quantities; 3. Be able to interpret the terms in the specification of pricing for bill of quantities; 4. Understand the relationship between different format tables of pricing for bill of quantities	4. The main content of specification of pricing for bill of quantities(★)	2	2. Exercise of calculation specification	1
			5. The format of pricing for bill of quantities	1		
3	Calculation specification of bill of quantities for building and decoration construction	1. Understand the content of calculation specification; 2. Be able to interpret the calculation rules, project features and their engineering content of those common projects in Appendix; 3. Be able to apply calculation specification and understand the difference from quota pricing rules	6. The main content of calculation rules	2	3. Exercise of calculation rules	1
			7. The common projects in the Appendix of Specification(★)	2		
4	Compiling the bill of quantities	1. Compile the bill of quantities upon five elements of different sections and different projects; 2. Apply the calculation rules to work out the bill of quantities; 3. Complete a set of compilation of bill of quantities of a unit project	8. The application of calculation rules in the Appendix(★)	2	4. Exercise of the application of calculation rules	2
			9. The calculation of engineering quantities(★)	2	5. The calculation of real cases of bill of quantities	2
			10. The compilation of bill of quantities(★)	3	6. The compilation of real case bills	1

续表

序号	教学情境/任务/项目/单元	教学目标	知识点	学时	实践项目	学时
5	招标控制价计算	1. 掌握清单项目与定额项目的划分与对应关系； 2. 掌握工料机单价的计算或确定方法； 3. 掌握综合单价编制方法； 4. 掌握招标控制价的计算程序； 5. 掌握各项费用的计算方法	11. 计价工程量计算（★）	3	7. 实例计价工程量计算	2
			12. 综合单价编制方法（★）	3	8. 实例综合单价计算	2
			13. 招标控制价费用计算程序及方法（★）	2	9. 实例招标控制价计算	1
6	招标控制价及投标报价的异同	1. 会计算建筑与装饰工程的投标报价； 2. 能掌握控制价与投标价的异同	14. 投标报价实例计算（★）	3	10. 实例投标报价计算	2
			15. 分析总结招标控制价及投标报价的异同	2	11. 绘制异同表格	1
	合计			32		16

备注：重要知识点标识★

五、教学实施建议

(一)教学参考资料

袁建新.《工程量清单计价》(第五版),中国建筑工业出版社,2020。

中华人民共和国住房和城市建设部.《建设工程工程量清单计价规范》(GB 50500—2013),中国计划出版社,2013。

中华人民共和国住房和城乡建设部.《房屋建筑与装饰工程工程量计算规范》(GB 50854—2013),中国计划出版社,2013。

四川省住房和城乡建设厅.《四川省清单计价定额》,2020。

(二)教师素质要求

教师具备中级以上职称,有相应的工程监理、建造师或造价职工程师业资格,或拥有工程相应专业背景,能够理论结合实际,对工程全过程管理有一定工作经验或学习经历。

continued

Item	Topic	Teaching Objectives	Teaching Contents and Teaching Hours			
			Lesson Contents	Hrs	Student Activities	Hrs
5	The calculation of tendering control price	1. Master the classification and relevance between the bill project and quota project; 2. The calculation and deciding methods of the unit price of labor, materials and machinery; 3. The compiling methods of comprehensive unit price; 4. The calculation procedure of tendering control price; 5. The calculation methods of various costs	11. Calculating the pricing for bill of quantities (★)	3	7. Real case calculation of pricing for bill of quantities	2
			12. The compiling methods of comprehensive unit price(★)	3	8. Real case calculation of comprehensive unit price	2
			13. The calculation procedure and methods of tendering control price(★)	2	9. Real case calculation of tendering control price	1
6	The difference between the tendering control price and the bidding offer	1. Calculate the bidding offer of building and decoration construction; 2. Master the similarity and difference between the control price and the bidding offer	14. Real case calculation of bidding offer ★	3	10. Real case bidding offer calculation	2
			15. Analysis and summary of the similarity and difference	2	11. Table drawing of similarity and difference	1
Total				32		16
Note	Mark the important points with ★					

5. Teaching Suggestions

5.1 Teaching Resources

Yuan Jianxin. *Pricing for Bill of Quantities.* 5th Edition. China Architecture & Building Press, 2020.

MoHURD. *Specification of Pricing for Bill of Quantities of Construction Engineering Press* (GB 50500—2013). China Planning Press, 2013.

MoHURD. *Calculation Specification of Pricing for Bill of Quantities of Building and Decoration Construction Engineering*(GB 50854—2013), China Planning Press, 2013.

Province. Ministry of Housing and Urban-Rural Development of Sichuan Province *The Quotas of Pricing for Bill of Quantities in Sichuan Province by Administrative Headquarter of Construction Cost Pricing in Sichuan*, 2020.

5.2 Teachers'Qualification

Teachers should have intermediate titles and above, with corresponding qualifications of engineering supervision, construction architect or pricing engineer; or they should have the corresponding professional background of construction engineering, and can apply their theoretic knowledge into practice, with certain learning or working experience in the whole process management of project.

(三)教学场地、设施要求

该课程为理论授课,普通教室。

六、考核评价

(一)基本概念及知识点

工程造价基本原理、《建设工程工程量清单计价规范》(GB 50500—2013)、《房屋建筑与装饰工程工程量计算规范》(GB 50854—2013)的主要知识点,以单选题、多选题及判断题的形式出题,分值为 30~40 分。

(二)计算题

分值为 60~70 分。

(1)计算给定图纸中某些常用项目清单工程量,编制清单计价表;

(2)计算计价工程量,编制工程量清单综合单价分析表;

(3)计算总价措施项目费、单价措施项目费;

(4)计算其他项目费;

(5)计算招标控制价或投标报价。

5.3   Teaching Facilities

This course can be delivered in common classrooms.

6. Evaluation

6.1   The Basic Concepts and Knowledge Points

In this part, students are tested on the grasp of basic knowledge of construction cost and the main knowledge points in the two reference books: *Specification of Pricing for Bill of Quantities of Construction Engineering* (GB 50500—2013), and *Calculation Specification of Pricing for Bill of Quantities of Building and Decoration Construction Engineering* (GB 50854—2013). The test questions are in the form of single-choice questions, multiple-choice questions and judgment questions. The part accounts for 30 – 40 points of the total score.

6.2   Calculation Questions

This part accounts for 60 ~ 70 points of the total score, including:

(1) Calculate the inventory quantity of some common projects in a given drawing, and prepare a list pricing form of bill of quantities;

(2) Calculate the price of bill of quantities, and compile the comprehensive unit price analysis table of bill of quantities;

(3) Calculate the total price of implementation project cost, and unit price of implementation project cost;

(4) Calculate other project costs;

(5) Calculate the tendering control price or the bidding offer.

四 川 建 筑 职 业 技 术 学 院

课 程 标 准

课程名称： __工程量清单计价实训__

课程代码： __500115__

课程学分： __2__

基本学时： __60__

适用专业： __工程造价（中英合作办学项目）__

执 笔 人： __袁 鹰__

编制单位： __建筑工程造价教研室__

审 核： __工程管理系（院）（签章）__

批 准： __教务处（盖章）__

Sichuan College of Architectural Technology

Curriculum Standard

Curriculum Name: Practical Training for Pricing for Bill of Quantities

Curriculum Code: 500115

Credits: 2

Teaching Hours: 60

Applicable Specialty: Construction Cost

(Sino-British Cooperative Program)

Compiled by: Yuan Ying

On Behalf of: Teaching & Research Section of Construction Cost

Reviewed by: The Department of Engineering Managment

Approved by: Teaching Affairs Office

《工程量清单计价实训》课程标准

一、课程概述

工程量清单计价实训是工程造价、工程造价实验班、建设工程管理、工程造价(中英合作办学项目)专业的一门专业实训课程,是核心理论课工程量清单计价课程的实训课程。课程2学分,2周60学时,在第4学期开设,其先修专业课为《工程造价概论》《建筑工程预算》,后续专业课为《装饰工程计量与计价》《工程造价控制》《工程结算》《BIM工程计价软件应用》。本实训课程重在培养学生的实际操作能力。通过本课程,学生能根据实际施工图、《建设工程工程量清单计价规范》(GB 50500—2013)及《房屋建筑与装饰工程工程量计算规范》(GB 50854—2013)、《四川省清单计价定额》及有关资料,编制工程量清单、编制招标控制价或清单投标报价。

二、课程设计思路

设置依据:《建筑工程工程量清单计价规范》《房屋建筑与装饰工程工程量计算规范》、四川建筑职业技术学院人才培养方案(2017版)。

课程框架的设计:教师给定任务书和指导书,辅导学生根据相关资料,自主编制工程量清单、招标控制价或投标报价。

教学目标的确定:学生能自主编制清单计价的相关造价文件。

教学内容的选取:教师提供指导书、任务书、图纸,学生能根据《建设工程工程量清单计价规范》(GB 50500—2013)、《房屋建筑与装饰工程工程量计算规范》(GB 50854—2013)的基本要求、计量计价方法,依据施工图纸及相应条件正确地编制工程量清单;能根据工程量清单、规范、定额、工料机价格编制招标控制价或投标报价。

三、课程目标

通过本课程,学生将具备基本的职业素质,具有高级造价职业人才所必需的思想基础、思维方式及职业道德,能掌握工程量清单计价的编制程序和方法,具备造价基本技能,具备在工程造价工作岗位及相关岗位上解决实际问题的能力。

(一)知识目标

(1)掌握工程量清单计价的原理及基本知识;

(2)掌握工、料、机单价的确定方法;

Practical Training for Pricing for Bill of Quantities **Curriculum Standard**

1. Curriculum Description

This is a comprehensive and practical course, and it is an important practical training for the course of *Pricing for Bill of Quantities*. It is a professional course for construction cost specialty (Sino-British Cooperative Program). It consists of 2 credits, 60 teaching hours of 2 weeks, and is delivered in the fourth semester. The prerequisite courses were *Introduction to Construction Cost*, *Building Construction Budget*, and the units followed are *Decoration Engineering Quantities and Pricing*, *Construction Cost Control*, *Project Settlement*, and *BIM Management* and *Construction Cost Software Application*. This course focuses on cultivating students'hands-on abilities. Through the training, students shall be able to prepare and compile bill of quantities, and work out the tendering control price and the bill-based bidding offer price skillfully according to the actual construction drawings and other references like *Pricing Specification of Bill of Quantities in Construction Engineering* (GB 50500—2013), *Calculation Specification of Bill of Quantities for Building and Decoration Construction* (GB 50854—2013), and *Quotas of Bill of Quantities in Sichuan Province*.

2. Curriculum Design Ideas

This course is designed according to the educational concept of construction cost specialty, vocational skills and comprehensive abilities required by construction industry and other enterprises. The setting is based on the professional training program and teaching plan of construction cost specialty. The design and selection of curriculum framework, teaching objectives and teaching contents are in accordance with the qualification and skills requirements of construction cost professionals, the test syllabus of cost engineers and so on. The teaching and learning resources include the resource library for construction cost in SCAT, the construction cost information in Sichuan province, etc.

3. Curriculum Objectives

Through this course, students should learn to acquire basic professional qualities, the ideological foundation, thinking style and professional ethics necessary for senior professional talents, to master the basic knowledge and skills of preparing and compiling pricing for bill of engineering quantity, and to cultivate the ability to solve practical problems in construction cost occupations and other relevant jobs.

3. 1 Knowledge Objectives

After learning this course, students will be able to

(1) master the principle and basic knowledge of pricing for bill of quantities;

(2) master the methods of deciding the unit price of labor, materials and machinery;

（3）熟悉《建设工程工程量清单计价规范》（GB 50500—2013）；

（4）熟悉《房屋建筑与装饰工程工程量计算规范》（GB 50500—2013）；

（5）熟悉清单计价编制程序和方法。

（二）素质目标

（1）具备举一反三、处理综合问题的思维能力；

（2）具有自我约束、自我管理的能力；

（3）仿真练习，培养爱岗敬业的思想、实事求是的工作作风和创新意识；

（4）熟悉工程造价工作的有关政策法规；

（5）加强职业道德的意识，认识工程造价人员的执业权限与基本要求。

（三）能力目标

（1）能正确使用《建设工程工程量清单计价规范》（GB 50500—2013）及《房屋建筑与装饰工程工程量清单计算规范》（GB 50854—2013）；

（2）能正确应用定额；

（3）会计算清单工程量和计价工程量；

（4）会计算或确定工料机单价；

（5）能熟练地编制综合单价；

（6）能编制工程量清单；

（7）能编制招标控制价或投标报价。

四、课程内容与学时分配

教学内容与学时安排表

序号	教学情境/任务/项目/单元	教学目标	教学内容及学时分配		
			知识点	实践项目	学时
1	编制工程量清单	1. 会识读施工图，并能将有问题处以图纸会审方式予以处理； 2. 根据常规施工方案，结合工程情况拟定该工程的施工方案； 3. 能编制工程量清单		1. 熟悉图纸，逐步完善图纸会审、施工方案	5
				2. 清单工程量计算	12
				3. 完成分部分项工程量清单	5
				4. 完成措施项目清单	4
				5. 完成其他项目清单	2
				6. 完成规费及税金项目清单	1
				7. 按要求装订成册	1

(3) be familiar with the reference book *Pricing Specification of Bill of Quantities in Construction Engineering* (GB 50500—2013);

(4) be familiar with the reference book *Calculation Specification of Bill of Quantities for Building and Decoration Construction* (GB 50854—2013);

(5) be familiar with the compiling procedure and methods of pricing for bill of quantities.

3.2 Quality Objectives

After learning this course, students will have the quality of

(1) dealing with comprehensive issues and being able to draw inferences about other cases from one instance;

(2) self-discipline and self-management;

(3) devoted, responsible and practical working style and innovative spirit;

(4) being familiar with relevant policies and regulations of construction cost work;

(5) strengthening the awareness of professional ethics and recognizing the responsibilities and basic requirements of construction cost professionals.

3.3 Ability Objectives

After learning this course, students will have the ability of

(1) using properly the pricing and calculation specification books for reference;

(2) using the quotas correctly;

(3) calculating the bill of quantities and its pricing;

(4) calculating or determine the unit price of labor, materials and machinery;

(5) preparing proficiently and compile comprehensive unit prices;

(6) compiling bill of quantities;

(7) compiling the tendering control price of bill of quantities and the bidding offer.

4. Curriculum Contents and Teaching Hours

Table of Teaching Contents and Teaching Hours

| Item | Topic | Teaching Objectives | Teaching Contents and Teaching Hours | | |
			Lesson Contents	Student Activities	Hrs
1	Prepare and compile bill of quantities	Be able to read the construction drawings and handle mistakes or problems in the form of joint review; 2. Work out the construction plan according to the commons and on-site situation; 3. Be able to compile bill of quantities		1. Be familiar with the drawing, complete the joint review and the construction plan	5
				2. Calculate the bill of quantities	12
				3. Complete the divided bills	5
				4. Complete the task plan bills	4
				5. Complete other project bills	2
				6. Complete the bill of charges and taxes	1
				7. Bound the bill book as required	1

序号	教学情境/任务/项目/单元	教学目标	教学内容及学时分配		
			知识点	实践项目	学时
2	编制招标控制价	1. 会划分清单项目的主项和附项,计算计价工程量; 2. 能根据相关资料计算计价工程量; 3. 能掌握综合单价的计算方法,正确熟练地计算; 4. 能计算各项目的费用,汇总形成招标控制价		1. 计算划分清单项目的主项和附项,计算计价工程量	8
				2. 计算综合单价	10
				3. 编制完成分部分项工程费	5
				4. 编制完成措施项目费	4
				5. 编制完成其他项目费	1
				6. 编制完成规费及税金	1
				7. 按要求装订成册	1
合计					60
备注:					

五、教学实施建议

(一)教学参考资料

袁建新.《工程量清单计价》(第五版),中国建筑工业出版社,2020。

中华人民共和国住房和城乡建设部.《建设工程工程量清单计价规范》(GB 50500—2013).中国计划出版社,2013。

中华人民共和国住房和城乡建设部.《房屋建筑与装饰工程工程量计算规范》(GB 50854—2013).中国计划出版社,2013。

四川省住房和城乡建设厅.《四川省清单计价定额》,2020。

(二)教师素质要求

教师具备中级以上职称,有相应的工程监理、建造师或造价职工程师业资格,或拥有工程相应专业背景,能够理论结合实际,对工程全过程管理有一定工作经验或学习经历。

(三)教学场地、设施要求

该课程为理论授课,普通教室。

continued

Item	Topic	Teaching Objectives	Teaching Contents and Teaching Hours		
			Lesson Contents	Student Activities	Hrs
2	Compile the tendering control price	1. Be able to classify the main items and sub-items in the bill project, and work out the pricing for bill of quantities; 2. Be able to calculate the pricing for bill of quantities according to references; 3. Master the calculation methods of comprehensive unit price and apply skillfully; 4. Be able to calculate the costs of divided projects and pool them to become the final tendering control price		1. Calculate the main items and sub-items in the bill project, and work out the pricing for bill of quantities	8
				2. Calculate the comprehensive unit price	10
				3. Compile and complete the construction costs for divided projects and sections	5
				4. Compile and complete the task plan cost	4
				5. Compile and complete other project costs	1
				6. Compile and complete charges and taxes	1
				7. Bound the bill book as required	1
Total					60
Note					

5. Teaching Suggestions

5.1 Teaching Resources

Yuan Jianxin. *Pricing for Bill of Quantities*. 5th Edition China Architecture & Building Press, 2020.

MoHURD. *Specification of Pricing for Bill of Quantities of Construction Engineering* (GB 50500—2013). China Planning Press, 2013.

MoHURD. *Calculation Specification of Pricing for Bill of Quantities of Building and Decoration Construction Engineering* (GB 50854—2013). China Planning Press, 2013.

Ministry of Housing and Urban-Rural Development of Sichuan Province. *The Quotas of Pricing for Bill of Quantities in Sichuan Province by Administrative Headquarter of Construction Cost Pricing in Sichuan Province*, 2020.

5.2 Teachers'Qualification

Teachers should have intermediate titles and above, with corresponding qualifications of engineering supervision, construction architect or pricing engineer; or they should have the corresponding professional background of construction engineering, and can apply their theoretic knowledge into practice, with certain learning or working experience in the whole process management of project.

5.3 Teaching Facilities

This course can be delivered in common classrooms.

六、考核评价

1. 考核内容及评分办法

（1）考核内容：实训成果的完整性、规范性；工程造价的合理性；

（2）评分办法：实训过程占20%，内容考核比重占30%；格式考核比重占10%，纪律占20%，面试占20%。

实训过程：是否独立完成实训任务，实训过程中协作能力、团队意识、沟通能力的体现，是否具有创新意识和开拓精神。

实训成果：内容的完整性、格式是否规范，卷面是否整洁，方法是否正确，计算结果是否正确合理。

实训纪律：是否遵守学校作息时间，有无无故缺席、迟到、早退现象。

其他：是否具备较好的职业能力，兼顾方法能力和社会能力。

2. 成果鉴定标准

根据《四川省学生学业考核方法》及《四川建筑职业技术学院课程设计、毕业设计、综合训练管理规定》，采取五级记分制，即优、良、中、及格、不及格。

（1）优：准时到班进行集中实训，有良好的团队意识和协作精神，能独立完成实训，实训成果完整，工程量清单和工程造价计算合理，具备分析问题、处理问题的能力，书写工整，格式规范，口试回答问题正确。

（2）良：准时到班进行集中实训，有良好的团队意识和协作精神，独立完成实训，实训成果完整，工程造价计算较合理，具有一定分析问题、处理问题的能力，书写较工整，格式规范，口试回答问题基本正确。

（3）中：准时到班进行集中实训，有较好的团队意识和协作精神，完成实训、实训成果完整、工程造价计算基本合理、书写基本工整，格式规范，口试回答问题基本正确。

（4）及格：能到班进行集中实训，有一定团队意识和协作精神，能完成实训，实训成果基本完整，格式基本规范，口试能回答一些问题。

（5）不及格：不到班进行集中实训，抄袭实训成果或实训成果不完整，不能回答口试问题。

6. Evaluation

6.1　The Content and the Grading

（1）The evaluation content includes the completeness and standardization of the training results, and the rationality of construction cost；

（2）Scoring methods：the training process accounts for 20%, content assessment accounts for 30%, the format assessment accounts for 10%, discipline accounts for 20%, and the interview accounts for 20%.

The training process：whether to complete the practical training task independently；the display of collaboration ability, teamwork consciousness and communicative ability in the course of practical training；whether to embody the innovative and pioneering spirit.

Practical training results：the completeness of the content；whether the format is standardized；whether the final presence is clean；whether the methods used are all correct；whether the calculation results are correct and reasonable.

Training discipline：whether to observe the school schedule；whether there is any absence, lateness, and early retreat without excuses.

Other requirements：whether to embody strong professional competences, together with strategic abilities and social abilities.

6.2　Assessment Standard for Training Achievement

According to the *Students Academic Evaluation Methods in Sichuan Province* and the *Curriculum Design*, *Graduation Design*, *Comprehensive Practical Training Management Regulations for Students in SCAT*, a five-level scoring system is adopted, that is excellent, good, mediocre, pass, failed.

（1）Excellent：perfect attendance；have a good sense of teamwork and collaboration；be able to independently complete the practical training；the completeness of the training workbooks and presentation；the calculation of bill of quantities and project cost is reasonable and correct；display the abilities of analyzing and solving all the problems；neat writing；format complied with specifications；be able to answer interview questions fully correctly.

（2）Good：attend the class without any delay；have a good sense of teamwork and collaboration；be able to independently complete the practical training；the completeness of the training workbooks and presentation；the calculation of bill of quantities and project cost is reasonable and correct；display certain abilities of analyzing and solving most problems；neat writing；format complied with specifications；be able to answer interview questions mostly but not fully correctly.

（3）Mediocre：attend the class without any delay；have a good sense of teamwork and collaboration；the completeness of the training workbooks and presentation；the calculation of bill of quantities and project cost is mostly reasonable；almost neat writing；format complied with specifications；be able to answer interview questions basically correctly.

（4）Pass：attend the training class with occasional delay；have certain sense of teamwork and collaboration；complete the practical training with help of others；the training results are basically correct；format basically complied with specifications；be able to answer some interview questions.

（5）Failed：lots of absence from class；copy others'training workbooks or the result presentation is very incomplete；unable to answer any interview questions correctly.

四 川 建 筑 职 业 技 术 学 院

课 程 标 准

课程名称：　　　工程造价软件应用

课程代码：　　　150187

课程学分：　　　　　3

基本学时：　　　　　48

适用专业：　　工程造价（中英合作办学项目）

执 笔 人：　　　　汪世亮

编制单位：　　　建筑工程造价教研室

审　　核：　　工程管理系（院）（签章）

批　　准：　　　教务处（盖章）

Sichuan College of Architectural Technology

Curriculum Standard

Curriculum Name: Application of Construction Cost Software

Curriculum Code: 150187

Credits: 3

Teaching Hours: 48

Applicable Specialty: Construction Cost

(Sino-British Cooperative Program)

Compiled by: Wang Shiliang

On Behalf of: Teaching & Research Section of Building

Construction Cost

Reviewed by: The Department of Engineering Management

Approved by: Teaching Affairs Office

《工程造价软件应用》课程标准

一、课程概述

《工程造价软件应用》课程是人才培养方案中非常重要的环节。《工程造价软件应用》课程是一门实践操作性强的专业课程,是将工程造价理论和方法计算机化的体现,同时也是工程造价工作的主要手段。对于工程造价(中英合作办学项目)专业,《工程造价软件应用》课程在第 5 学期开设。该课程的先修课程是《建筑构造》《建筑施工工艺》《建筑材料》《建筑工程预算》《钢筋工程量计算》以及《工程量清单计价》等。

二、课程设计思路

工程造价软件熟练应用是用人单位对工程造价专业毕业生的基本素质要求。《工程造价软件应用》课程采用的是项目教学法,课程以使用工程造价软件完成一个中等体量建筑工程项目的工程量计算为目标,选取现今工程造价行业主要软件(广联达软件或清华斯维尔软件)为使用工具,形成完整的项目工程量计算表。该课程教学内容为建筑行业现行的主要建筑结构(例如砖混结构、框架结构、框剪结构等)的软件建模与工程量统计分析,该课程需在专用的计算机房进行。

三、课程目标

本课程的目标是:使学生具有使用工程造价软件完成中等体量现行常见建筑结构工程项目的工程量计算,并形成完整工程量计算书的能力;使学生具有在工作中使用手工计算工程量与软件计算工程量互相印证、补充的能力;使学生掌握软件计算工程量的原理知识、软件建模方法、算量设置技巧,并进一步巩固工程量计算规则;使学生养成在工作中面对不同地区、不同性质建筑物时选择正确的工程设置、适用的建模方法进行完整的工程量计算的素质,养成依据国家法规、规范、标准采用正确方法工作的素质。

Application of Construction Cost Software Curriculum Standard

1. Curriculum Description

Application of Construction Cost Software is a very important part in the professional training program. *Application of Construction Cost Software* is a practical and operable professional course, which reflects the computerization of construction cost theory and method. At the same time, construction cost software is also the main means of construction cost work. For the specialty of construction cost (Sino-British cooperative program), *Application of Construction Cost Software* is offered in the fifth semester. The seprevious delivered courses are *Building Structure*, *Building Construction Technology*, *Building Materials*, *Building Construction Budget*, *Calculation of Reinforcement Quantity* and *Pricing for Bills of Quantities*, etc.

2. Curriculum Design Ideas

Skillful application of construction cost software is the basic requirement for graduates of the specialty of construction cost. In this course, the teaching method is the project-based pedagogy. This course takes the construction cost software as the target to complete the quantity calculation of medium-sized construction projects. The main software in the construction cost industry (guanglianda software or Tsinghua sware software) is selected as the tool to form a complete project quantity calculation table. The content of this course is the software modeling and engineering quantity statistical analysis of the current main building structures in the construction industry (such as brick-concrete structure, frame structure, frame-shear structure, etc.). This course needs to be carried out in a special computer room.

3. Curriculum Objectives

The objectives of this course are to enable students to use construction cost software to complete the volume calculation of engineering projects of medium volume and current common building structures, and to form a complete volume calculation book; to let students have the ability to verify and supplement each other by using manual calculation and software calculation in work; to enable students to master the principle knowledge, software modeling method and calculation setting skills of software calculation quantity, and to further consolidate the calculation rules of engineering quantity; to have students develop the ability of selecting the correct engineering setting and the applicable modeling method to complete the quantity calculation in face of buildings of different regions and different natures, and to adopt the correct method to work according to the regulations, norms and standards in construction industry in China.

(一)知识目标

(1)掌握软件工程设置方法；

(2)掌握轴网的绘制方法；

(3)掌握基础、柱、梁、板、墙及所有钢筋的绘制方法；

(4)掌握门、窗绘制方法；

(5)掌握楼梯、台阶、散水等其他构件的绘制方法；

(6)掌握异形构件(包含异形构件的钢筋)的绘制方法；

(7)掌握建筑装饰、建筑面积以及脚手架的布置方法；

(8)掌握软件识别CAD图纸生成模型的方法；

(9)掌握对构件挂接做法并生成工程量清单的方法；

(10)掌握分析、汇总、查询工程量并生成工程量计算表的方法。

(二)素质目标

(1)养成不断自我提高、不断更新知识的素质；

(2)养成面对地区差异、行业差异、建筑结构差异等复杂情况采用正确方法算量的素质；

(3)养成必须依据国家法规、规范、标准工作的素质。

(三)能力目标

(1)具有正确设置工程项目信息的能力；

(2)具有正确设置工程量计算规则和钢筋算量的能力；

(3)具有按照施工图纸定义构件和布置构件以及完成建模的能力；

(4)具有分析、汇总和查询工程量的能力。

四、课程内容与学时分配

教学内容与学时安排表

序号	教学情境/任务/项目/单元	教学目标	教学内容及学时分配			
			知识点	学时	实践项目	学时
1	软件简介与工程设置	掌握软件界面特点与工程设置方法	1. 软件简介 2. 工程信息 3. 层高设置 4. 材料设置 5. 算量设置(★) 6. 钢筋设置(★)	2	1. 工程信息 2. 层高设置 3. 材料设置 4. 算量设置(★) 5. 钢筋设置(★)	1

3.1　Knowledge Objectives

After learning this course, students will be able to

(1) master the setting method of software engineering;

(2) master the drawing method of shaft network;

(3) master the drawing method of foundation, column, beam, board, wall and all rebar;

(4) master painting methods of doors and windows;

(5) master the drawing methods of stairs, steps, and other components;

(6) master the drawing method of special-shaped components (including steel reinforcement of special-shaped components);

(7) master the method of building decoration, building area and scaffold layout;

(8) master the method of software identification CAD drawing generation model;

(9) master the method of component hooking and generating bill of quantities;

(10) master the method of analyzing, summarizing, querying the project amount and generating the project amount calculation table.

3.2　Quality Objectives

After learning this course, students will have the quality of

(1) improving myself and updating knowledge constantly;

(2) using correct methods to calculate quantities in face of regional differences, industrial differences, architectural structure differences and other complex situations;

(3) obeying national laws, regulations and standards.

3.3　Ability Objectives

After learning this course, students will have the ability of

(1) setting project information correctly;

(2) setting the calculation rules of engineering quantity and reinforcement algorithm correctly;

(3) defining and arranging components and completing modeling according to construction drawings;

(4) analyzing, summarizing and inquiring project quantities.

4. Curriculum Contents and Teaching Hours

Table of Teaching Contents and Teaching Hours

Item	Topic	Teaching Objectives	Teaching Contents and Teaching Hours			
			Lesson Contents	Hrs	Student Activities	Hrs
1	Software profile with engineering setup	Master software interface features and engineering setting methods	1. Software introduction	2		1
			2. Engineering information		1. Engineering information	
			3. Layer height setting		2. Layer height setting	
			4. Material setting		3. Material setting	
			5. Calculation settings(★)		4. Calculation settings(★)	
			6. Steel bar setting(★)		5. Steel bar setting(★)	

序号	教学情境/任务/项目/单元	教学目标	知识点	学时	实践项目	学时
			教学内容及学时分配			
2	轴网与基础、基础钢筋绘制	掌握轴网、基础与基础钢筋的绘制方法	7. 正交轴网与弧形轴网的绘制(★)	3	6. 正交轴网与弧形轴网的绘制(★)	1
			8. 独立基础与独立基础钢筋的布置(★)		7. 独立基础与独立基础钢筋的布置(★)	
			9. 条形基础与条形基础钢筋的布置		8. 条形基础与条形基础钢筋的布置	
3	柱、柱筋与梁、梁筋的布置	掌握柱、柱筋与梁、梁筋的绘制方法	10. 柱与柱筋的布置(★)	4	9. 柱与柱筋的布置(★)	2
			11. 梁与梁筋的布置(★)		10. 梁与梁筋的布置(★)	
4	板与板钢筋的布置	掌握板、板筋的绘制方法	12. 板与板筋的布置(★)	4	11. 板与板筋的布置(★)	2
5	剪力墙与剪力墙钢筋的布置	掌握剪力墙与墙钢筋的绘制方法	13. 剪力墙与墙筋的布置(★)	4	12. 剪力墙与墙筋的布置(★)	2
6	砌体墙与门窗、过梁的布置	掌握砌体墙与门窗、过梁的布置方法	14. 砌体墙的布置	2	13. 砌体墙的布置	1
			15. 门窗的布置		14. 门窗的布置	
			16. 过梁的布置		15. 过梁的布置	
7	楼梯、台阶、散水等其他构件的布置	掌握楼梯、散水、台阶等其他构件的布置方法	17. 楼梯的布置	2	16. 楼梯的布置	1
			18. 散水和台阶的布置		17. 散水和台阶的布置	
			19. 其他构件的布置		18. 其他构件的布置	
8	装饰构件与建筑面积、脚手架的布置	掌握装饰构件与建筑面积、脚手架的布置	20. 装饰构件的布置(★)	1	19. 装饰构件的布置(★)	1
			21. 建筑面积的布置		20. 建筑面积的布置	
			22. 脚手架的布置		21. 脚手架的布置	

continued

Item	Topic	Teaching Objectives	Teaching Contents and Teaching Hours			
			Lesson Contents	Hrs	Student Activities	Hrs
2	Axis net and foundation, foundation reinforcement drawing	Master the drawing method of shaft mesh, foundation and foundation reinforcement	7. Drawing of orthogonal shaft net and arc shaft net(★)	3	6. Drawing of orthogonal shaft net and arc shaft net(★)	1
			8. Layout of steel reinforcement of independent foundation and independent foundation(★)		7. Layout of steel reinforcement of independent foundation and independent foundation(★)	
			9. Layout of bar foundation and bar foundation reinforcement		8. Layout of bar foundation and bar foundation reinforcement	
3	Column, column reinforcement and beam, beam reinforcement layout	Master the drawing method of column, column bar, beam and beam bar	10. Column and reinforcement layout(★)	4	9. Column and reinforcement layout(★)	2
			11. Beam and beam reinforcement layout(★)		10. Beam and beam reinforcement layout(★)	
4	Layout of plate and plate reinforcement	Master the drawing method of board and rib	12. Layout of boards and stiffeners(★)	4	11. Layout of boards and stiffeners(★)	2
5	Arrangement of shear wall and reinforcement of shear wall	Master the drawing method of shear wall and reinforcement	13. Arrangement of shear wall and wall reinforcement(★)	4	12. Arrangement of shear wall and wall reinforcement(★)	2
6	Arrangement of masonry walls, doors, Windows and lintels	Master the arrangement of masonry walls, doors, windows and lintels	14. Arrangement of masonry walls	2	13. Arrangement of masonry walls	1
			15. Arrangement of doors and windows		14. Arrangement of doors and windows	
			16. Arrangement of lintel		15. Arrangement of lintel	
7	The arrangement of stairs, stairs and other components	Master the layout method of stairs, water dispersing, steps and other components	17. The layout of the stairs	2	16. The layout of the stairs	1
			18. The layout of water and steps		17. The layout of water and steps	
			19. Arrangement of other components		18. Arrangement of other components	
8	Arrangement of decorative elements, building area and scaffolding	Master the layout of decorative components, building area and scaffolding	20. Layout of decorative components(★)	1	19. Layout of decorative components(★)	1
			21. Layout of building area		20. Layout of building area	
			22. Arrangement of scaffolding		21. Arrangement of scaffolding	

序号	教学情境/任务/项目/单元	教学目标	教学内容及学时分配			
			知识点	学时	实践项目	学时
9	识别 CAD 图纸建模	掌握识别 CAD 图纸建模的方法	23. 导入 CAD 图纸 24. 识别构件 25. 识别钢筋(★)	8	22. 导入 CAD 图纸 23. 识别构件 24. 识别钢筋(★)	4
10	构件挂接做法并分析、汇总、查询工程量	掌握构件挂接做法并分析、汇总、查询工程量的方法	26. 构件挂接做法 27. 分析、汇总、查询工程量 28. 生成工程量清单	2	25. 构件挂接做法 26. 分析、汇总、查询工程量 27. 生成工程量清单	1
	合计			32		16

备注:重要知识点标注★

五、教学实施建议

(一)教学参考资料

张晓敏、李社生.《建筑工程造价软件应用:广联达系列软件》,中国建筑工业出版社,2013。

(二)教师素质要求

任课教师为工程造价专业教师,熟练使用广联达或者清华斯维尔软件且为双师型教师。

(三)教学场地、设施要求

教学场地为计算机机房,为满足教学要求需要设 3 间计算机机房,每间容纳 50 个学生;电脑配置要求 i3 及以上处理器,内存 8 G 以上,硬盘要求 500 G 以上;软件要求广联达图形算量与钢筋算量软件 3 套,每套 50 节点,斯维尔三维算量软件 3 套,每套 50 节点。

六、考核评价

本课程考核方式为实际操作,考核内容为在固定时间(100 分钟)完成一定规模建筑的软件计算工程量工作,教师考核的内容包括工程量输出的准确性以及模型的完整性。成绩构成为平时成绩占 30%,考试成绩占 70%。

continued

Item	Topic	Teaching Objectives	Teaching Contents and Teaching Hours			
			Lesson Contents	Hrs	Student Activities	Hrs
9	Identifying CAD drawings for modeling	Master the method of identifying CAD drawing modeling	23. Import CAD drawings	8	22. Importing CAD drawings	4
			24. Identification components		23. Identification components	
			25. Steel reinforcement identification(★)		24. Steel reinforcement identification(★)	
10	Component hooking practice and analysis, summary, query the amount of work	Master the method of component hooking and analyze, summarize and inquire the project quantity	26. Component connection method	2	25. Component connection method	1
			27. Analyzed, summarized and inquired the project amount		26. Analyzing, summarizing and inquiring the project amount	
			28. Generate the bill of quantities		27. Generating the bills of quantities	
	Total			32		16
Note:Mark important points with ★						

5. Teaching Suggestions

5.1 Teaching Resources

Zhang Xiaomin, Li Shesheng. *Application of Construction Cost Software*: *Guanglianda Series Software*, China Architecture & Building Press, 2013.

5.2 Teachers'Qualification

The teachers of this course should be majored in construction cost and should be proficient in using Guanglianda or Tsinghua Sware software.

5.3 Teaching Facilities

Computer rooms are required for the course. In order to meet the teaching requirements, three computer rooms are necessary, with each room accommodating 50 students. Computer configuration requirement for processors is i3 and above, internal memory should be more than 8 G, and hard disk capacity should be more than 500 G. The software requires three sets of Guanglianda graphic calculation software and steel bar calculation software, each set of 50 nodes, and three sets of Sware 3d calculation software, each set of 50 nodes.

6. Evaluation

The assessment method of this course is practical operation, and the assessment content is to complete the software calculation work of a certain scale building in a fixed time(100 minutes). The content of the assessment includes the accuracy of the output of the project quantity and the integrity of the model. Daily performance accounts for 30% of the total score, and the final examination accounts for 70%.

四 川 建 筑 职 业 技 术 学 院

课 程 标 准

课程名称： 工程造价软件应用实训

课程代码： 150189

课程学分： 1

基本学时： 30

适用专业： 工程造价（中英合作办学项目）

执 笔 人： 汪世亮

编制单位： 建筑工程造价教研室

审　　核： 工程管理系（院）（签章）

批　　准： 教务处（盖章）

Sichuan College of Architectural Technology

Curriculum Standard

Curriculum Name: Construction Cost Software Application Training

Curriculum Code: 150189

Credits: 1

Teaching Hours: 30

Applicable Specialty: Construction Cost

(Sino-British Cooperative Program)

Compiled by: Wang Shiliang

On Behalf of: Teaching & Research Section of Building

Construction Cost

Reviewed by: The Department of Engineering Management

Approved by: Teaching Affairs Office

《工程造价软件应用实训》课程标准

一、课程概述

《工程造价软件应用实训》课程是人才培养方案中非常重要的环节。《工程造价软件应用实训》是《工程造价软件应用》课程的实训环节。对于工程造价专业(中英合作办学项目),《工程造价软件应用实训》课程在第5学期开设。该课程的先修课程是《建筑构造》《建筑施工工艺》《建筑材料》《建筑工程预算》《钢筋工程量计算》《工程量清单计价》以及《工程造价软件应用》等。

二、课程设计思路

工程造价软件熟练应用是用人单位对工程造价专业毕业生的基本素质要求。《工程造价软件应用实训》课程采用的是实际操作训练方法,以使用工程造价软件完成一个中等体量建筑工程项目的工程量计算为目标,选取现今工程造价行业主要软件(广联达软件或清华斯维尔软件)为使用工具,在规定时间内完成建模并形成完整的项目工程量计算表。该实训内容为建筑行业现行的主要建筑结构(例如砖混结构、框架结构、框剪结构等)的软件建模与工程量统计分析,需在专用的计算机房进行。

三、课程目标

该课程的目标是使学生具有使用工程造价软件完成中等体量现行常见建筑结构的工程项目的工程量计算,并形成完整工程量计算书的能力;使学生具有在工作中使用手工计算工程量与软件计算工程量互相印证、补充的能力;使学生掌握软件计算工程量的原理知识、软件建模方法、算量设置技巧,并进一步巩固工程量计算规则;使学生养成在工作中面对不同地区、不同性质建筑物时选择正确的工程设置、适用的建模方法进行完整的工程量计算的素质,养成依据国家法规、规范、标准采用正确方法工作的素质。

(一)知识目标

(1)掌握软件工程设置方法;

Construction Cost Software Application Training Curriculum Standard

1. Curriculum Description

Construction Cost Software Application Training is a very important part in the professional training program. *Construction Cost Software Application Training* is the practical training of the course *Application of Construction Cost Software*. For the specialty of construction cost (Sino-British cooperative program) , *Construction Cost Software Application Training* is offered in the fifth semester. The delivered courses are *Building Structure*, *Building Construction Technology*, *Building Materials*, *Building Construction Budget*, *Calculation of Reinforcement Quantity*, *Pricing for Bills of Quantities* and *Application of Construction Cost Software*, etc.

2. Curriculum Design Ideas

Skillful application of construction cost software is the basic requirement for graduates of the specialty of construction cost. The teaching method for this course is practical training with the purpose of completing the calculation of engineering quantity of medium size construction engineering project with the selected software in construction cost industry(such as Guanglianda software or Tsinghua Sware software) within the prescribed time and forming a complete project quantity calculation sheet. The practical training content is the software modeling and statistical analysis of the current main building structures in the construction industry(such as brick-concrete structure, frame structure, frame-shear structure, etc.) , which shall be carried out in a special computer room.

3. Curriculum Objectives

The objectives of this course are to enable students to use construction cost software to complete the volume calculation of engineering projects of medium volume and current common building structures, and to form a complete volume calculation book; to let students have the ability to verify and supplement each other by using manual calculation and software calculation in work; to enable students to master the principle knowledge, software modeling method and calculation setting skills of software calculation quantity, and further consolidate the calculation rules of engineering quantity; to have students develop the ability of selecting the correct engineering setting and the applicable modeling method to complete the quantity calculation in face of buildings of different regions and different natures, and to adopt the correct method to work according to the regulations, norms and standards in construction industry in China.

3.1 Knowledge Objectives

After learning this course, students will be able to

(1) master the setting method of software engineering;

(2)掌握轴网的绘制方法;

(3)掌握基础、柱、梁、板、墙及所有钢筋的绘制方法;

(4)掌握门、窗绘制方法;

(5)掌握楼梯、台阶、散水等其他构件的绘制方法;

(6)掌握异形构件(包含异形构件的钢筋)的绘制方法;

(7)掌握建筑装饰、建筑面积以及脚手架的布置方法;

(8)掌握软件识别CAD图纸生成模型的方法;

(9)掌握对构件挂接做法并生成工程量清单的方法;

(10)掌握分析、汇总、查询工程量并生成工程量计算表的方法。

(二)素质目标

(1)养成不断自我提高、不断更新知识的素质;

(2)养成面对地区差异、行业差异、建筑结构差异等复杂情况,采用正确方法算量的素质;

(3)养成必须依据国家法规、规范、标准工作的素质。

(三)能力目标

(1)具有正确设置工程项目信息的能力;

(2)具有正确设置工程量计算规则和钢筋算量的能力;

(3)具有按照施工图纸定义构件和布置构件以及完成建模的能力;

(4)具有分析、汇总和查询工程量的能力。

四、课程内容与学时分配

教学内容与学时安排表

序号	教学情境/任务/项目/单元	教学目标	教学内容及学时分配	
			实践项目	学时
1	工程设置,轴网与基础、基础钢筋绘制,柱、柱筋绘制	熟练完成工程设置,轴网与础、基础钢筋绘制,柱、柱筋绘制	1. 工程设置	6
			2. 轴网绘制	
			3. 基础与基础钢筋绘制(★)	
			4. 柱与柱钢筋绘制(★)	
2	梁、梁筋的布置	熟练完成梁、梁筋的布置	5. 梁、梁筋的布置(★)	6
3	板与板钢筋的布置、剪力墙与剪力墙钢筋的布置	熟练完成板与板钢筋的布置、剪力墙与剪力墙钢筋的布置	6. 板与板筋的布置、剪力墙与剪力墙钢筋的布置(★)	6

(2) master the drawing method of shaft network;

(3) master the drawing method of foundation, column, beam, board, wall and all rebar;

(4) master painting methods of doors and windows;

(5) master the drawing methods of stairs, steps, and other components;

(6) master the drawing method of special-shaped components (including steel reinforcement of special-shaped components);

(7) master the method of building decoration, building area and scaffold layout;

(8) master the method of software identification CAD drawing generation model;

(9) master the method of component hooking and generating bill of quantities;

(10) master the method of analyzing, summarizing, querying the project amount and generating the project amount calculation table.

3.2 Quality Objectives

After learning this course, students will have the quality of

(1) constantly improving oneself and constantly updating knowledge;

(2) using correct methods to calculate quantities in face of regional differences, industrial differences, architectural structure differences and other complex situations;

(3) obeying national laws, regulations and standards.

3.3 Ability Objectives

After learning this course, students will have the ability of

(1) setting project information correctly;

(2) setting the calculation rules of engineering quantity and reinforcement algorithm correctly;

(3) defining and arranging components and completing modeling according to construction drawings;

(4) analyzing, summarizing and inquiring project quantities.

4. Curriculum Contents and Teaching Hours

Table of Teaching Contents and Teaching Hours

Item	Topic	Teaching Objectives	Teaching Contents and Teaching Hours	
			Student Activities	Hrs
1	Engineering setting, drawing axis net, foundation and foundation reinforcement, drawing column and column reinforcement	Skilled completion of engineering Settings, axis network and foundation, foundation reinforcement drawing, column, column drawing	1. Project setting 2. Axis network drawing 3. Drawing of foundation and foundation reinforcement(★) 4. Column and column reinforcement drawing(★)	6
2	Arrangement of beam and beam reinforcement	Skilled completion of beam, beam reinforcement layout	5. beam, beam reinforcement layout(★)	6
3	The layout of plate and plate reinforcement, the layout of shear wall and shear wall reinforcement	Proficient in the layout of plates and steel bars, and the layout of shear walls and steel bars of shear walls	6. Layout of plates and reinforcement, and layout of shear walls and reinforcement of shear walls(★)	6

序号	教学情境/任务/项目/单元	教学目标	教学内容及学时分配	
			实践项目	学时
4	砌体墙与门窗、过梁、楼梯、散水、台阶、装饰等构件的布置	熟练完成砌体墙与门窗、过梁、楼梯、散水、台阶、装饰等构件的布置	7. 砌体墙的布置	6
			8. 门窗的布置	
			9. 过梁的布置	
			10. 楼梯的布置	
			11. 散水和台阶的布置	
			12. 装饰构件以及其他构件的布置	
5	构件挂接做法并分析、汇总、查询工程量	掌握构件挂接做法并分析、汇总、查询工程量的方法	13. 构件挂接做法（★）	6
			14. 分析、汇总、查询工程量（★）	
			15. 生成工程量清单、打印、成果装订	
合计				30

备注:重要知识点标注★

五、教学实施建议

(一)教学参考资料

张晓敏、李社生.《建筑工程造价软件应用:广联达系列软件》,中国建筑工业出版社,2013。

(二)教师素质要求

任课教师为工程造价专业教师,熟练使用广联达或者清华斯维尔软件且为双师型教师。

(三)教学场地、设施要求

教学场地为计算机机房,为满足教学要求需要设 3 间计算机机房,每间容纳 50 个学生;电脑配置要求 i3 及以上处理器,内存 8G 以上,硬盘要求 500G 以上;软件要求广联达图形算量与钢筋算量软件 3 套,每套 50 节点,斯维尔三维算量软件 3 套,每套 50 节点。

六、考核评价

本课程考核方式为实际操作,教师考核的内容包括工程量输出的准确性以及模型的完整性。成绩构成为:平时成绩占 30%,作业成果成绩占 70%。

continued

Item	Topic	Teaching Objectives	Teaching Contents and Teaching Hours	
			Student Activities	Hrs
4	Arrangement of masonry walls, doors and windows, lintels, stairs, sprinklers, steps, decorations, etc	Skilled completion of masonry walls and doors and windows, lintel, stairs, water, steps, decoration and other components of the layout	7. Arrangement of masonry walls	6
			8. Arrangement of doors and windows	
			9. Arrangement of lintel	
			10. the layout of the stairs	
			11. the layout of water and steps	
			12. Arrangement of decorative components and other components	
5	Component hooking practice and analysis, summary, query the amount of work	Master the method of component hooking and analyze, summarize and inquire the project quantity	13. Component hooking method(★)	6
			14. Analyzing, summarizing and inquiring project quantity(★)	
			15. Generated the bill of quantities, printed and bound the results	
Total				
Note: Mark the important points with ★				

5. Teaching Suggestions

5.1 Teaching Resources

Zhang Xiaomin, Li Shesheng. *Application of construction cost software: guanglianda series software*, China Architecture & Building Press, 2013.

5.2 Teachers'Qualification

The teachers of this curriculum are construction cost professional teachers, proficient in using Guanglianda or Tsinghua Sware software, and double-qualified teachers.

5.3 Teaching Facilities

The teaching site is the computer room. In order to meet the teaching requirements, you need 3 computer rooms, each room can accommodate 50 students. Computer configuration requirements i3 and above processors, memory of more than 8G, hard disk requirements of more than 500G; The software requires three sets of Guanglianda graphic calculation software and steel bar calculation software, each set of 50 nodes, and three sets of Sware 3d calculation software, each set of 50 nodes.

6. Evaluation

The assessment method of this curriculum is practical operation, and the content of teacher assessment includes the accuracy of engineering output and the integrity of the model. Score composition is daily performance accounts for 30% of the total score, and the final examination accounts for 70%.

四 川 建 筑 职 业 技 术 学 院

课 程 标 准

课程名称：　　　水电安装工程预算

课程代码：　　　150281

课程学分：　　　4

基本学时：　　　48

适用专业：　工程造价（中英合作办学项目）

执 笔 人：　　　刘　渊

编制单位：　　安装工程造价教研室

审　　核：　工程管理系（院）（签章）

批　　准：　　　教务处（盖章）

Sichuan College of Architectural Technology

Curriculum Standard

Curriculum Name: Project Budget of Water and Electricity Installation

Curriculum Code: 150281

Credits: 3

Teaching Hours: 48

Applicable Specialty: Construction Cost

(Sino-British Cooperative Program)

Compiled by: Liu Yuan

On Behalf of: Teaching & Research Section of Installation Project Cost

Reviewed by: The Department of Engineering Management

Approved by: Teaching Affairs Office

《水电安装工程预算》课程标准

一、课程概述

《水电安装工程预算》是工程造价(中英合作办学项目)专业的一门操作应用型专业必修课程。本课程主要研究水电安装工程定额与预算的基本理论和水电安装工程预算实际方法及水电安装工程造价文件的编制方法。

本课程的教学任务是:阐述水电安装工程预算与建筑工程预算、装饰工程预算的联系与区别以及通用安装工程计价定额的组成及应用;说明水电安装预算的编制原理与方法;指导学生通过综合练习,具备独立编制水电安装工程预算的能力。

《水电安装工程预算》在第5学期开设,是在学生学完《房屋建筑工程及装饰工程预算》《建筑电气安装工艺与识图》《建筑水暖设备安装工艺与识图》以后的后续课程。

二、课程设计思路

(一)设置依据

工程造价(中英合作办学项目)人才培养方案以及水电安装工程预算实际工程过程需求。

(二)课程框架设计

水电安装工程预算						
定额		施工图预算的编制		安装工程量计算		
定额基础知识	通用安装工程定额组成及应用	安装工程费用组成与计算	施工图预算编制依据及程序	给水排水、采暖及燃气工程计量	消防工程计量	建筑电气安装工程计量

Project Budget of Water and Electricity Installation Curriculum Standard

1. Curriculum Description

This course is a compulsory professional applied one in construction cost specialty (Sino-British Cooperative Program). This course focuses on the basic theory of quotas and budgets for water and electricity installation project, and also the compiling methods for the installation budgets and the cost documents.

This course focuses on the illustration of the differences and connections between project budget of water and electricity installation and building construction budget, decoration construction budget, the composition and application of general installation project quotas, the calculation theory and methods of budget of water and electricity installation. In this course, students are trained to have the comprehensive ability in making the project budget of water and electricity installation.

This course is delivered in the fifth semester after students complete those prerequisite courses — *Building Construction and Decoration Project Budget*, *Construction Electrical Installation Processand Diagram Reading*, and *Construction Plumbing Equipment Installation Process and Diagram Reading*.

2. Curriculum Design Ideas

2.1 Design Basis

This course is based on the professional training program of construction cost (Sino-British Cooperative Program) and the requirement for the practical works in calculation the budget of water and electricity installation.

2.2 Curriculum Design Framework

Project Budget of Water and Electricity Installation						
Quotas		Budget Preparation on Installation Project Diagrams		Calculation of Installation Project Cost		
The basics of quotas	The composition and application of general installation project quotas	The compositions and calculation of installation project cost	The basis and procedure of budget preparation on installation project diagrams	Quantities calculation of water supply and drainage, heating and gas engineering	Quantities of fire control engineering	Quantities of building electrical installation engineering

三、课程目标

以职业能力培养为核心的课程设计,在重视学生专业能力培养的同时,重视方法能力与社会能力的培养。通过学习,学生能在定额计价模式下完成各种造价文件的编制。

(一)知识目标

(1)熟悉安装工程定额的内容,了解安装工程定额的发展变化趋势;理解水电安装预算的编制原理和计价改革发展方向。

(2)掌握安装工程预算定额的组成、应用方法(包括定额套用、换算及定额补充的要点)。

(3)熟悉工程费用的组成及各项费用的计算方法,以及水电安装工程施工图预算编制依据及程序。

(4)掌握安装工程工程量的计算方法,熟悉水电安装工程预算的编制方法及步骤。

(二)素质目标

(1)培养学生热爱建筑业,具有爱岗敬业和奉献精神;

(2)教育学生了解、熟悉行业规范,并且熟悉本专业的相关法规、政策;

(3)培养学生树立为社会主义市场经济发展应具有的开拓进取精神以及客观公正、勇于奉献、廉洁自律的工作作风和职业素养。

(三)能力目标

(1)综合应用各门课程相关知识的能力,学会熟练使用通用安装工程计价定额;

(2)正确列制其分项工程项目名称、正确计算其工程量的能力;

(3)能合理、正确地编制安装工程施工图预算,具备独立编制水电安装工程预算的能力;

(4)通过在校期间的培养,使学生为今后考取相应执业资格证打下一定的基础。

3. Curriculum Objectives

With the vocational-oriented curricular design, this course aims to cultivate students'professional competences, method-applied ability and social ability. Through the course, students should learn to complete the preparation and compilation of various cost documents with the quota-pricing model.

3. 1　Knowledge Objectives

After learning this course, students will be able to

(1) be familiar with the contents of the installation project quota, and understand the development trend of the installation project quota; understand the principle of preparation of the water and electricity installation budget and the direction of pricing reform.

(2) master the composition and application methods of the installation project budget quota, including key points of quota application, conversion and supplement.

(3) be familiar with the composition and calculation methods of project costs of the costs, and the basis and procedures of budge preparation on the water and electricity installation project diagram.

(4) master the calculation method of the quantities of installation engineering, and be familiar with the budget preparation methods and steps for the water and electricity installation project.

3. 2　Quality Objectives

After learning this course, students will have the quality of

(1) continuous interest and love for the industry of architecture, and devoted, responsible and committed spirits;

(2) understanding and being familiar with the industry regulations and the relevant rules and policies;

(3) good spirits of entrepreneurship, and working styles and professional qualification of fare justice, dedication, integrity and self-discipline for the development of socialism market economy.

3. 3　Ability Objectives

After learning this course, students will have the ability of

(1) using relevant know-hows comprehensively, esp. be skillful of applying the commonly used installation project pricing quotas;

(2) makingcorrectlylists of sub-projects titles and calculate the relevant engineering quantities;

(3) preparing the budget on the installation construction diagram rationally and correctly, and independently compile the budge of water and electricity installation project;

(4) laying a solid foundation for obtaining the corresponding qualifications in the future.

四、课程内容与学时分配

教学内容与学时安排表

序号	教学情境/任务/项目/单元	教学目标	教学内容及学时分配			
			知识点	学时	实践项目	学时
1	绪论	知识回顾、新课程引入及教学要求	1. 本课程的研究对象与任务	2		
			2. 本课程学习内容			
			3. 本课程特点			
2	第一章	定额	4. 定额基础知识回顾	2		
			5. 通用安装工程定额组成及应用			
3	第二章	安装工程施工图预算	6. 安装工程费用组成与计算	2	定额套用及换算练习	
			7. 安装工程施工图预算编制依据及程序			
4	第三章	给水排水、采暖及燃气工程计量	8. 基础知识	10	安装工程材料价格市场调查	6
			9. 定额规则解释及计算方法		工程量计算练习	
5	第四章	消防工程计量	10. 基础知识	4	工程量计算练习	2
			11. 定额规则解释及计算方法			
6	第五章	建筑电气安装工程计量	12. 基础知识	12	实训模型参观	2
			13. 定额规则解释及计算方法		工程量计算练习	6
合计				32		16

4. Curriculum Contents and Teaching Hours

Table of Teaching Contents and Teaching Hours

Item	Topic	Teaching Objectives	Teaching Contents and Teaching Hours			
			Lesson Contents	Hrs	Student Activities	Hrs
1	Introduction	Knowledge review, introduction to the new unit and learning requirement	1. Subjects and tasks of this unit	2		
			2. Learning contents			
			3. Features of this unit			
2	Chapter 1	Quotas	4. Review of the foundation knowledge of quotas	2		
			5. The composition and application of general installation project quotas			
3	Chapter 2	Budget on installation project diagram	6. The compositions and calculation of installation project cost	2	Exercise of quotas application and conversion	
			7. The basis and procedure of budget preparation on installation project diagrams			
4	Chapter 3	Quantities calculation of water supply and drainage, heating and gas engineering	8. Foundation knowledge	10	Market survey of the prices of installation project materials	6
			9. Quota rules interpretation and application		Exercise of quantities calculation	
5	Chapter 4	Quantities of fire control engineering	10. Foundation Knowledge	4	Exercise of quantities calculation	2
			11. Quota rules interpretation and application			
6	Chapter 5	Quantities of building electrical installation engineering	12. Foundation knowledge	12	Practical training model visit	2
			13. Quota rules interpretation and application		Exercise of quantities calculation	6
Total				32		16

五、教学实施建议

(一)教学参考资料

刘渊、袁媛.《安装工程计量与计价》,东南大学出版社,2017。

其他辅助资料:各地造价信息资料和各地建设工程造价网、地区计价定额、通用安装工程工程量清单计价定额、建设工程工程量清单计价规范等。

(二)教师素质要求

担任该门课程的老师是专任教师资格,具备水电安装工程识图及工程造价的专业背景,具有一定实践经验及必备的教学能力。

(三)教学场地、设施要求

一般教室或多媒体教室。

六、考核评价

(一)考核方法

考试形式:闭卷结合平时考勤及操作练习进行综合考核。

题型:包括单项选择、多项选择、判断、案例分析计算等。

(二)考核说明

考试重点:安装工程工程量的计算作为考核重点,考核的知识点应覆盖该门课程各个章节的知识内容。

总评成绩的构成比例:闭卷考试占70%,平时考勤及操作练习占30%。

考核原则:考核应坚持客观、科学和公正的原则。

5. Teaching Suggestions

5.1   Teaching Resources

Liu Yuan, Yuan Yuan. *Quantities and Pricing of Installation Construction Engineering.* Southeast University Press, 2017.

Other auxiliary materials: local pricing information materials and local construction cost websites, regional pricing quota, general quota of bill of quantities of installation engineering, regulations of pricing for bill of quantities of construction engineering, etc.

5.2   Teachers'Qualification

Teachers should have the professional ethics and the teacher qualification certificate; they should have the education background of water and electricity installation engineering diagram reading and construction cost, and have the relative qualification certificates and certain social or professional practices; they should have some professional teaching experience.

5.3   Teaching Facilities

This course can be delivered in either common classrooms or multimedia classrooms.

6. Evaluation

6.1   The Method of Assessment

Comprehensive evaluation combines closed-book examination, the usual attendance and operation practice. The final examination includes single-choice selection, multiple-choice selection, judgment section, case analysis and calculation, etc.

6.2   Assessment Requirement

The evaluation of this course is in accordance with the *Regulation of Academic Evaluation for Students in SCAT*: the total score is consisted of the daily performance(accounts for 30%) and the final examination(accounts for 70%).

The test covers the knowledge contents of all the chapters in this course, with highlights on the calculation of installation engineering quantities.

The evaluation will follow the principle of objectivity, science and impartiality.

四川建筑职业技术学院

课程标准

课程名称：<u>水电安装工程预算实训</u>

课程代码：<u>500243</u>

课程学分：<u>1</u>

基本学时：<u>30</u>

适用专业：<u>工程造价（中英合作办学项目）</u>

执笔人：<u>刘　渊</u>

编制单位：<u>安装工程造价教研室</u>

审　核：<u>工程管理系（院）（签章）</u>

批　准：<u>教务处（盖章）</u>

Sichuan College of Architectural Technology

Curriculum Standard

Curriculum Name: Practical Training for Project Budget of Water and

Electricity Installation

Curriculum Code: 500243

Credits: 1

Teaching Hours: 30

Applicable Specialty: Construction Cost

(Sino-British Cooperative Program)

Compiled by: Liu Yuan

On Behalf of: Teaching & Research Section of Installation Project Cost

Reviewed by: The Department of Engineering Management

Approved by: Teaching Affairs Office

《水电安装工程预算实训》课程标准

一、课程概述

《水电安装工程预算实训》是工程造价(中英合作办学项目)专业的一个操作应用型专业实践环节。本课程的教学任务:指导学生通过综合实训,具备独立编制水电安装工程施工图预算的能力。

《水电安装工程预算实训》在第 5 学期开设,是在学生学完《水电安装工程预算》课程以后的后续实践环节。

二、课程设计思路

(1)设置依据:最新人才培养方案以及水电安装预算实际工程过程需求。

(2)课程框架的设计:

水电安装工程预算实训			
工程计量		工程计价	
识图列项	计算定额工程量	套用定额、确定综合单价	计算各项费用,完成施工图预算的编制

(3)教学目标的确定:通过实训,学生能完成一套水电安装工程施工图预算的编制。

(4)理论与实践比例:32:16。

三、课程目标

以职业能力培养为核心的课程设计,在重视学生专业能力培养的同时,重视方法能力与社会能力的培养。通过实训,学生能完成水电安装工程造价的计算和相关造价文件的编制。

(一)知识目标

(1)识读指定的图纸,熟悉相关资料;

(2)设置定额项目名称并计算工程量,填制工程量计算表;

Practical Training for Project Budget of Water and Electricity Installation Curriculum Standard

1. Curriculum Description

This is a comprehensive and practical course, and it is an important practical training for construction cost specialty(Sino-British Cooperative Program). Through this course: students can further familiarize and master the independent preparation and compilation of budget on water and electricity installation engineering diagram.

This course is delivered in the fifth semester after students complete the prerequisite courses *Project Budget of Water and Electricity Installation Engineering*.

2. Curriculum Design Ideas

(1) This course is designed based on professional training program of the major of engineering cost (Sino-British Cooperative Program) and the practical needs of the works related with water and electricity installation.

(2) The curriculum framework of this course is as follows:

Practical Training for Project Budget of Water and Electricity Installation			
Engineering Quantities		Calculation of Engineering Cost	
Construction drawings reading & quotas projects listing	Quotas projects engineering quantities calculating	Apply the quotas & decide the comprehensive unit price	Calculate the cost of varied items & complete the preparation of construction drawing budget

(3) The design highlights the professionalism, practice and openness of specialty courses.

(4) The ratio of theory courses to practice is 32 to 16.

3. Curriculum Objectives

With the vocational-oriented curricular design, this course aims to cultivate students'professional competences, method-applied ability and social ability. Through the course, students should learn to complete the preparation and compilation of various cost documents with the quota-pricing model.

3.1 Knowledge Objectives

After learning this course, students will be able to

(1) read the specified drawings and be familiar with the relevant materials;

(2) set the titles of quotas projects and calculate the quantity, and fill in the calculation table of the engineering quantity;

（3）套用定额，确定综合单价；

（4）计算分部分项工程费及单价措施费；

（5）根据有关资料计算工程造价；

（6）填写编制说明、封面等。

（二）素质目标

（1）培养学生热爱建筑业，具有爱岗敬业和奉献精神；

（2）教育学生了解、熟悉行业规范，并且熟悉本专业的定额、相关法规、政策；

（3）培养学生树立为社会主义市场经济发展应具有的开拓进取精神以及客观公正、勇于奉献、廉洁自律的工作作风和职业素养。

（三）能力目标

（1）培养学生综合应用各门相关知识的能力，学会熟练使用通用安装工程工程量清单计价定额；

（2）具备能较为熟练地编制水电安装施工图预算的编制能力；

（3）通过在校期间的培养，为学生今后考取相应执业资格证书打下一定的基础。

四、课程内容与学时分配

教学内容与学时安排表

序号	教学情境/任务/项目/单元	教学目标	教学内容及学时分配		
			知识点	实践项目	学时
1	1-1	工程计量		1. 识读施工图纸	21
				2. 设置定额项目名称（★）	
				3. 计算工程量（★）	
2	1-2	确定综合单价		4. 套用定额	3
				5. 单价换算	
3	1-3	分部分项工程费及单价措施项目费计算		6. 未计价材料分析计算	3
				7. 计算分部分项工程和单价措施项目费	
4	1-4	计算造价、完成相关表格填制		8. 计算总价措施项目费及其他项目费	3
				9. 计算、汇总工程造价	
				10. 填制表格	
合计					30
备注：					

(3) use the quota to determine the comprehensive unit price;

(4) calculate the sub-project and divided sections'costs and the unit price of operation;

(5) be able to calculate the construction cost in accordance with relevant materials;

(6) be able to write the preparation instructions and settle the cover and so on.

3.2 Quality Objectives

After learning this course, students will have the quality of

(1) continuous interest and love for the industry of architecture, and cultivate their devoted, responsible and committed spirits;

(2) understanding and being familiar with the industry regulations and the relevant rules and policies;

(3) good spirits of entrepreneurship, and working styles and professional qualification of fare justice, dedication, integrity and self-discipline for the development of socialism market economy.

3.3 Ability Objectives

After learning this course, students will have the ability of

(1) using relevant know-hows comprehensively, esp. mastering the general quotas of pricing for bill of quantities of installation engineering;

(2) preparing the budget on the water and electricity installation engineering diagram skillfully;

(3) laying a solid foundation for obtaining the corresponding qualifications in the future.

4. Curriculum Contents and Teaching Hours

Table of Teaching Contents and Teaching Hours

Item	Topic	Teaching Objectives	Teaching Contents and Teaching Hours		
			Lesson Contents	Student Activities	Hrs
1	1-1	Engineering quantities		1. Read the construction drawings	21
				2. Set the titles of quotas projects(★)	
				3. Calculate the engineering quantities(★)	
2	1-2	Deciding the comprehensive unit price		4. Apply the quotas	3
				5. Unit price conversion	
3	1-3	Calculating the sub-project and divided sections construction cost and unit price for operation		6. Analyze and Calculate the unnominated materials	3
				7. Calculate the subs cost and the unit price for operation	
4	1-4	Calculating the construction cost and complete the filling-in of relevant tables		8. Calculate the total costs and other fees	3
				9. Summarize the construction cost	
				10. Complete tables	
Total					30
Note:					

五、教学实施建议

(一)教学参考资料

(1)刘渊、袁媛.《安装工程计量与计价》,东南大学出版社,2017。

(2)其他辅助资料:当地造价信息资料和当地建设工程造价网、地区计价定额、建设工程工程量清单计价规范及当地有关工程造价文件、政策。

(3)指导老师指定的施工图、造价信息资料、实训任务书及指导书等。

(二)教师素质要求

担任该门课程的老师是专任教师资格,具备水电安装工程识图及工程造价的专业背景,具有一定实践经验及必备的教学能力。

(三)教学场地、设施要求

一般教室。

六、考核评价

(一)成绩评定标准

《水电安装工程预算实训》成绩按优、良、中、及格、不及格五个等级评定:

(1)优:内容完善、计算正确,字迹工整、清晰,完全由学生自己独立完成;

(2)良:内容完善、计算基本正确,字迹工整,完全由学生自己独立完成;

(3)中:内容完善、计算有小部分错误,字迹工整,完全由学生自己独立完成;

(4)及格:内容完善、计算部分有错,大部分由学生自己独立完成;

(5)不及格:内容不完善,计算错误很大,有大量抄袭内容。

(二)评定方法

首先要通过口试答辩检查,并以编制内容的质量、格式的应用是否正确和书写是否工整、清晰,实训出勤等为评定依据与考核内容。

(三)考核的比重

口试答辩占40%,书面编制内容占40%,实训出勤占20%。

5. Teaching Suggestions

5.1 Teaching Resources

Liu Yuan, Yuan Yuan. *Quantities and Pricing of Installation Construction Engineering.* Southeast University Press, 2017.

Required auxiliary materials: construction drawings specified by the tutor, pricing information materials, task book and guidelines for practical training, etc.

Other auxiliary materials: local pricing information materials and local construction cost websites, regional pricing quota, general quota of bill of quantities of installation project, regulations of pricing for bill of quantities of construction engineering, etc.

5.2 Teachers'Qualification

Teachers should have the professional ethics and the teacher qualification certificate; they should have the education background of water and electricity installation engineering diagram reading and construction cost, and have the relative qualification certificates and certain social or professional practices; they should have some professional teaching experience.

5.3 Teaching Facilities

This course can be delivered in common classrooms.

6. Evaluation

6.1 The Standard of Assessment

The score is graded into five level system of excellent, good, mediocre, pass, and failed.

(1)Excellent: complete content, correct calculation, neat and clear handwriting, completed by the student him/herself;

(2)Good: complete content, basically correct calculation, neat handwriting, completed by the student him/herself;

(3)Mediocre: complete content, small errors in calculation, neat handwriting, completed by the student him/herself;

(4)Pass: complete content, partially wrong calculation, completed mostly by the student;

(5)Failed: uncompleted content, big mistakes in calculation, lots of plagiarism.

6.2 The Method of Assessment

Students shall pass the oral test first, and then be checked and examined by the standard of assessment above and the usual attendance.

6.3 Assessment Requirement

The evaluation of this course is in accordance with the *Regulation of Academic Evaluation for Students in SCAT*; the total score is consisted of the oral Q&A(accounts for 40%), the usual attendance (accounts for 20%)and the compiled preparation book(accounts for 40%).

四川建筑职业技术学院

课程标准

课程名称：_____定额原理_____

课程代码：_____150359_____

课程学分：_____1.5_____

基本学时：_____24_____

适用专业：___工程造价（中英合作办学项目）___

执 笔 人：_____曹碧清_____

编制单位：___建筑工程造价教研室___

审　　核：_____工程管理系（签章）_____

批　　准：_____教务处（盖章）_____

Sichuan College of Architectural Technology

Curriculum Standard

Curriculum Name: _____Quota Principle_____

Curriculum Code: _____150359_____

Credits: _____1. 5_____

Teaching Hours: _____24_____

Applicable Specialty: _____Construction Cost_____

_____(Sino-British Cooperative Program)_____

Compiled by: _____Cao Biqing_____

On Behalf of: Teaching & Research Section of Construction Cost

Reviewed by: _____The Department of Engineering Management_____

Approved by: _____Teaching Affairs Office_____

《定额原理》课程标准

一、课程概述

《定额原理》是建筑工程造价专业(中英合作办学项目)的一门专业必修理论课程(无实践环节),于大二上学期开设。在人才培养方案中,《工程造价概论》《建筑工程预算》为该门课程的先修课程;《工程量清单计价》《工程造价职业资格实务培训》《工程造价控制》《工程造价软件应用》《工程结算》为其后续课程。

二、课程设计思路

以培养与国际行业标准接轨的工程造价专业一线技术技能人才为宗旨进行教学设计,构建适应工程造价专业(中英合作办学项目)要求的课程框架,以学会编写与之相适应的企业定额为教学目标,充分利用现行《四川省建筑工程工程量清单计价定额》(房屋建筑与装饰分册)、成都市工程造价信息、德阳市工程造价信息等教学资源,紧密结合行业发展动向和学生现有知识储备进行课堂教学。

三、课程目标

通过课程学习,学生掌握建筑工程定额的种类、施工过程和工作时间研究、定额的各种具体编制方法等知识点,形成编写人工定额、材料定额和机械台班定额的技能,养成熟练应用定额进而独立编写定额的素质。

(一)知识目标

(1)掌握建筑工程定额的种类;

(2)掌握施工过程的构成因素、影响因素以及组成明细;

(3)掌握技术测定法中测时法、工作日写实法、写实记录法和简易测定法的适用范围、优缺点以及记录时间的具体方法;

(4)掌握理论计算法中标准砖墙体材料用量、装饰材料用量和半成品配合比材料用量的计算方法;

Quota Principle Curriculum Standard

1. Curriculum Description

Quota Principle is a compulsory theoretical course(without practice)for the specialty of construction cost(Sino-British Cooperative Program), which is delivered in the third semester. In the professional training program, the delivered courses are *Introduction to Construction Cost* and *Building Construction Budgets*, while the courses followed are *Pricing for Bills of Quantities*, *Professional Qualification Training of Construction Cost*, *Construction Cost Control*, *Application of Construction Cost Software* and *Engineering Settlement*.

2. Curriculum Design Ideas

The teaching design is based on the objective of training students into front-line technical and skilled talents in construction cost and approaching international industry standards, and the curriculum framework are in accordance with the requirements of construction cost(Sino-British Cooperative Program). With the teaching goal of compiling relative enterprise quotas, we should make full use of teaching resources such as *The Quota-based Valuation of Bill Quantity of Construction Works in Sichuan* (housing construction and decoration booklets), construction cost information in Chengdu, and construction cost information in Deyang. Classroom teaching should be closely combined with industry development trends and students'knowledge background.

3. Curriculum Objectives

Through this course, students can master the types of construction engineering quotas, construction process and working time research, various specific methods of quota preparation and other knowledge points, which forms the skills of compiling artificial quotas, material quotas and mechanical quotas, and cultivates the quality of applying quotas and then compiling it independently.

3. 1 Knowledge Objectives

After learning this course, students will be able to

(1) master the types of construction quotas;

(2) master the component factors, influencing factors and detailed composition of the construction process;

(3) master the application scope, advantages and disadvantages of time chronometry, detailed record of work date, realistic recording and the simple measurement in the technical measuring method, as well as the specific methods of recording time;

(4) master the theoretical calculation method of standard brick wall material consumption, decorative material consumption and mix proportion consumption of semi-finished production;

（5）掌握经验估计法中加权平均法、概率与经验公式计算方法；

（6）掌握统计分析法和比较类推法的编写方法。

（二）素质目标

（1）养成灵活运用科学方法编写企业定额的素质；

（2）养成实事求是、刻苦钻研的职业操守。

（三）能力目标

（1）能够独立运用技术测定法编写人工定额；

（2）能够独立运用理论计算法编写材料消耗定额；

（3）能够独立运用技术测定法、理论计算法等编写机械台班消耗定额；

（4）能够独立运用费用文件和给定资料编写企业管理费和利润；

（5）能够依据现场第一手统计资料编写出制定项目的基价。

四、课程内容与学时分配

教学内容与学时安排表

序号	教学情境/任务/项目/单元	教学目标	教学内容及学时分配			
			知识点	学时	实践项目	学时
1	概述	介绍课程学习目的和教学内容架构	1. 企业定额的概念； 2. 企业定额的作用； 3. 企业定额的种类(★)； 4. 企业定额的编制方法	2		
2	施工过程和工作时间研究	讲授施工过程的分解、构成要素和影响因素；分析定额时间和非定额时间的构成	5. 施工过程研究； 6. 工作时间研究(★)	2		
3	技术测定法	讲授测时法、写实记录法、工作日写实法和简易制定法记录时间的具体操作步骤以及它们的优缺点，总结人工定额编制的要点	7. 概述； 8. 测时法(★)； 9. 写实记录法； 10. 工作日写实法； 11. 简易制定法	4		

(5) master the weighted average method, probability and empirical formula calculation method in experience estimating method;

(6) master the compiling methods of statistical analysis and comparative analogy.

3.2 Quality Objectives

After learning this course, students will have the quality of

(1) compiling enterprise quotas flexibly and scientifically;

(2) taking the professional ethics of seeking truth from facts and studying hard.

3.3 Ability Objectives

After learning this course, students will have the ability of

(1) using the technical measurement method to compile the artificial quota independently;

(2) using theoretical calculation method to compile material consumption quota by themselves;

(3) using technical measurement method and theoretical calculation method to compile mechanical consumption quota by themselves;

(4) using cost documents and given information to compile enterprise management fees and profits independently;

(5) compiling the base price of the project according to the first-hand statistics on the spot.

4. Curriculum Contents and Teaching Hours

Table of Teaching Contents and Teaching Hours

| Item | Topic | Teaching Objectives | Teaching Contents and Teaching Hours | | | |
			Lesson Contents	Hrs	Student Activities	Hrs
1	Introduction	Introduce the learning purpose and contents structure of the course	1. The concept of enterprise quota; 2. The role of enterprise quota; 3. Types of enterprise quotas (★); 4. The method of compiling enterprise quota	2		
2	Study on construction process and working time	Explain the decomposition, constitutive elements and influencing factors of construction process; analyze the composition of quota time and non-quota time	5. Study on construction process; 6. Study on working time			
3	Technical measurement method	Explain the specific steps of chronometry, realistic recording, detailed record of work date and simple measurement, as well as their advantages and disadvantages, and summarize the main points of the artificial quota compilation	7. Introduction; 8. Chronometry (★); 9. Realistic recording; 10. Detailed record of work date; 11. Simple measurement	4		

序号	教学情境/任务/项目/单元	教学目标	教学内容及学时分配			
			知识点	学时	实践项目	学时
4	理论计算法	讲授材料消耗定额的主要编制方法,包括砌体材料用量、装饰块料用量、半成品配合比用量的计算方法	12. 砌体材料用量计算(★); 13. 装饰块料用量计算; 14. 半成品配合比用量的计算(★)	4		
5	定额制定简易方法	讲授制定人工定额的简易方法的具体操作步骤:经验估计、统计分析、比较类推	15. 经验估计法(★); 16. 统计分析法(★); 17. 比较类推法	2		
6	定额编制方案	讲解编制定额的具体实施方案	18. 拟订适用范围(★); 19. 拟订结构形式; 20. 确定定额水平(★)	2		
7	人工定额的拟订	结合定额时间研究、技术测定法以及定额制定简易方法的基础知识讲授人工定额的拟订步骤	21. 概述(时间定额和产量定额)(★); 22. 具体拟订步骤(★)	2		
8	材料定额的拟订	结合理论计算法的知识讲授材料消耗定额的拟订方法	23. 概述(净用量、消耗量、损耗量)(★); 24. 直接性材料消耗定额的编制方法(★); 25. 周转性材料消耗定额的编制方法	2		
9	机械台班定额的拟订	结合机械定额时间和非定额时间的基础知识讲解机械台定额的编制方法	26. 概述(时间定额和产量定额)(★); 27. 具体拟订步骤(★)	2		

continued

Item	Topic	Teaching Objectives	Teaching Contents and Teaching Hours			
			Lesson Contents	Hrs	Student Activities	Hrs
4	Theoretical calculation method	Introduce the main compilation methods of material consumption quota, including the calculation methods of masonry material consumption, decorative block material consumption and semifinished product mix design	12. Calculation of masonry material consumption(★); 13. Calculating the consumption of decorative blocks; 14. Calculation of mix design of semifinished products(★)	4		
5	Simple methods of quota setting	Explain the specific operation steps of the simple method of compiling artificial quota: experience estimation, statistical analysis, comparative analogy	15. Experience of estimating method(★); 16. Statistical analysis(★); 17. Comparative analogy	2		
6	Quota preparation scheme	Explain the specific implementation plan of quota compilation	18. Drafting the scope of application(★); 19. Formulation of structure; 20. Determining the level of quota (★)	2		
7	Drafting of artificial quota	Combine the basic knowledge of quota time study, technical measurement method and simple method of quota formulation to teach the steps of formulating artificial quota	21. Introduction (time quota and production quota) (★); 22. Specific formulation steps(★)	2		
8	Formulation of material quota	Explain the method of material consumption quota based on the knowledge of theoretical calculation	23. Introduction (net consumption, loss) (★); 24. The method of compiling direct material consumption quota(★); 25. The method of compiling consumption quota of turnover materials	2		
9	Formulation of mechanical Quota	Explain the method of mechanical quota compilation based on the basic knowledge of mechanical quota time and non-quota time	26. Introduction (time quota and production quota) (★); 27. Specific formulation steps(★)	2		

续表

序号	教学情境/任务/项目/单元	教学目标	教学内容及学时分配			
			知识点	学时	实践项目	学时
10	企业定额编制与实务	结合定额编制方案、人工定额、材料消耗定额和机械台班定额的基础知识以及实例讲解企业定额的编制方法	28. 编制企业定额的基础工作; 29. 企业定额的编制步骤(★); 30. 企业定额编制实例	2		
	合计			24		
备注:重要知识点标注★						

五、教学实施建议

(一)教学参考资料

袁建新.《企业定额编制原理与实务》,中国建筑工业出版社,2003。

(二)教师素质要求

具有教师资格,热爱教学岗位,熟知本地区现行定额的基本架构。

(三)教学场地、设施要求

该课程为纯理论课程,普通教室即可。

六、考核评价

该课程成绩设置为总分100分,期末考核和平时成绩比例为7∶3。以闭卷考核的方式考查学生对基础知识的掌握程度,要求学生能依据材料独立编写指定项目的基价。

continued

Item	Topic	Teaching Objectives	Teaching Contents and Teaching Hours			
			Lesson Contents	Hrs	Student Activities	Hrs
10	Compilation and Practice of Enterprise Quota	The method of compiling enterprise quota is explained with the basic knowledge of quota compilation scheme, artificial quota, material consumption quota, mechanical quota, and examples	28. Basic work of compiling enterprise quota; 29. The procedure of compiling enterprise quota(★); 30. Examples of enterprise quota compilation	2		
Total				24		
Note: Mark the important points with ★						

5. Teaching Suggestions

5.1 Teaching Resources

Yuan Jianxin. *Principle and Practice of Enterprise Quota Compilation*, China Architecture & Building Press, 2003.

5.2 Teachers'Qualification

Teachers should have the teachers'qualification certificate, love teaching, and be familiar with the basic structure of the current quota in their region.

5.3 Teaching Facilities

This course is quite theoretical, so it can be delivered in common classrooms.

6. Evaluation

The daily performance accounts for 30% of the total score and the final examination accounts for 70%. The exam is taken without reference materials. Students are required to compile the base price of a specified project independently based on the materials to test their basic knowledge of this course.

四川建筑职业技术学院

课 程 标 准

课程名称： 工程造价控制

课程代码： 150029

课程学分： 2

基本学时： 32

适用专业： 工程造价（中英合作办学项目）

执 笔 人： 袁 鹰

编制单位： 建筑工程造价教研室

审 核： 工程管理系（院）（签章）

批 准： 教务处（盖章）

Sichuan College of Architectural Technology

Curriculum Standard

Curriculum Name: Construction Cost Control

Curriculum Code: 150029

Credits: 2

Teaching Hours: 32

Applicable Specialty: Construction Cost

(Sino-British Cooperative Program)

Compiled by: Yuan Ying

On Behalf of: Teaching & Research Section of

Engineering Management

Reviewed by: The Department of Engineering Management

Approved by: Teaching Affairs Office

《工程造价控制》课程标准

一、课程概述

工程造价控制是工程造价(中英合作办学项目)专业的一门专业必修课程,课程 2 学分,32 学时,在第四学期开设,其前行专业课为建筑工程预算、建设项目管理。课程主要研究建设项目各个阶段工程造价控制的理论和方法,是一门综合性、实践性较强的应用型课程。

二、课程设计思路

课程框架的设计:建设项目全过程控制的理论和方法。

教学目标的确定:具备工程造价控制的理论,熟悉工程造价控制的方法,具备工程造价控制的基本能力。

理论与实践比例:理论教学,无实践环节。

教学内容的选取:以方法的学习和应用为主,理论为方法服务。

教学资源:选取造价控制各阶段的实例进行教学。

三、课程目标

该课程的目标是使学生具备基本的职业素质、具有工程管理高级职业人才所必需的思想基础、思维方式及职业素质,能掌握工程造价控制的基本知识和基本技能,基本具备在工程管理相关岗位上解决实际问题的能力。

(一)知识目标

(1)能明确建设项目各阶段工程造价的控制内容、控制重点;

(2)在建设项目各阶段,会应用工程造价的控制方法进行造价管理。

(二)素质目标

(1)树立爱岗敬业的思想,自觉遵守职业道德及行业规范;

(2)培养全过程造价管理的思维方式;

(3)培养学生专业兴趣和工作热情。

(三)能力目标

能在工程造价管理工作中运用全过程控制的理论和方法达成对工程造价各阶段进行控制的目标。

Construction Cost Control **Curriculum Standard**

1. Curriculum Description

Construction Cost Control is a compulsory course of the specialty of construction cost in Sino-British Cooperative Program. There are 2 credits and 32 class hours for this course. It is offered in the fourth semester. This course mainly studies the theory and method of construction cost control in each stage of construction project.

2. Curriculum Design Ideas

The frame of curriculum design is the theory and method of the whole process of construction cost control.

The setting of teaching objectives is based on cultivating students'theory of construction cost control, method of construction cost control and ability of construction cost control.

This course is quite theoretical and there is no practical training program in the teaching process.

The contents of the course focus on the mastering and applying the methods in construction cost control and the theories serve for the methods.

Practical cases in each stage of cost control are selected for teaching resource.

3. Curriculum Objectives

The objective of this course is to enable students to have basic professional quality, the ideological basis, thinking mode and professional quality that are necessary for senior professional talents in engineering management; to enable students master the basic knowledge and skills of construction cost control, and have the ability to solve practical problems in posts related with engineering management.

3.1   Knowledge Objectives

After learning this course, students will be able to

(1) define the control content and key points of project cost at each stage of the construction project;

(2) use proper cost control method in each stage of construction project to manage it.

3.2   Quality Objectives

After learning this course, students will have the quality of

(1) loving and dedication to work, and consciously abiding professional ethics and industry norms;

(2) taking the thinking mode of whole-process cost management;

(3) taking professional interest and work enthusiasm.

3.3   Ability Objectives

After learning this course, students will have the ability of controlling the construction cost with the theory and method of whole process control method in each stage of the construction project.

四、课程内容与学时分配

教学内容与学时安排表

序号	教学情境/任务/项目/单元	教学目标	教学内容及学时分配			
			知识点	学时	实践项目	学时
1	工程造价控制基本理论	（1）掌握工程造价的有关概念：工程造价、工程造价控制、静态投资、动态投资(★)、建设项目投资估算、设计概算、施工图预算等(★)；熟悉工程造价的特点和计价特征(★)； （2）了解工程造价控制的基本内容；了解工程造价依据的种类及内容； （3）了解我国造价工程师执业资格制度的建立；掌握造价工程师的权利和义务(★)； （4）了解工程造价咨询服务业管理制度	工程造价基本理论；工程造价控制的基本理论	2		
			工程造价计价依据；造价工程师执业资格制度；工程造价咨询服务业管理制度	3		
2	决策阶段的工程造价控制	（1）了解决策阶段控制工程造价的意义和影响工程造价的因素； （2）熟悉可行性研究报告的作用、主要内容和审批程序(★)； （3）掌握投资估算的内容(★)； （4）掌握财务评价基本报表的项目组成，各类数据的分析与评价(★)； （5）掌握财务评价指标的含义、体系、评价方法及准则(★)	决策阶段控制工程造价的意义；影响工程造价的因素；决策阶段工程造价控制的内容	2		
			决策阶段控制工程造价的方法；建设项目财务评价	4		

4. Curriculum Contents and Teaching Hours

Table of Teaching Contents and Teaching Hours

Item	Topic	Teaching Objectives	Teaching Contents and Teaching Hours			
			Lesson Contents	Hrs	Student Activities	Hrs
1	Basic theory of construction cost control	(1) master the related concepts of project cost: project cost, project cost control, static investment(★), dynamic investment, construction project investment estimation, design budget, construction drawing budget, etc(★); be familiar with the characteristics of construction cost and valuation characteristics(★); (2) Understand the basic content of project cost control; Understand the types and contents of construction cost basis; (3) Understand the establishment of the professional qualification system for cost engineers in China; Master the rights and obligations of cost engineers(★); (4) Understand the management system of construction cost consultation service industry	Basic theory of construction cost; The basic theory of construction cost control	2		
			Basis for project cost valuation; Professional qualification system of cost engineer; Management system of construction cost consultation service industry	3		
2	Project cost control in the decision-making stage	(1) Understand the significance of controlling project cost in the decision-making stage and the factors affecting project cost; (2) Be familiar with the role, main contents and approval procedures of the feasibility study report(★); (3) Master the content of investment estimation(★); (4) Master the project composition of the basic financial evaluation statements, and analyze and evaluate all kinds of data(★); (5) Grasp the meaning, system, evaluation method and criterion of financial evaluation index(★)	The significance of controlling project cost in decision-making stage; Factors affecting project cost; The content of the project cost control in the decision-making stage	2		
			The method of controlling project cost in decisionmaking stage; Financial evaluation of construction projects	4		

序号	教学情境/任务/项目/单元	教学目标	教学内容及学时分配			
			知识点	学时	实践项目	学时
3	设计阶段的工程造价控制	(1)了解设计的内容和程序,熟悉工程设计与工程造价的关系; (2)掌握设计方案优选的概念和设计方案优选的方式(★);掌握运用价值工程优化设计的途径(★);掌握设计方案的技术经济评价(★); (3)掌握设计概算的审查内容和审查方法(★); (4)掌握施工图预算的审查方法(★); (5)了解推行限额设计的意义;掌握设计目标的设置方法(★);掌握限额设计的控制方法(★)	设计阶段工程造价控制的内容和程序;工程设计与工程造价的关系;工程设计方案的优选	3		
			设计概算的审查方法;施工图预算的审查方法;限额设计方法	3		
4	招投标阶段的工程造价控制	(1)了解招标投标阶段控制工程造价的目的、意义及招标投标程序; (2)了解建设工程施工招标投标的意义,熟悉投标报价的控制方法; (3)了解设备、材料采购招标投标的意义、方法; (4)了解开标、评标和定标的程序,掌握评标方法	概述;建设工程施工招标投标;设备、材料采购招标投标	2		
			建设工程开标、评标和定标	2		

continued

Item	Topic	Teaching Objectives	Teaching Contents and Teaching Hours			
			Lesson Contents	Hrs	Student Activities	Hrs
3	Design phase of the project cost control	(1) Understand the content and procedure of design, and be familiar with the relationship between engineering design and construction cost; (2) Master the concept of design scheme optimization and design scheme optimization method (★); Grasp the application of value engineering optimization design approach(★); Master the technical and economic evaluation of the design scheme(★);	The content and procedure of construction cost control in the design stage; The relationship between engineering design and construction cost; Optimization of engineering design scheme	3		
		(3) Master the review content and review method of design budget estimates(★); (4) Master the construction drawing budget review method(★); (5) Understand the significance of introducing quota design; Master the method of setting design goals (★); Master the control method of quota design(★)	Methodology for the review of design budget proposals; The examination method of construction drawing budget; Quota design method	3		
4	Project cost control in the bidding stage	(1) to understand the purpose, significance and bidding procedures of controlling project cost at the bidding stage; (2) understand the significance of construction bidding and be familiar with the control method of bidding quotation; (3) understand the significance and methods of bidding for equipment and materials procurement;	Overview; Tendering and bidding for construction projects; Tendering and bidding for equipment and materials procurement	2		
		(4) understand the procedures of bid opening, evaluation and selection, and master the evaluation methods	Bid opening, evaluation and selection of construction projects	2		

序号	教学情境/任务/项目/单元	教学目标	教学内容及学时分配			
			知识点	学时	实践项目	学时
5	工程实施阶段的工程造价控制	(1)了解工程变更的概念及产生的原因;掌握工程变更的确认及处理程序(★);掌握工程变更价款的计算及控制方法(★); (2)了解索赔的概念、分类;掌握索赔的程序及处理原则(★);掌握索赔的内容与特点;掌握业主反索赔的内容与特点(★);掌握索赔费用的计算及控制方法(★); (3)掌握工程价款结算的计算方法及控制方法(★); (4)了解 FIDIC 合同条件下工程价款的支付与结算方法; (5)熟悉资金使用计划的编制与控制方法;掌握投资偏差的分析方法及纠正措施(★)	工程变更与合同价调整;工程索赔	2		
			工程价款结算的控制	2		
			FIDIC 合同条件下工程价款的支付与结算;资金使用计划	2		
6	工程竣工阶段的工程造价控制	(1)了解竣工验收的概念、依据和验收程序;了解竣工决算的内容及编制步骤; (2)掌握保修费用的处理方法(★)	竣工验收	3		
			竣工决算;保修费用的处理	2		
合计				32		
备注:重要知识点标注★						

五、教学实施建议

(一)教学参考资料

尹贻林 .《工程造价计价与控制》,中国计划出版社,2010。

continued

Item	Topic	Teaching Objectives	Teaching Contents and Teaching Hours			
			Lesson Contents	Hrs	Student Activities	Hrs
5	Project cost control during the implementation stage	(1) Understand the concept and causes of engineering change; Master the confirmation and processing procedures of engineering changes (★); Master the calculation and control method of engineering change price(★); (2) Understand the concept and classification of claims; Master the claim procedures and processing principles; Master the content and characteristics of the claim (★); Master the content and characteristics of the owners'counterclaim (★), master the calculation and control method of claim cost (★); (3) Master the calculation method and control method of project price settlement(★);	Engineering change and contract price adjustment; Engineering claims	2		
			Control of settlement of project price	2		
		(4) Understand the payment and settlement method of project price under the conditions of FIDIC contract; (5) Be familiar with the preparation and control methods of fund use plan; Master the analysis method of investment deviation and corrective measures(★)	Payment and settlement of the project price under the conditions of the FIDIC contract; Fund use plan	2		
6	Project cost control at the completion stage	(1) Understand the concept, basis and acceptance procedures of completion acceptance; To understand the contents and preparation steps of the final accounts for completion; (2) Master the handling method of warranty cost(★)	Completion inspection and acceptance	3		
			Final accounts for completion; Handling of warranty costs	2		
		Total		32		
Note: Mark the important points with ★						

5. Teaching Suggestions

5.1 Teaching Resources

Yin Yilin. *Construction Cost Valuation and Control.* China Planning Press, 2011.

(二)教师素质要求

教师具备中级以上职称,有相应的工程监理、建造师或造价职工程师业资格,或拥有工程相应专业背景,能够理论结合实际,对工程全过程管理有一定工作经验或学习经历。

(三)教学场地、设施要求

该课程为理论授课,普通教室。

六、考核评价

(一)基本概念及知识点(分值为 30 ~ 40 分)

工程造价控制基本理论的主要知识点,以单选题、多选题及判断题的形式出题。

(二)计算题(分值为 60 ~ 70 分)

(1)财务评价指标计算题;

(2)价值工程计算题;

(3)设计方案优选计算题;

(4)工程变更与合同价款调整计算题;

(5)工程竣工结算计算题。

5. 2 Teachers'Qualification

Teachers should be lecturers or higher titles of this specialty and they should have the qualification of engineering supervision, construction engineer or cost engineer and have the corresponding professional background of engineering management. Teachers should be able to combine theory with practice with certain working experience or learning experience in the whole process of engineering management.

5. 3 Teaching Facilities

This course can be delivered in common classrooms.

6. Evaluation

6. 1 Basic concepts and knowledge points(with a score of 30 ~ 40 points)

The main knowledge points of the basic theory of construction cost control are presented in the form of single choice questions, multiple choice questions and T/F questions.

6. 2 Application of the concepts and knowledge points(with a score of 60 ~ 70 points)

(1) Financial evaluation indexes;

(2) Value engineering;

(3) Design scheme optimization;

(4) Engineering change and contract price adjustment;

(5) Project completion and settlement.

四川建筑职业技术学院

课 程 标 准

课程名称：　　　英国建筑工程技术

课程代码：　　　210212

课程学分：　　　4

基本学时：　　　48

适用专业：　　　工程造价（中英合作办学项目）

执 笔 人：　　　托尼·夏洛克

编制单位：　　　林肯学院

审　　核：　　　瑞克·隆

批　　准：　　　詹姆斯·福斯特

Sichuan College of Architectural Technology

Curriculum Standard

Curriculum Name: Construction Technology in the U. K.

Curriculum Code: 210212

Credits: 4

Teaching Hours: 48

Applicable Specialty: Construction Cost

    (Sino-British Cooperative Program)

Compiled by: Tony Sherlock

On Behalf of: Lincoln College

Reviewed by: Rick Long

Approved by: James Foster

《英国建筑工程技术》课程标准

一、课程概述

本课程旨在帮助学习者了解常见形式的低层建筑所使用的技术,包括其子结构的设计和构造,以及上层建筑施工中使用的技术。

二、课程设计思路

(1)了解目前英国国内建筑中各种形式的低层建筑;

(2)了解与各种形式与低层建筑相关的基础设计和施工;

(3)了解英国低层住宅上部结构的技术;

(4)了解对建筑施工产生影响的问题和限制。

三、教学目标

(一)了解目前英国国内建筑中各种形式的低层建筑

低层建筑的形式:预制件包括木框架,钢框架,混凝土框架,承重,非承重;单层或有两到三层楼的低层;独立式住宅;排屋;尖屋顶;平屋顶;短跨结构;中等跨度结构;施工方法的差异;每种方法的优点和局限性。

建筑:家庭住宅。

(二)了解与各种形式的低层建筑相关的基础设计和施工

心土调查:现场调查和心土调查(区域地质、岩性、地下水);记录和分析结果;土壤分类。

地基设计:设计原则;地基的选择(条形地基、垫式地基、筏形地基和桩地基);结构要求;地基收缩的影响和预防措施;基坑底部隆起;不均匀沉降(沉降差)。

方法:挖掘;施工。

Construction Technology in the U. K. Curriculum Standard

1. Curriculum Description

This module aims to provide the learners with the opportunity to gain an understanding of the technology used in common forms of low-rise construction, including the design and construction of their sub-structure, the techniques used in Construction of the superstructure.

2. Curriculum Design Ideas

(1) Understand the various forms of low-rise construction currently used for domestic building in the U. K. ;

(2) Understand foundation design and construction associated with the various forms of low-rise construction;

(3) Understand the techniques used in the construction of superstructures for low-rise domestic buildings;

(4) Understand the implications of issues and constraints on building construction.

3. Curriculum Objectives

3. 1 Understand the various forms of low-rise construction currently used for domestic building in the U. K.

Forms of low-rise construction: prefabricated including timber frame, steel frame, concrete frame, load bearing, non-load bearing; single storey and low-rise of two to three storeys; detached; terraced; pitched roofs; flat roofs; short span; medium span; differences in construction methods; advantages and limitations of each method.

Buildings: houses.

3. 2 Understand foundation design and construction associated with the various forms of low-rise construction

Subsoil investigation: site survey and subsoil investigation (regional geology, lithology, ground water); recording and interpretation of results; classification of soils.

Foundation design: principles of design; choice of foundations (strip, pad, raft and pile foundations); structural requirements; effects of and precautions against subsoil shrinkage; ground heave; differential settlement.

Methods: excavation; construction.

挖掘:挖掘深度达 5 m;除水;地基处理;壕沟中的临时支持以及相关的健康和安全问题;各类挖掘和运土机械。

施工:用于条、垫、筏、桩和梁基础的施工技术;材料选择;选用方法对造价的影响;场地和设备要求;健康与安全问题;环境问题;法律限制。

(三) 了解英国低层住宅上部结构的技术

上部结构设计原则:设计原则和影响主要和次要元素选择的因素(地板、墙壁、屋顶、楼梯、窗户、门)

上部建筑:(地板、墙壁、屋顶、楼梯、窗户和门);材料的选择;场地和设备要求。

上部结构完工:影响内外部装饰面选择的因素;可用装饰面的类型及其应用方法。

(四) 了解建筑施工必须考虑的环境和法律问题

环境问题:建筑施工所使用的材料和方法造成的对环境的影响。

相关法规:《建筑条例》概述;《工作健康和安全法》(1974 年);《施工健康、安全和福利条例》(1996 年);《建筑设计与管理条例》(2015 年);《残疾歧视法案》(1995 年);《设备使用和安全防护规章》;《健康有害物控制规程》;《个人防护装备》;《瑞德法》;《城乡规划法》。

基础设施:施工场地(特点、用途);为传统和现代工程提供建筑材料;预制部件;系统建筑。

四、教学内容和学时安排表

根据《学习成果量表》和相关规范,本课程将开展各种教学活动和并做评估。任课教师根据学生的学习进度以及学生对英语的理解和内容的掌握规划调整教学内容。因此,具体的课堂活动应在《学习成果量表》的指导下开展。

五、教学实施建议

因为本课程实践活动居多,因此教学内容必须有所限定(上课时应提供教学内容指南)。

(一) 教师素质要求

上课教师应具备专业背景以及建筑行业认证和教师资格证。具体聘用视个人简历而定。

Excavation: excavation up to five meters depth; water elimination; ground improvement; temporary supports in trenches and associated health and safety issues; various types of excavation and earth moving plant.

Construction: construction techniques used for strip, pad, raft, pile and beam foundations; selection of materials; economic implications of methods used; plant requirements; health and safety issues; environmental issues; legislative constraints.

3.3　Understand the techniques used in the construction of superstructure for low-rise domestic buildings in the U. K.

Principles of superstructure design: principles of design and factors affecting choice of primary and secondary elements(floors, walls, roofs, stairs, windows, doors)

Superstructure construction: (floors, walls, roofs, stairs, windows and doors); selection of materials; plant and equipment requirements.

Superstructure finishes: factors affecting the choice of internal and external finishes; types of finish available and methods used in their application.

3.4　Understand the environmental and legislative issues that must be considered when constructing buildings

Environmental issues: environmental impact resulting from materials and methods used in the construction of buildings.

Legislative constraints: A brief overview of *Building Regulations*; *Health and Safety at Work Act* (1974); *Construction Health, Safety and Welfare Regulations* (1996); *Construction Design and Management Regulations* (2015); *DDA* (1995); *PUWER*; *COSHH*; *PPE*; *RIDOR*; *Town and Country Planning legislation.*

Infrastructure: construction plant(characteristics, uses); supply of building materials for traditional and modern projects; prefabricated components; system building.

4. Curriculum Contents and Teaching Hours

Various activities will be delivered and assessed as per the *Learning Outcomes* and specification. Resources will be planned and delivered by individual teacher, depending on the pace of the students, understanding of English and grasp of the content. Therefore specific class activities will be directed by the *Learning Outcomes.*

5. Teaching Suggestions

Limited resources are necessary for this module, as most activities will be undertaken practically. (Instruction to resources needed will be given at the time of delivery).

5.1　Qualifications of the Teacher

Professional Chartered(Construction)qualifications and Teaching Quals.

(二) 教学场所和设施要求

教师应在四川建院分配的教室进行授课。

设施要求:多媒体、投影仪、书写板、画板、书写笔。

六、考核评价

总评成绩由平时课堂表现、上课问答以及作业成绩三部分组成。

5. 2　Teaching Places & Facility Requirements

Teaching undertaken in allocated classrooms at SCAT.

Teaching Resources: PowerPoint/OHP/Chalk board/drawing boards/writing materials.

6. Evaluation/Examination

Assessment of activities conducted by observation/question & answer/assessment of course work.

四川建筑职业技术学院

课程标准

课程名称：　　　　英国建筑材料

课程代码：　　　　210217

课程学分：　　　　2

基本学时：　　　　24

适用专业：　　　工程造价（中英合作办学项目）

执笔人：　　　　托尼·夏洛克

编制单位：　　　　林肯学院

审　核：　　　　瑞克·隆

批　准：　　　詹姆斯·福斯特

Sichuan College of Architectural Technology

Curriculum Standard

Curriculum Name: Material Technology in the U. K.

Curriculum Code: 210217

Credits: 2

Teaching Hours: 24

Applicable Specialty: Construction Cost

    (Sino-British Cooperative Program)

Compiled by: Tony Sherlock

On Behalf of: Lincoln College

Reviewed by: Rick Long

Approved by: James Foster

《英国建筑材料》课程标准

一、课程概述

本课程旨在让学习者了解英国低层民用建筑所用的各类型建筑材料。

本课程引导学生学习与建筑材料、建筑环境相关的学科内容,让学习者对建筑材料的基本性质和用途有一个初步理解。设计本课程的目的就在于让学生学习英国低层民用建筑所用建筑材料的相关知识。

二、课程设计思路

本课程经过英国伦敦城市行业协会的评估认可,其实施方案能保证课程质量的完成。

(1)了解建筑材料的性质与用途。

(2)了解建筑材料的结构性能。

(3)通过建筑材料了解现实环境中的可持续性要求。

模块形式,例如授课形式/资源等。

该门课程以课堂讲授为基础,并开展多种形式的教学活动。这些活动都要逐一评估,确保整个教学模块的教学质量。

三、课程目标

(一)了解建筑材料的性质和用途

建筑材料:金属与合金,如铁、钢、锌、铜、黄铜、铝、铅;木材和木材制品;烧土制品,如砖、瓷砖;陶瓷和混凝土;塑料和其他合成材料;涂料和表面材料,如油漆、透明涂层、木材处理剂。

材料的性质:该领域内所涉及的性质,如强度、弹性、孔隙率、吸水性、湿热环境中的位移、导电导热性能、保温绝缘性能、耐久性、可加工性、密度、比热容、黏性。

材料的用途:建筑、可持续性、节能、环境问题、可再生资源的利用。

(二)了解建筑材料的结构性能

结构性能:用途与性能的关系;荷载形式;结构材料的固有特性(木材、钢、钢筋混凝土);组成结构构件(如梁、柱、架、基座、基础)的材料性能。

Material Technology in the U. K. Curriculum Standard

1. Curriculum Description

This curriculum provides learners with an understanding of materials used in various types of construction in low-rise domestic buildings in the U. K.

This curriculum introduces the student to the relevant study of construction materials and the built environment and provides learners with a basic understanding of the properties and use of construction materials. This unit has been designed to enable learners studying materials used in the construction of various low-rise domestic properties in the U. K.

2. Curriculum Design Ideas

The quality of the curriculum units is assessed by City & Guilds and accredited for meeting its quality assurance practices. The design of the curriculum based on motivating students to

(1) understand the properties and use of construction materials;

(2) understand the structural behaviour of construction materials;

(3) understand the need for sustainability in the context on construction materials.

Master the format of module i. e. delivery/resources.

The delivery is classroom based and will take the format of various activities, which will be individually assessed, leading to a module qualification.

3. Curriculum Objectives

3. 1　Understand the properties and use of construction materials

Materials: metals and alloys e. g. iron, steel, zinc, copper, brass, aluminium, lead; timber and timber products; clay products e. g. bricks, tiles; cements and concretes; plastics and other artificial materials; coatings and finishes e. g. paints, clear finishes, wood treatments.

Properties of materials: as appropriate to field of study e. g. strength, elasticity, porosity and water absorption, thermal and moisture movement, thermal and electrical conductivity/resistivity, durability, workability, density, specific heat capacity, viscosity.

Uses of materials: construction; sustainability; energy efficiency; environmental issues; use of renewable resources.

3. 2　Understand the structural behaviour of construction materials

Structural behaviour: relationship between behaviour and use; forms of loading; inherent properties of structural materials(timber, steel, reinforced concrete); behaviour of structural materials when formed into structural members e. g. beams, columns, frames, pads, bases.

(三)通过建筑材料了解现实环境中的可持续性要求

可持续性问题:现代施工方法和实践的效果;英国的可持续发展政策;建筑学相关内容。

四、课程内容与学时分配

教学过程中设计了针对每个学习目标和具体内容的各种实践活动,并对每项活动进行评估。每位老师可根据学生的学习进度、英文水平、对内容的掌握程度,自行规划、使用教学资源。学习目标决定了具体的教学活动实施情况。

五、教学实施建议

(一)教学资源

PPT、在线资源、个人研究成果。

(二)教师素质

建筑专业资格证书、教师资格证书。

简历中的适当素质。

(三)教学场地

教学活动将在四川建院的指定教室内展开。

(四)教学条件

PPT、投影仪、黑板、画板、文具。

六、考核评价

教学活动的评估将根据教师对学生的观察、课题问答情况、课程作业评分来确定。

3.3   Understand the need for sustainability in the context on construction materials

Sustainable issue;Effect on modern construction methods and practices; sustainable policies within the U. K. ; and architectural input.

4. Curriculum Contents and Teaching Hours

Various activities will be delivered and assessed as per the learning outcomes and specification. Resources will be planned and delivered by individual teacher, depending on the pace of the students, understanding of English and grasp of the content. Therefore, specific class activities will be directed by the Learning Outcomes.

5. Teaching Suggestions

5.1   Teaching Resources

PowerPoint, On-Line research, Individual study.

5.2   Qualifications of the Teacher

Professional Chartered(Construction)qualifications and Teaching Quals.
See CV's.

5.3   Teaching Places

Teaching undertake at Sichuan College in classrooms allocated.

5.4   Teaching Facilities

PowerPoint/OHP/Chalk board/drawing boards/writing materials.

6. Evaluation/Examination

Assessment of activities undertaken by observation/question & answer/assessment of course work.

四 川 建 筑 职 业 技 术 学 院

课 程 标 准

课程名称：　　　英国建筑工程计量

课程代码：　　　210218

课程学分：　　　4

基本学时：　　　60

适用专业：　　　工程造价

执 笔 人：　　　托尼·夏洛克

编制单位：　　　林肯学院

审　　核：　　　瑞克·隆

批　　准：　　　詹姆斯·福斯特

Sichuan College of Architectural Technology

Curriculum Standard

Curriculum Name: Measurement Processes in

Construction in the U. K.

Curriculum Code: 210218

Credits: 4

Teaching Hours: 60

Applicable Specialty: Construction Cost

(Sino-British Cooperative Program)

Compiled by: Tony Sherlock

On Behalf of: Lincoln College

Reviewed by: Rick Long

Approved by: James Foster

《英国建筑工程计量》课程标准

一、课程概述

本课程旨在帮助学生掌握建筑工程测量技术以及在不同工程阶段的应用。学生不仅要从格式、编码和测量规则三方面了解测量的方法,还要学会运用适当的测量技术来编制工程量清单的技能。通过本课程的学习,学生除了学会工程开工,还要能计算地基、下部结构和上部结构的工程量,并掌握编制工程量清单的技能。

二、课程设计思路

课程质量得到了"英国城市和行业协会"机构的评估和认证,以确保教学质量。

(1)了解测量技术;

(2)能够计算工程量;

(3)能够编制工程量清单。

授课以课堂教学为主,并开展各种实践活动。所有的课程教学活动都会一一测评,从而保证该模块的学习质量。

三、教学目标

(一)了解建筑工程测量技术

技术:使用公认的标准方法,以标准形式将建筑图纸上的工程量摘抄记录到尺寸坐标图纸中。

标准测量方法:用于建筑工程施工使用的格式、编码系统、测量规则和尺寸表。

(二)能够计算工程量

计算工程量:开工程序,测量程序,工程量计算。

应用:地基和下部结构;上部结构如外墙、内墙、房顶、内饰和部件(门、窗、楼梯、踢脚线、框缘和抹灰)。

(三)能够编制工程量清单

测量技术和步骤:技术如传统的相关工作步骤。

合同文件:其他文件、主要费用和暂定款(施工费用)。

应用:用于建筑工程(简单工程项目、商贸工程项目)。

Measurement Processes in Construction in the U. K. Curriculum Standard

1. Curriculum Description

This course provides the learner with an understanding of measurement techniques and their uses in various project stages. Learners will gain an understanding of methods of measurement in terms of format, coding and measurement rules. Learners will develop the skills needed to produce quantities by applying appropriate mensuration techniques. Learners will be able to perform take-off and produce quantities for foundation, substructure and superstructure. Learners will develop skills to produce bills of quantities.

2. Curriculum Design Ideas

The quality of the courses is assessed by City & Guilds and accredited for meeting its quality assurance practices. The design of the curriculum based on motivating students to

(1) understand measurement techniques.

(2) be able to produce quantities.

(3) be able to produce bills of quantities.

The delivery is classroom based and will take the format of various activities, which will be individually assessed, leading to a module qualification.

3. Curriculum Objectives

3. 1 Understanding measurement techniques

Techniques: extraction of quantities from drawings in a standard form onto dimension paper using recognized and standard methods.

Standard methods of measurement: For construction using format, coding systems, measurement rules and dimension sheets.

3. 2 Be able to produce quantities

Producing quantities: Take off procedures, mensuration procedures, computation of quantities.

Applications: Foundations and substructure; superstructure eg external walls, internal walls, roof, internal finishing and components(doors, windows, staircases, skirting , architrave and plastering).

3. 3 Be able to produce bills of quantities

Measurement techniques and processes: techniques e. g. traditional associated working up processes.

Contract documents: Other documentation, prime cost and provisional sums.

Applications: to construction projects(simple work sections, trade sections).

四、课程内容和学时安排

根据《学习成果量表》和相关规范,本课程将开展各种教学活动和并做评估。任课教师根据学生的学习进度以及学生对英语的理解和内容的掌握规划调整教学内容。因此,具体的课堂活动应在《学习成果量表》的指导下开展。

五、教学实施建议

(一)教学用具

圆规、铅笔、计算器、尺寸坐标图纸。

(二)教师素质要求

上课教师应具备专业背景以及建筑行业认证和教师资格证。

具体聘用视个人简历而定。

(三)教学场所和设施要求

教师应在四川建筑职业技术学院分配的教室进行授课。

设施要求:多媒体、投影仪、书写板、画板、书写笔。

六、考核评价

总评成绩由平时课堂表现、上课问答以及作业成绩三部分组成。

4. Curriculum Contents and Teaching Hours

Various activities will be delivered and assessed as per the *Learning Outcomes* and specification. Resources will be planned and delivered by the individual teacher, depending on the pace of the students, understanding of English and grasp of the content. Therefore specific class activities will be directed by the *Learning Outcomes*.

5. Teaching Suggestions

5.1 Resources

Scale rule, pencils, calculator and Dimension paper.

5.2 Qualifications of the Teacher

Professional Chartered(Construction)qualifications and Teaching Quals.
See CV's.

5.3 Teaching Places & Facility Requirements

Teaching undertaken in allocated classrooms at SCAT.

5.4 Teaching Resources

PowerPoint/OHP/Chalk board/drawing boards/writing materials.

6. Evaluation/Examination

Assessment of activities undertaken by observation/question & answer/assessment of course work.

四川建筑职业技术学院

课 程 标 准

课程名称：   英国建筑工程计价

课程代码：   210219

课程学分：   4

基本学时：   48

适用专业：   工程造价（中英合作办学项目）

执 笔 人：   托尼·夏洛克

编制单位：   林肯学院

审   核：   瑞克·隆

批   准：   詹姆斯·福斯特

Sichuan College of Architectural Technology

Curriculum Standard

Curriculum Name: Construction Estimating in the U. K.

Curriculum Code: 210219

Credits: 4

Teaching Hours: 48

Applicable Specialty: Construction Cost

(Sino-British Cooperative Program)

Compiled by: Tony Sherlock

On Behalf of: Lincoln College

Reviewed by: Rick Long

Approved by: James Foster

《英国建筑工程计价》课程标准

一、课程概述

本课程让学生对工程估算的过程和技巧有所了解。它让学生有机会学习推导决算工程量,计算综合单价,确定建设项目的大致价值等技能。

估算是承包商对自己所开展的建设工作确立成本的过程。工程建设合同投标的过程通常包括承包商对所要开展的工作进行精确计量,并用这个计量结果来估算成本、编制投标书。

招标是对发包人提出的工作进行定价的过程。实施的方式是发布文件以便准备进行估算,挑选参与投标的承包商,评估收到的价格以便确定合同的归属。标书需要经过准确性评估,并对数字进行检查,确保在最终定标和向胜出的承包商递交的合同文件中没有任何错误。

学生将有机会从工程图纸和其他文件中获得详细的工程量,估算各种建设工作的成本,将估算结果转化成标书,为即将投标的工程项目提供成本估算大纲。

二、课程设计思路

本课程经过英国伦敦城市行业协会的评估认可,其实施方案能保证课程质量的完成。

(1)了解预算的目的,以及工程定价的常用技巧。

(2)了解招标过程(与下一模块相关联)。

本课程以课堂讲授为基础,并开展多种形式的教学活动。这些活动都要逐一评估,确保整个教学模块的教学质量。

三、课程目标

(一)了解预算的目的,以及工程定价的常用技巧

预算的目的:估算净成本;初步定价;利润与一般间接费;工程量和价值对选择预算方法的影响。

Construction Estimating in the U. K. **Curriculum Standard**

1. Curriculum Description

This unit gives learners an understanding of the processes and techniques involved in estimating. It also gives learners an opportunity to develop skills in producing final quantities, calculating all-in rates and determining the approximate value of building projects.

Estimating is concerned with the processes used by contractors to establish the cost to them of carrying out construction work. The process of bidding for a construction contract normally involves the contractor measuring the works required accurately and using the outcomes to estimate the costs and compile the tender.

Tendering is the process of obtaining a price for the client's work. This is done by issuing the documents needed to prepare the estimate, selecting contractors to bid, and evaluating the prices received in order to award the contract. Tenders are then evaluated for accuracy and checked numerically to ensure that no mistakes have been made before formal acceptance and the issue of contract documents to the successful contractor.

Learners will have the opportunity to obtain detailed quantities from drawings and other documents, estimate the cost of a variety of construction works, convert the estimate into a tender and provide outline cost estimates for proposed construction projects.

2. Curriculum Design Ideas

The quality of the course is assessed by City & Guilds and accredited for meeting its quality assurance practices. The design of the curriculum based on motivating students to

(1) understand the purpose of estimating and the common techniques used to price construction work

(2) understand the process of tendering(linked to the next module).

The delivery is classroom based and will take the format of various activities, which will be individually assessed, leading to a module qualification.

3. Curriculum Objectives

3. 1 Understand the purpose of estimating and the common techniques used to price construction work

Purposes of estimating: estimating net cost; pricing of preliminaries; profit and general overheads; the effects of quantity and value on the chosen method of estimating.

预算技巧：人工、设备和材料、计量单位的单位费率、标准账面费率、输出表、历史费率、工程研究。

文件：工程估价规程。

(二) 了解招标过程

常用投标方法：依据工程的规模、大小和价值确定的招标方法；各种(单一可选标段、两标段可选、开放式、系列)建设工程项目的类型(建筑工程、土木工程或建筑装饰工程)；目标成本；实测工程量；投标费用。

文件：图纸、施工计划、技术说明、施工进度表、工程量清单、活动进度、与总承包人相关的投标过程规范、分包项目、供应分包。

四、课程内容与学时分配

教学过程中设计了针对每个学习目标和具体内容的各种实践活动，并对每项活动进行评估。每位老师可根据学生的学习进度、英文水平、对内容的掌握程度，自行规划、使用教学资源。学习目标决定了具体的教学活动实施情况。

五、教学实施建议

(一) 教学资源

直尺、固定尺寸的纸、建筑工程标准计量规则(SMM7 或 NRM2)、铅笔、橡皮、计算器和绘图纸

(二) 教师素质

建筑专业资格证书、教师资格证书。

具体聘用视个人简历而定。

(三) 教学场地

教学活动将在四川建筑职业技术学院指定的教室内展开。

(四) 教学条件

PPT、投影仪、黑板、绘图板、书写材料。

六、考核评价

教学活动的评估将根据教师对学生的观察、日常问答情况、课程作业评分来确定。

Estimating techniques: labour, plant and materials; rates per unit of measurement; standard price book rates; output tables; historical rates; work study.

Documentation: Code of Estimating Practice.

3.2 Understand the process of tendering

Common methods of tendering: methods of tendering relevant to the scale, size and value of the construction works; type of work (building, civil engineering or building services work) for a range of construction works (eg single stage selective, two stage selective, open, serial); target cost; measured term; fee bidding.

Documentation: e. g. drawings, schedules, specifications, schedules of work, bills of quantities, activity schedules; codes of procedure for tendering relevant to main and principal contractors, subcontract packages and supply packages.

4. Curriculum Contents and Teaching Hours

Various activities will be delivered and assessed as per the Learning Outcomes and specification. Resources will be planned and delivered by individual teacher, depending on the pace of the students, understanding of English and grasp of the content. Therefore specific class activities will be directed by the Learning Outcomes.

5. Teaching Suggestions

5.1 Teaching Resources

Scale rules, dimension paper, SMM7/NRM2, pencil, eraser, calculator and drawing paper.

5.2 Qualifications of the Teacher

Professional Chartered (Construction) Qualifications and Teaching Quals.
See CV's.

5.3 Teaching Places

Teaching undertake at Sichuan College of Architectural Technology in classrooms allocated.

5.4 Teaching Facilities

PowerPoint/OHP/Chalk board/drawing boards/writing materials.

6. Evaluation/Examination

Assessment of activities undertaken by observation/question & answer/assessment of course work.

四 川 建 筑 职 业 技 术 学 院

课 程 标 准

课程名称：　　　英国工程招投标

课程代码：　　　210220

课程学分：　　　2

基本学时：　　　32

适用专业：工程造价（中英合作办学项目）

执 笔 人：　　托尼·夏洛克

编制单位：　　　林肯学院

审　　核：　　　瑞克·隆

批　　准：　詹姆斯·福斯特

Sichuan College of Architectural Technology

Curriculum Standard

Curriculum Name: <u>Construction Tendering in the U. K.</u>

Curriculum Code: <u>210220</u>

Credits: <u>2</u>

Teaching Hours: <u>32</u>

Applicable Specialty: <u>Construction Cost</u>

<u>(Sino-British Cooperative Program)</u>

Compiled by: <u>Tony Sherlock</u>

On Behalf of: <u>Lincoln College</u>

Reviewed by: <u>Rick Long</u>

Approved by: <u>James Foster</u>

《英国工程招投标》课程标准

一、课程概述

本课程让学习者有机会了解编制投标书所需的信息,理解如何使用不同类型的标书文件,掌握工程量清单中分项工程或单个标段的费率计算,掌握标书的编制方法。

成功盈利的建筑企业要求预算员懂商业、能力强、重细节。预算员对项目成本进行预估,以便公司在确定项目利润率的基础上递交投标书。

项目利润率的确定要基于公司要求的回报率,同时要考虑当前的工作量,前期已预计的工作量,项目相关的风险水平,当前和未来的市场状况,以及投标该项目的其他竞争者的已有工作量和预计工作量。

学习者将了解从标书文件的接收到递交最终投标书整个投标过程。这个过程包括了文件初审、实地视察、风险分析、对材料和分包项目的询盘、综合费率计算、工程量清单计价、评标会。学习者将注意到工作中对准确性的高度需求,任何错误将会导致金钱损失或投标失败。

估价是投标过程的有机组成部分。这个工作领域内波动极大,要求预算员适应紧张的工作日程,在递交投标书截止日期之前完成工作。预算员会因自己预算的准确性和对市场状况的准确评估而获得满足感和骄傲。

预算员和估算师都需要相同的知识和技能,小型建筑企业通常把两者的角色合二为一。学完本课程后,学生将有能力确定综合费率,并将其应用于标书文件当中,在考虑自身的经营决策后,形成投标总价,以便针对工程施工编制标书。

二、课程设计思路

本课程经过英国伦敦城市行业协会的评估认可,其实施方案能保证课程质量的完成。

(1)了解标书编制所需的基本信息;

(2)了解如何运用不同的标书文件;

Construction Tendering in the U. K. Curriculum Standard

1. Curriculum Description

This curriculum will give learners the opportunity to develop knowledge of the information needed to produce a tender, an understanding of how different types of tender documentation are used, and to develop skills needed to calculate unit rates for an element or trade section of a bill of quantities and produce a tender.

Successful and profitable construction companies require estimators who are commercially aware, highly skilled and pay close attention to detail. The estimator produces an estimate of project cost to enable the company to submit a tender after the decision has been made on the amount of profit to add to the project.

This decision is based on the company's required return whilst taking into account their current workload and advance order book, level of risk associated with the project, the current and future market conditions, and the perceived workload or current order book of competitors who may also tender for the project.

Learners will develop an understanding of tendering procedures from receipt of the tender documentation through to submission of the final tender. This encompasses the initial inspection of the documents, inspection of the site, risk analysis, materials and sub-contract enquiries, calculation of analytical unit rates, pricing of the bill of quantities and the tender adjudication meeting. Learners will become aware of the need to work with great accuracy as any errors could lead to financial losses or an unsuccessful tender.

Estimating is an integral part of the tendering process. It is an extremely vibrant and dynamic field of work that requires the estimator to work to tight deadlines as the tender submission date approaches. Estimators gain great satisfaction and pride from the accuracy of their estimates and their assessment of commercial market conditions.

Estimators and quantity surveyors require similar skills and knowledge and, in smaller construction companies, the roles are often combined. After completing this unit learners will be able to build up analytical unit rates and apply them to tender documentation in order to produce a tender for construction work, taking into account the commercial decisions to be made in arriving at a tender sum.

2. Curriculum Design Ideas

The quality of the units is assessed by City & Guilds and accredited for meeting its quality assurance practices. The design of the curriculum based on motivating students to

(1) know the basic information needed to produce a tender;

(2) understand how to use different types of tender documentation;

(3)能进行工程量清单中分项工程或单个标段的费率计算;

(4)能针对特定的施工项目或分项工作编制标书。

该门课程以课堂讲授为基础,并开展多种形式的教学活动。这些活动都要逐一评估,确保整个教学模块的教学质量。

三、课程目标

(一)了解标书编制所需的基本信息

相关信息:标书文件;合同条件和要求,例如:合同格式,合同有效期,开工日期、约定的和已明确的损失赔偿,标书文件的类型,图纸与技术规范,单价表;图纸会审;实地勘察;实地视察所需收集的信息;材料和分包项目询盘;人工成本;覆盖率;计量标准(NRM)覆盖规则;全包工价;公司提留;预期回报率;风险因素。

(二)了解如何运用不同的标书文件

标书文件:类型、标段、计量方法。

类型:工程量清单、图纸与技术说明、设计与建设、单价表。

标段:初期、前期、核定工作量、日工作量、总成本、备用款、临时费。

(三)针对特定的施工项目或分项工作编制标书

单价:材料价格、运输和装卸成本、覆盖率、垃圾处理费、厂房租金、人工成本和全包工价、合理的管理费加成比例、复核程序,ICT预算软件的运用(业余版和行业专用版)。

(四)能针对特定的施工项目或分项工作编制标书

最终标书:工程量清单已计量工作的计价;初步计价;付款凭证和备用款到位;影响预算转化为投标总价的因素和相关策略;投标后和合同授予前期手续。

(3) calculate unit rates for an element or trade section of a bill of quantities;

(4) produce a tender for a specific construction trade or element.

The delivery is classroom based and will take the format of various activities, which will be individually assessed, leading to a module qualification.

3. Curriculum Objectives

3.1　Know the basic information needed to produce a tender

Information: tender documentation; contract terms and conditions eg form of contract, contract period, commencement date, liquidated and ascertained damages, type of documentation, drawings and specification, schedule of rates; inspection of drawings; site investigations; information to collect during site visit; materials and sub-contract enquiries; labour constants; coverage rates; standard method of measurement(NRM) coverage rules; all in labour rates; company overheads; desired return; risk factors.

3.2　Understand how to use different types of tender documentation

Tender documentation: types; sections; methods of measurement.

Types: bill of quantities; drawings and specification; design and build; schedule of rates.

Sections: preliminaries; preambles; measured work; day works; prime cost sums; provisional sums and contingencies.

3.3　Be able to calculate unit rates for an element or trade section of a bill of quantities

Unit rates: material price, delivery or offloading costs; coverage rates; waste allowance; plant hire charges; labour constants and all in labour rates; the addition of the appropriate overhead percentage; checking procedures; use of ICT-based estimating packages(both non-specialist and industry specific).

3.4　Be able to produce a tender for a specific construction trade or element

Final tender: pricing the bill of quantities measured work sections; pricing of preliminaries; completion of PC and provisional sums; risk, commercial factors and strategies to convert the estimate into a tender sum; procedures used in receiving and opening tenders; post-tender pre-contract procedures.

四、课程内容与评估标准

学习项目 学完本课程,学生将能够:	评估合格标准 学生能够:
了解标书编制所需的基本信息。	在各种活动中学会标书编制所需的各种文件和信息。 设计一次模拟招标投标,标明开标和截止日期、合同期限、单个分项工程的工作进度表,未完工的损失赔偿。上一模块涉及的工地勘察信息和图纸会审;工程量清单的定义。 编制人工总价、核算覆盖率;对管理费,利润和风险进行解释
了解如何运用不同的标书文件	通过课堂活动了解工程量清单的标段
能进行工程量清单中分项工程或单个标段的费率计算。	设计课堂活动加深确立费率的相关知识

五、教学实施建议

(一)教学资源

绘图工具、计算器、SMM7/NRM2、固定尺寸的纸。

(二)教师素质

建筑专业资格证书、教师资格证书。

(三)教学场地及设备要求

教学活动将在四川建筑职业技术学院的指定教室内展开。

(四)教学条件

PPT、投影仪、黑板、画板、文具。

六、考核评价

教学活动的评估将根据教师对学生的观察、日常问答情况、课程作业评分来确定。

4. Curriculum contents and assessment criteria

Learning outcomes On successful completion of this unit a learner will:	Assessment criteria for pass The learner can:
Know the basic information needed to produce a tender	Produce through various activities the documents and information needed to produce a tender. Produce a model tender including start and completion dates, contract period, schedule of works for one construction element, damages for non-completion. Site visit information and inspection of drawings from a previous module. What is a bill of quantities. Activity to produce labour constants and coverage rates. Explain overheads, profit and risk
Understand how to use different types of tender documentation	Examine through activity the sections of the bill of bill of quantities
Be able to calculate unit rates for an element or trade section of a bill of quantities	Produce activities that develop knowledge of the build-up of unit rates

5. Suggestions for the Teaching:

5.1 Teaching Resources

Scale Rule, Calculator, SMM7/NRM2, Dimension paper.

5.2 Qualifications of the Teacher

Professional Chartered(Construction)qualifications and Teaching Quals.

5.3 Teaching Place

Teaching undertake at Sichuan College in classrooms allocated.

5.4 Teaching Facilities

PowerPoint/OHP/Chalk board/drawing boards/writing materials.

6. Evaluation/Examination

Assessment of activities undertaken by observation/question & answer/assessment of course work.

四 川 建 筑 职 业 技 术 学 院

课 程 标 准

课程名称：　　　　　毕业设计

课程代码：　　　　　150344

课程学分：　　　　　8

基本学时：　　　　　8周

适用专业：　　　　　工程造价

执 笔 人：　　　　　李剑心

编制单位：　　　　　建筑工程造价教研室

审　　核：　　　　　工程管理系(院)(签章)

批　　准：　　　　　教务处(盖章)

Sichuan College of Architectural Technology

Curriculum Standard

Curriculum Name: Graduation Design Project

Curriculum Code: 150344

Credits: 8

Teaching Hours: 8 Weeks

Applicable Specialty: Construction Cost

(Sino-British Cooperative Program)

Compiled by: Li Jianxin

On Behalf of: Teaching & Research Section of Construction Cost

Reviewed by: The Department of Engineering Management

Approved by: Teaching Affairs Office

《毕业设计》课程标准

一、课程概述

《毕业设计》是工程造价专业(中英合作办学项目)的实践必修课程。毕业设计是使学生在大学所学各类课程的基础上,对所学课程系统总结复习,并进行综合实践运用的过程。工程造价的计价与确定是一个技术性、经济性、政策性极强的复杂综合工作,通过毕业设计,学生要熟练掌握工程量清单的编制、工程量清单报价书的编制,熟悉与工程量清单计价有关的各种条件;熟悉相关施工方案及工艺要求,正确使用计价规范、计价定额、标准图集、建筑构造手册、建筑规范标准等,掌握工程造价指标分析方法,了解工程造价动态控制及管理的主要影响因素,增强实际工作能力。该课程属于专业课,在第6学期开设,先修课程为《工程造价概论》《建筑工程预算》《工程量清单计价》。

二、课程设计思路

该课程根据国家"十三五"规划、行指委的《工程造价教学基本要求》《四川建院工程造价人才培养方案》,以及社会需求进行设置。每一名学生必须独立完成一幢完整的房屋建筑(框架结构)的工程量清单和工程量清单报价书的编制工作。包括建筑工程、装饰装修工程及各单位工程对应的措施项目、其他项目等的工程量清单和清单报价的编制,并进行综合单价分析及造价指标分析。实训时间为8周,实训场地为普通教室。

三、课程目标

学生能够独立编制招标工程量清单、招标控制价、投标报价和结算价。

(一)知识目标

(1)建筑工程量计算;
(2)装饰工程量计算;
(3)工程量清单及相关知识;
(4)工程量清单计价及相关知识。

Graduation Design Project Curriculum Standard

1. Curriculum Description

This is a compulsory and practical course for construction cost specialty students (Sino-British Cooperative Program). Graduation project enables students to systematically review and summarize the units they have learned in colleges, and apply them in comprehensive practice. The valuation and determination of construction cost is a complex and comprehensive work that is subject to technicality, economy and policy. Through this course, students are required to master the preparation of bill of quantities, the compilation of quotation of bill of quantities, be familiar with various conditions related to the pricing for bill of quantities, and corresponding construction plans and techniques. Students shall be able to correctly and skillfully use valuation norms, valuation quotas, standard atlas, building construction manuals, building standards, etc. Students shall master the analysis method of construction cost indicators, understand the main factors affecting the dynamic control and management of construction cost, and improve their actual working ability. This course is a professional one, which is delivered in the sixth semester. The prerequisite courses were *Construction Cost*, *Budget of Construction Project* and *Pricing for Bill of Quantities*.

2. Curriculum Design Ideas

This course is designed in accordance with the 13*th Five-Year Plan of the State*, the *Basic Teaching Requirements of Construction Cost* by the Industry Educational Committee, the *Cultivation Plans of Construction Cost Talents in SCAT*, and the social needs analysis. Each student must independently complete the preparation of a complete bill of quantities for a building(frame structure) and a compilation of quotation for the bill of quantities, which encompasses works for construction project, building and decoration projects, as well as the corresponding tasks for each unit project and other relevant projects and engineering. The training time is 8 weeks and the venue is ordinary classrooms.

3. Curriculum Objectives

This course aims to help studentsindependently prepare and compile bill of quantities for tendering works, tender control prices, bidding quotations and settlement prices.

3. 1   Knowledge Objectives

After learning this course, students will be able to

(1)construction engineering works calculation;

(2)decoration works calculation;

(3)bill of quantities and related knowledge;

(4)pricing for bill of quantities and related knowledge.

(二)素质目标

(1)具备一定的辩证思维的能力;

(2)具有爱岗敬业的思想,实事求是的工作作风和创新意识;

(3)熟悉工程造价工作的有关政策法规;

(4)加强职业道德的意识,认识工程造价人员的执业权限与基本要求。

(三)能力目标

(1)计算建筑工程量;

(2)计算装饰工程量;

(3)编制工程量清单;

(4)编制招标控制价;

(5)编制投标报价;

(6)编制结算价。

四、课程内容与学时分配

教学内容与学时安排表

序号	教学情境/任务/项目/单元	教学目标	教学内容及学时分配		学时/周
			知识点	实践项目	
1	编制招标控制价、招标工程量清单	掌握编制招标控制价、招标工程量清单		1. 工程量计算(★)	2.5
				2. 软件计算工程量(★)	2
				3. 编制招标控制价(★)	1
				4. 编制招标工程量清单(★)	
2	编制投标报价环节	掌握编制投标报价		5. 提取计价工程量(★)	0.5
				6. 编制投标报价(★)	
3	编制工程结算	掌握编制结算报价		7. 编制工程结算造价(★)	0.5
4	分析总结			8. 分析总结(★)	1
5	成果验收			9. 成果验收(★)	0.5
合计					8
备注:重要知识点标注★					

3. 2 Quality Objectives

After learning this course, students will have the quality of

(1) critical thinking;

(2) taking devoted, responsible, practical work style and innovative consciousness;

(3) being familiar with the policies and laws for construction cost work;

(4) strengthened professional ethics, and understanding the responsibilities and permission and basic requirements for construction cost professionals.

3. 3 Ability Objectives

After learning this course, students will have the ability of

(1) calculating the construction project works;

(2) calculating the decoration project works;

(3) preparing and compile the bill of quantities;

(4) preparing and compile the tender control prices;

(5) preparing and compile the bidding quotations;

(6) preparing and compile the settlement prices.

4. Curriculum Contents and Teaching Hours

Table of Teaching Contents and Teaching Hours

Item	Topic	Teaching Objectives	Teaching Contents and Teaching Hours		
			Lesson Contents	Student Activities	Wks/ W
1	Prepare and compile tender control price and the corresponding bill of quantities	Master the preparation and compilation of tender control price and the corresponding bill of quantities		1. Calculating the engineering works(★)	2. 5
				2. Use software to calculate the works(★)	
				3. Prepare and compile the tender control price(★)	2
				4. Prepare and compile the bill of quantities for tendering(★)	1
2	Prepare and compile bidding quotations	Master the preparation and compilation of the bidding quotations		5. Collecting bill of quantities for pricing	0. 5
				6. Prepare and compile the bidding quotations(★)	
3	Prepare and compile project settlement and clearance	Master the preparation and compilation of settlement quotation		7. Prepare and compile the pricing for project settlement and clearance(★)	0. 5
4	Analysis and summary			8. Analysis and summary(★)	1
5	Project results are checked and accepted			9. Project results are checked and accepted (★)	0. 5
Total					8
Note: Mark the important points with ★					

五、教学实施建议

(一)教学参考资料

(1)四川省建设工程造价管理总站,四川省住房与城乡建设厅.《四川省建设工程工程量清单计价定额》,2020年。其中定额为:建筑与装饰工程(一)(二)、附录三本。

(2)中华人民共和国住房与城乡建设部.《建设工程工程量清单计价规范》(GB 50500—2013),中国计划出版社,2013。

(3)中华人民共和国住房与城乡建设部.《建筑与装饰工程工程量计算规范》(GB 50854—2013),中国计划出版社,2013。

(4)建安工程费组成文件、毕业设计相关资料。

(二)教师素质要求

任课老师应当具有工程造价专业知识,有造价专业执业证书者优先。

(三)教学场地、设施要求

普通教室。

六、考核评价

考核内容:毕业设计作业验收、毕业设计答辩。

考核方式:实际操作。

成绩构成:成绩构成比例原则按《四川建筑职业技术学院学生学业考核办法》执行。

5. Teaching Suggestions

5.1 Teaching Resources

(1) Ministry of Housing and Urban-Rural Development of Sichuan Province. *Building and Decoration Construction Engineering Part I, Part II and Appendix. The Quotas of Pricing for Bill of Quantities of Construction Engineering in Sichuan Province by Administrative Headquarter of Construction Cost Pricing in Sichuan Province*, 2020.

(2) MoHURD. *Specifications for Pricing for Bill of Quantities* (GB 50500—2013). China Planning Press, 2013.

(3) MoHURD. *Specifications for Bill of Quantity of Building and Decoration Engineering* (GB 50854—2013). China Planning Press, 2013.

(4) component documents for building and installation works, and relevant materials for graduation project

5.2 Teachers'Qualification

Teachers must have teacher qualification certificate, with construction cost major background and relevant professional qualification certificate and practical working and teaching experience.

5.3 Teaching Facilities

This course can be delivered in common classrooms.

6. Evaluation

The evaluation of this course includes the compiled files and the oral presentation, highlighting the practical operation. The total score is graded in accordance with the *Regulation of Academic Evaluation for Students in SCAT*.

四川建筑职业技术学院

课程标准

课程名称：　　　　　顶岗实习

课程代码：　　　　　150344

课程学分：　　　　　8

基本学时：　　　　　240

适用专业：　　工程造价(中英合作办学项目)

执笔人：　　　　　戴明元

编制单位：　　中外合作办学项目教研室

审　核：　　国际技术教育学院(签章)

批　准：　　　　教务处(盖章)

Sichuan College of Architectural Technology

Curriculum Standard

Curriculum Name: Post Practice

Curriculum Code: 150344

Credits: 8

Teaching Hours: 240

Applicable Specialty: Construction Cost

(Sino-British Cooperative Program)

Compiled by: Dai Mingyuan

On Behalf of: Teaching & Research Section of International

Joint Program

Reviewed by: The Department of ISTE

Approved by: Teaching Affairs Office

《顶岗实习》课程标准

一、课程概述

顶岗实习(毕业实习)是该专业的必修课程,在第6学期9~16周开设。毕业实习之前,学生已学完该专业设计的所有课程,该课程是学生在该专业的最后学习阶段。学生通过毕业实习(顶岗实习),将拓展加深所学专业知识,在建设工程预结算、建设工程招标、建筑工程投标报价、工程造价控制、工程造价审计等方面得到锻炼,为毕业后从事该专业相关工作打下坚实基础。

二、课程设计思路

该课程为实践课,依据该专业人才培养方案而制定。学生完成理论课程学习和校内实习及毕业设计即可到企业顶岗实习。实习分为两个阶段:熟悉情况阶段和具体工作阶段。熟悉情况阶段:在实习单位的工程技术人员和实习指导教师的安排与指导下,学生了解企业情况、工程情况、工作情况,包括了解企业文化,工程施工方案,施工组织设计,工程造价软件应用等。具体工作阶段:在熟悉情况的基础上,学生参与部分具体工作,如编制施工预算,施工图预算,工程量计算,参与编制施工组织设计,编制材料、制品供应计划,工程投标报价,整理交工资料以及编制计划和检查计划完成情况等。

三、课程目标

通过毕业实习,能将所学知识用于建设工程预结算、建设工程招标、建筑工程投标报价、工程造价控制、工程造价审计等工作。

(一)知识目标

(1)具有编制工程预算、计算工程量等相关知识;
(2)具有编制工程量清单、工程量清单投标报价等相关知识;
(3)具有涉外工程资料管理、涉外工程计量计价等相关知识;

Post Practice **Curriculum Standard**

1. Curriculum Description

Post Practice is the compulsory course for the specialty of construction cost (Sino-British Cooperative Program). It is delivered in the ninth to sixteenth week in the sixth semester. Before this course, students have finished all the courses for them, and this course is the last one for them before graduation. Through this course, students can further their study and put the knowledge they have learned in the college into practice, such as, building construction budget and settlement, construction bidding and pricing, construction pricing control and pricing for bills of quantities, construction cost audit and so on. This course will lay a strong foundation for the students to work on posts related with this specialty.

2. Curriculum Design Ideas

This course is designed based on the requirement of professional training program and it is a course of practice. After the students finish the theoretical courses and on-campus practice and graduation project, they can go to some companies to finish the post practice. It can be divided into two phases: one is to be familiar with the work situation and the other one is the specific work phase. In the familarity phase, students can understand the enterprise situation, the engineering situation, the work situation, including understanding the enterprise culture, the construction plan, the construction organization design, the application of the construction cost software, etc. under the arrangement and guidance of the engineering technicians or trainee instructors in the internship units. In the specific work phase, students can participate in some specific work, such as the preparation of construction budget, construction drawing budget, project volume calculation and they can participate in preparation of construction organization design, materials, product supply plans, project bidding quotations, sorting of materials and preparation plans and checking the completion of the plan based on what they have learned in the first phase.

3. Curriculum Objectives

The objectives of this course is to help students put the knowledge that they have learned in college into practice, such as, building construction budget and settlement, construction bidding and pricing, construction pricing control and pricing for bills of quantities, construction cost audit and so on.

3. 1 **Knowledge Objectives**

After learning this course, students will be able to

(1) have the basic knowledge of construction budge, construction volume calculation and so on;

(2) have the basic knowledge of bills of quantities, project bidding quotations and so on;

(3) have the basic knowledge of managing the overseas project data and pricing the overseas projects;

(4)具有编制招标控制价、编制工程结算等相关知识。

(二)素质目标

(1)具有强烈的爱国主义精神、社会责任感及良好的思想品德、社会公德和职业道德；

(2)具有求实创新的科学精神、刻苦钻研的实干精神及较强的团队协作意识；

(3)具有一定的审美情趣、艺术修养和文化品位，有较高的人文、科学素养；

(4)具有健康的身心素质、健全的人格、坚强的意志和乐观向上的精神风貌。

(三)能力目标

(1)有编制工程预算、计算工程量的能力；

(2)有编制工程量清单、具备工程量清单投标报价的能力；

(3)有编制招标控制价、编制工程结算的能力；

(4)有涉外工程资料管理、涉外工程计量计价的能力；

(5)有学习发展的能力，能根据工作和时代发展，不断更新知识和技能。

四、课程内容与学时分配

(一)课程内容

工程造价专业毕业生可能涉及的工作内容，具体工作内容由企业根据实际工作情况和工作岗位而定，但须服从实习单位工作安排。顶岗实习的基本任务是：在实习单位工程技术人员和指导教师的指导下，了解工程造价专业所涉及的工作内容，参与工程造价部分具体工作，实习结束时具备独立完成该岗位工作的专业技术能力。

(二)学时分配

240 学时(共 8 周)，由实习单位安排。

五、教学实施建议

(一)实习成果及质量要求

(1)实习日志：每天记录所见所感，工作过程、内容、质量要点及疑难问题。要求字迹工整清晰，每天内容翔实，图、文、表并茂，有体会有收获。(日志不少于 200 字)

(2)实习周志：每周记录所见所感，工作过程、内容、质量要点及疑难问题。内容翔实，图、文、表并茂，有体会有收获。(按照学院要求，周志在网上提交，不少于 500 字)

(4) have the basic knowledge of biding control price, project settlement and so on.

3. 2 Quality Objectives

After learning this course, students will have the quality of

(1) strong spirit of patriotism, social responsibility and good ideological character, social morality and professional ethics;

(2) having innovative and scientific spirit, hard-working spirit and strong sense of teamwork;

(3) having certain aesthetic taste, artistic cultivation and cultural taste, and having higher humanistic and scientific literacy;

(4) having a healthy physical and mental quality, a sound personality, strong will and optimistic spirit.

3. 3 Ability Objectives

After learning this course, students will have the ability of

(1) preparing construction budget and calculating engineering quantities;

(2) preparing bills of quantities and pricing or bidding of bills of quantities;

(3) preparing bidding price and project settlement;

(4) managing the foreign related project data and pricing the foreign related projects;

(5) learning and developing continuously and constantly updating knowledge and skills based on work experience and the development of the times.

4. Curriculum Contents and Teaching Hours

4. 1 Curriculum Contents

This course includes any work related with the specialty of construction cost. As for the specific work, it is arranged by the internship unit based on the practical situation or position. The students should follow the guidance or rules of the internship unit. The basic task of the post practice includes: be familiar with the work related with their major under the guidance of engineering technicians or trainee instructors; be involved in specific work related with their major and finally be able to finish the work independently.

4. 2 Teaching Hours

240 teaching hours in eight weeks arranged by internship unit.

5. Teaching Suggestions

5. 1 Practice Results and Quality Requirement

(1) Practice diary: Record what you see and feel, and work process, content, key points, and difficulties you meet every day. The handwriting should be neat and clean, and daily content should be specific with graphics and pictures. There should be a comment or a summary of your gain. (with not less than 200 words)

(2) Weekly summary: Record what you see and feel, and work process, content, key points, and difficulties you meet every week. There should be a comment or a summary of your gain. (According to the requirement of SCAT, the note should be submitted to a specific website with not less than 500 words)

(3)月总结:撰写本月实习情况,要求有工作岗位描述,所承担的具体工作,完成情况,完成质量及存在的问题,实习体会或收获,下一步的工作目标。内容翔实,图、文、表并茂,不少于 2 000字。(按照学院要求,月总结在网上提交,没有网络的地方提交纸质版)

(4)实习总结:对实习内容进行整理,要求有单位基本信息,实习岗位描述,任务完成情况,专业知识、社会能力、方法能力的提升,重点工作内容阐述,或对一些专门问题进行探讨。总结还应包括毕业实习是否到达预期期望,哪些地方需要改进,如何改进等。要求字迹工整美观、内容详略得当,重点突出、体会深刻、有独到见解。(实习总结不少于 3 000 字,可以网上提交,也可以提交纸质版)

(二)学院指导老师的主要任务

(1)保持与实习学生的信息沟通,每周必须联系学生,了解实习学生的工作生活情况,解决工作中的实际困难和技术问题。

(2)对实习学生的日志、周志、月总结、实习总结等及时批改,对发现的问题及时处理。

(3)根据国际学院的安排,不定期到实习单位检查实习学生实习情况。

(4)综合评价实习成绩,按规定时间提交教务处。

(三)毕业实习纪律要求

(1)实习地点在德阳的学生早上必须按项目上班时间到项目、按上课作息时间归寝室,否则按夜不归寝处理,毕业实习无周末。如需请假,假条需有项目领导、指导教师、辅导员三方签字。由系上安排的学生中途基本不允许自行再联系实习单位,除非所联系的实习单位是就业单位,并要求就业单位与实习指导教师取得联系确认,并上交实习申请表到系教务办公室备案。由单位安排的毕业实习学生服从单位安排。

(2)实习申请表、实习承诺书、实习工作册以班级为单位到国际学院学生办公室领取。实习日记每天一篇。上述三项缺少任何一样,其毕业实习成绩均直接评定为不合格。

(3) Monthly summary: Summarize the work in the month, and it should include position description, description of the specific tasks you have finished, the result of the task, any problems you have found or difficulty you have overcame, the thought about the internship or any gain you have got from the internship and the objective of next month. The content should be specific with graphics and pictures with no less than 2,000 words. (According to the requirement of SCAT, the summary should be submitted to a specific website. If there is no way to access to Internet, the written form is also acceptable)

(4) Internship summary: Summarize the process of internship and list the name of the internship unit, the internship position, the tasks that have been finished, the improvement of professional knowledge, social ability and working methods, the illustration of the main task or comment on some specific issues. Students should comment on the result of the internship, showing the conclusion about the expectation, the feedback about the things to be improved or the methods in improving themselves. The handwriting should be neat and clean and the content should contain summary or specific details. The focus should be obvious; the reflection should be profound and the comment should be initiative. (The summary should be no less than 3,000 words and it can delivered to the website or hand in the written form)

5.2　The main tasks of the instructors

(1) Maintain communication with students and contact students every week to understand the work or the life, solve practical difficulties and technical problems at work;

(2) Make timely revision for the students'internship dairy, weekly summary, monthly summary, internship summary and deal with problems on time;

(3) Make visit to the internship units from time to time to check the work of the internship students under the arrangement of the ISTE in SCAT;

(4) Make proper evaluation for the students'performance in the internship practice and submit the scores to teaching affairs office on time.

5.3　Graduate internship discipline requirements

(1) For students choosing internship units in Deyang, they must go to the units according to the project working hours. After work, they should go back to dormitory just as the normal schedule in dormitory requirement. Otherwise, it will be reported as the violation of the dormitory requirement and be dealt as absence from the college. There is no weekend during the internship weeks and if they ask for leave, it should be permitted by the leader of the project, the instructor and the college counselor. Generally speaking, the students in the internship units arranged by ISTE are not allowed to contact new internship unit by themselves unless the the internship unit will be the unit for their future work. In that case, the internship unit should contact the instructor and confirm this information, and at the same time they should submit an internship application to the teaching affairs office in ISTE. The students should follow the arrangement of the internship units.

(2) Internship application form, internship commitment, internship work book should be collected from the students office of ISTE by class. If any of the above three missing, their graduation internship are directly assessed as unqualified.

（3）所有实习学生每周必须通过短信、QQ、电子邮件等方式与实习指导教师汇报一周情况，否则按旷课一周处理。汇报内容必须包括学生姓名、学号、实习城市、实习项目部名称、项目部联系电话、本人联系电话、实习情况是否正常等。累计 3 周不汇报，按照相关文件精神要求，毕业实习成绩均直接评定为不及格。

（4）实习期间学生除遵守学校有关规定外，必须遵守所在实习单位的各项规章制度，包括作息制度、保密制度、资料和机具保管制度等。

（5）在实习期间请实习学生所在实习单位对学生考勤。实习期间原则上不请假。如有特殊情况需经实习单位批准。未经批准不得擅自离开岗位。学生请病假必须有医生的病假证明。学生实习期间必须住在实习单位安排的宿舍。每晚 10 点以前学生必须返回住宿地点。

（6）学生在实习期间发扬艰苦朴素的生活作风，衣着发式应符合"文明、安全、方便操作"的要求，学生在实习期间，严禁酗酒、闹事、赌博以及其他不文明礼貌的活动，严禁参加违反"四项基本原则"、破坏安定团结的非正常活动；不准进舞厅、卡拉 OK 厅、网吧和其他各类不适合学生的场所。

（7）学生在实习期间对工作应认真负责、谦虚谨慎。所从事的工作必须在工地工长和技术人员指导下进行。每项工作成果要请有关人员审核。不得擅自处理重大技术问题，防止质量事故和安全事故的发生。

（8）根据现场条件，因地制宜地开展"两个文明建设"的有关活动和文体活动，但必须在工作之余有组织地进行，不能影响工作的正常进行，任何情况下严禁下河游泳。

以上纪律，实习学生必须严格遵守。严重违反者，取消其毕业实习资格，立即遣送回学校处理。

六、考核评价

《毕业实习》是一门独立的实践教学课程，成绩评定根据以下多种因素采用五级制记分，分为优、良、中、合格、不合格五个层次，由企业考核成绩和学院考核成绩构成，各占 50%。任何一方考核不合格总成绩为不合格。

（1）实习出勤情况：由实习单位评定，出勤率低于三分之一者按照学院相关文件取消毕业实习资格，请实习单位有关部门或领导考勤；实习结束时，在实习报告考勤表上注明事假、病假、旷工、迟到、早退次数。

(3) All internship students must make a report to the internship instructor through Text message, QQ, e-mail and other means every week to summarize their life or work; otherwise it will be recorded as absence from work for one week. The report content must include: name, school number, internship city, internship project department name, project department contact number, students'contact number and the description of the internship. If there are in all 3 weeks without report, their graduation internship are directly assessed as unqualified in accordance with certain rules of SCAT.

(4) In addition to complying with the relevant rules of our college, students must abide the rules and regulations of their internship units, including the work time, confidentiality, information and equipment storage, etc.

(5) The internship units should check on their work attendance during the internship. Basically, students should not ask for leave from the internship units. If there is something necessary for asking for leave, the internship units should give permission. Students should not leave the internship position without getting permission. The sick leave should be testified by the certificate from hospital. Students should log in the dormitory of the internship units and should go back to the dormitory before 10 p. m. everyday.

(6) During the internship, students should carry forward the hard and simple style of life. The clothing should meet the requirements of "civilized, safe, convenient at work". Students are strictly prohibited from drinking, rioting, gambling and other uncivilized activities, and they are strictly prohibited from participating in activities against the "four basic principles" and violating the stability and unity of the society. Students are strictly prohibited from ballrooms, Karaoke rooms, Internet cafes and other types of places that are not suitable for students.

(7) Students should be responsible, modest and prudent in their work during their internships. The work performed must be carried out under the guidance of the site foreman and technicians. Each work outcome should be reviewed by the person concerned. No major technical problems shall be handled without authorization to prevent the occurrence of quality accidents and safety accidents.

(8) Carry out the "two civilization construction" related activities based on the condition of project department, but the activities must be organized without influencing their work. Students are strictly prohibited from swimming in the river.

Students must strictly abide the above discipline. Students violating them seriously will be called off from the internship unit and be settled by the college immediately.

6. Evaluation

Graduation internship is an independent practical course, so it is graded into five levels: excellent, good, medium, qualified, unqualified based on some factors. The grade consists of two parts (50% for each): the evaluation from the internship unit and from the college. If any of the evaluation is assessed as unqualified, the final grade will be marked as unqualified.

(1) Internship attendance: It is assessed by the internship unit. If the attendance rate is less than one-third, students will be called off from the internship unit based on the relevant requirement of the college. The internship unit should indicate the number of asking for leave, sick leave, absence, lateness, and early departure on the internship attendance report form.

(2)实习日记:字迹是否工整清晰,每天内容是否翔实、图、文、表并茂,是否有体会有收获。

(3)实习报告:字迹是否工整美观、内容详略是否得当(有单位概况、有专题内容),文笔是否流畅、重点是否突出、体会是否深刻、有无独到见解。

(4)实习单位鉴定意见(企业考核成绩):由企业填写,根据实习生的思想表现、专业能力、工作态度综合评定。企业指导教师签名,单位盖章。

(5)学院鉴定意见(学院考核成绩):由学院指导教师根据学生提交的资料(日志、周志、月总结、实习总结),每周交流沟通情况,不定期巡查记录,单位鉴定意见等综合评定。学院指导教师签名。

(2) Internship Diary: It is assessed by whether the handwriting is clear, whether the content of the day is specific with graphics or pictures and whether there is a gain in the process.

(3) Internship summary: It is assessed by whether the handwriting is clear, whether the content contains summary and details, such as the introduction to the internship unit or the comment on some specific issue, whether the writing is easy to read, whether it contains clear main idea, deep thought and initiative perspective.

(4) Internship unit assessment: It is comprehensively assessed by the internship unit based on students'performance, professional ability and work attitude. The assessment should be signed by the trainee instructor and sealed by the internship unit.

(5) College assessment: It is assessed by the instructor based on the documents they have submitted (internship diary, weekly summary, monthly summary and internship summary), the daily communication they have made with the instructor, the inspection records, the assessment of the internship unit, etc. The instructor should sign their name on the assessment form.

四川建筑职业技术学院

课 程 标 准

课程名称： FIDIC 合同条件与索赔

课程代码： 150353

课程学分： 2

基本学时： 32

适用专业： 工程造价（中英合作办学项目）

执 笔 人： 黄恒振

编制单位： 建筑工程管理教研室

审　　核： 工程管理系（院）（签章）

批　　准： 教务处（盖章）

Sichuan College of Architectural Technology

Curriculum Standard

Curriculum Name: FIDIC Contract Conditions and Claims

Curriculum Code: 150353

Credits: 2

Teaching Hours: 32

Applicable Specialty: Construction Cost

(Sino-British Cooperative Program)

Compiled by: Huang Hengzhen

On Behalf of: Teaching & Research Section of

Engineering Management

Reviewed by: The Department of Engineering Management

Approved by: Teaching Affairs Office

《FIDIC 合同条件与索赔》课程标准

一、课程概述

《FIDIC 合同条件与索赔》是为工程造价专业(中英合作办学项目)开设的一门重要的专业基础限选课程。本课程的主要任务是使学生掌握 FIDIC 合同条件,特别是施工合同条件的具体内容,了解基于 FIDIC 的合同管理的基本框架,并应用所学开展合同条件编制、工程变更及索赔等合同管理工作。

先修课程:《招投标与合同管理》《项目管理》等。

本课程需要安排在第 4 学期开设。

二、课程设计思路

采用国际知名的合同条件是中国建筑企业进入国际市场和国外建筑企业在中国开展工程建设的通行做法。随着建筑工程管理的国际化和规范化,掌握国际工程合同条件并加以应用成为建筑工程管理人才的基本要求。基于此背景,本课程的总体设计思路是将传统的以知识为主线构建的学科型课程模式,转变为以能力为主线的、面向应用的课程模式,充分考虑学生的学习规律和职业生涯发展,通过贯穿全课程的工程实践案例,让学生对国际合同管理的基本内容、FIDIC 合同条件的索赔条款、国际工程惯例及索赔工作有深入的认识,注重课程的知识性、实用性和发展性。

三、课程目标

(一) 知识目标

通过本课程的学习,学生可以较为全面地掌握 FIDIC 合同条件的具体规定,掌握合同管理的相关理论知识,基于 FIDIC 的国际工程招标投标的基本程序、环节和关键节点;掌握索赔的基本理论知识及 FIDIC 合同条件中的索赔条款的内容。

(二) 素质目标

(1)培养学生的专业素养、自觉遵守行业规范,能正确领会 FIDIC 合同条件所蕴含的工程合同管理的国际理念,并在实践工作中遵行;

FIDIC Contract Conditions and Claims Curriculum Standard

1. Curriculum Description

FIDIC Contract Conditions and Claims is an important basic elective course for the specialty of construction cost(Sino-British Cooperative Program). The main task of this course is to enable students to master FIDIC contract conditions, especially the specific content of conditions of contact for construction, understand the basic framework of contract management based on FIDIC, and apply what they have learned into carrying out contract management work, such as contract conditions preparation, engineering changes and claims.

The delivered courses are *Tendering and Contract Management*, *Project Management* and so on.

This course is delivered in the fourth semester.

2. Curriculum Design Ideas

Adopting internationally renowned contract conditions is a common practice for Chinese construction enterprises to enter the international market and foreign construction enterprises to carry out construction projects. With the internationalization and standardization of construction engineering management, mastering the conditions of international engineering contracts and applying them have become the basic requirements for management talents in construction engineering. Based on this, the overall design idea of this course is to transform the traditional knowledge-based curriculum model into a competency-based and application-oriented curriculum model, fully considering students' learning habits and career development. Through practicing cases in this course, students can have a deep understanding of the basic content of international contract management, FIDIC claim clause, international engineering practices and claims work, and pay attention to the knowledge, practicability and development of the course.

3. Curriculum Objectives

3.1 Knowledge Objectives

After learning this course, students will be able to master the specific provisions of FIDIC contract conditions, the relevant theory of contract management, the basic procedures, links and key points of international project bidding based on FIDIC, the basic theoretical knowledge of claims and the content of claim clauses in FIDIC contract conditions.

3.2 Quality Objectives

After learning this course, students will have the quality of

(1) good professional spirit, industry norms, and understanding and abiding the international concepts of engineering contract management contained in FIDIC contract conditions;

（2）通人情、明事理，能恰当处理合同与索赔管理工作中的人际关系；

（3）培养讲诚信、重承诺、肯吃苦、勇于负责的道德品质和爱岗敬业的工作态度；

（4）具备良好的人文和心理素质，愿与他人合作，在合同管理岗位工作中，能独立思考，有不断创新的精神。

（三）能力目标

通过本课程的学习，学生应初步具备工程合同管理的国际视野，熟悉国际上通行的工程合同模式，能够运用 FIDIC 合同条件的基本思想和方法处理实践中的合同与索赔管理问题。

四、课程内容与学时分配

教学内容与学时安排表

序号	教学情境/任务/项目/单元	教学目标	教学内容及学时分配			
			知识点	学时	实践项目	学时
1	FIDIC 合同综述	了解 FIDIC 的发展历史及重要性；理解 FIDIC 合同的独有特点；掌握 FIDIC 合同的最大特点	1. FIDIC 合同的发展历史和地位	1		
			2. FIDIC 合同独有的特点			
2	FIDIC 合同与 ICE 合同	了解 FIDIC 与 ICE 的历史，法律体系，合同文件优先次序；理解 FIDIC 合同与 ICE 合同的本质区别；掌握相关条款（语言和法律、延期赔偿费等）	3. FIDIC 合同与 ICE 合同的关联	2		
			4. FIDIC 合同与 ICE 合同的不同之处			
3	FIDIC 合同的宗旨	了解默示的含义和内容，合同内容，施工合同的原则，对实际项目合同的调整；理解业主的资金安排，现场作业和施工方法等内容；掌握承包商的违约因素，终止日的估价，终止后的付款，业主的违约事件终止时的付款，承包商暂停工作的权力及复工等	5. FIDIC 合同的宗旨	3		
			6. FIDIC 合同独有的特点和内容			

(2) being sociable and reasonable, and being able to properly handle the interpersonal relationship in maintain contract and claim management;

(3) integrity, commitment, willingness to endure hardships and undertaking responsibility and dedication to work;

(4) strong basis in humanistic and psychological knowledge, cooperating with others willingly, thinking independently and continuous innovation in the work of contract management positions.

3.3 Ability Objectives

After learning this course, students will have the international vision of engineering contract management, be familiar with the international common engineering contract mode, and be able to use the basic ideas and methods of FIDIC contract conditions to deal with the contract and claim management problems in practice.

4. Curriculum Contents and Teaching Hours

Table of Teaching Contents and Teaching Hours

Item	Topic	Teaching Objectives	Lesson Contents	Hrs	Student Activities	Hrs
1	Overview of FIDIC contract	Understand the history and importance of FIDIC; Understand the unique characteristics of FIDIC contract; Master the greatest characteristics of FIDIC contract	1. The development history and status of FIDIC contract	1		
			2. The unique characteristics of FIDIC contract			
2	FIDIC contract and ICE contract	Understand the history, legal system and contract documents of FIDIC and ICE; Understand the essential difference between FIDIC contract and ICE contract; Master relevant clauses (language and law, deferred compensation, etc.)	3. The relationship between FIDIC contract and ICE contract	2		
			4. Differences between FIDIC contract and ICE contract			
3	Purpose of FIDIC contract	Understand the implied meaning and content, contract content, construction contract principles, and adjustment to the actual project contract; Understand the employer's financial arrangements, site operations and construction methods, etc; Master the factors of contractor's default, the valuation at date of termination, the payment after termination, the payment at date of the employer' default termination, the contractor's right to suspend work and resume work, etc.	5. Purpose of FIDIC contract	3		
			6. The unique characteristics and contents of FIDIC contract			

序号	教学情境/任务/项目/单元	教学目标	教学内容及学时分配			
			知识点	学时	实践项目	学时
4	咨询工程师的地位和作用	了解咨询工程师代表的组成,任命助理,图纸和文件的保管与提供,现场要保留的一套图纸,工程进展中断,图纸误期和误期的费用,其他各种费用的承担情况; 理解 FIDIC 合同里对咨询工程师的职责和权力,咨询工程师的权力委托,变更及示履行的义务,咨询工程师的责任; 掌握业主选择设计者时的原则,咨询工程师应具有的思想品质	7. 业主选择设计者时的原则 8. 咨询工程师的责任和义务	2		
5	业主的义务	了解业主(项目)在工程承包业发展中的作用,咨询工程师易出的错误; 理解施工进度; 掌握业主在合同中的主要义务	9. 业主的义务	1		
6	承包商的权利和义务	了解投票阶段、施工阶段和竣工阶段的相关规定; 理解中标书不宜滥用的原因,承包商在中标通知书与开工令之间这段时间应做的工作,FIDIC 对工程的检验和保修的要求; 掌握承包商的权力和义务包含的三个阶段	10. 承包商的权利和义务的三个阶段	2		

continued

Item	Topic	Teaching Objectives	Teaching Contents and Teaching Hours			
			Lesson Contents	Hrs	Student Activities	Hrs
4	The status and role of consulting engineers	Understand the composition of representatives of consulting engineers, appoint assistants, custody and supply of drawings and documents, a set of drawings to be retained on site, interruption of project progress, delayed drawings and its cost, and other expenses; Understand the responsibilities and rights of consulting engineers in FIDIC contracts, Consulting Engineer's Power Entrustment, Change and Obligation; Master the principle of choosing designer for employer, and ideological qualities consulting engineer should have	7. Principles for employers to choose designers; 8. Responsibilities and obligations of consulting engineers	2		
5	Obligations of employer	Understand the role of employers (projects) in the development of engineering contracting industry and the mistakes easily made by consulting engineers; Understand construction progress; Master the main obligations of the employer in the contract	9. Obligations of employer			
6	Contractor's rights and obligations	Understand the relevant provisions of the voting stage, the construction stage and the completion stage. Understand the reasons for the inappropriate abuse of the letter of acceptance , what the contractor should do during the period between letter of acceptance and the order of commencement, and FIDIC's requirements for the inspection and maintenance of the project. Master three stages contained in contractor's rights and obligations	10. Three stages of contractor's rights and obligations	2		

序号	教学情境/任务/项目/单元	教学目标	教学内容及学时分配			
			知识点	学时	实践项目	学时
7	资质审查	了解资质审查的重要性及与公开招标的相似性； 理解资质审查的两种形式； 掌握预审和后审,资审人员的素质要求	11. 资质审查的重要性 12. 资质审查的两种形式	1		
8	B. Q. 单	了解 B. Q. 单的解释结构形式； 理解 FIDIC 合同最大特点,FIDIC 合同的签约总价； 掌握 B. Q. 单对于承包商的重要性	13. B. Q. 单的定义及意义 14. B. Q. 单的构成	1		
9	FIDIC 合同是单价合同	了解对 FIDIC 合同是单价合同的解释； 理解 B. Q. 单支付方式的特点,B. Q. 单在合同中的主要作用,什么是工料数； 掌握合同单价的地位高于一切	15. FIDIC 合同是单价合同	2		
10	投标技巧——不平衡报价	了解承包商怎样获得项目,承包商报价人员的水平； 理解承包商在投标报价的技术性,投标报价的关键,什么是早收钱； 掌握报价技巧	16. 投标技巧	1		
11	费用分析与成本控制	了解直接费所包含的项目,间接费及利润； 理解工费、料费、机械设备折旧费、工程开办费、管理费、自有资金占用费； 掌握利润与风险的关系	17. 费用分析与成本控制	1		

continued

Item	Topic	Teaching Objectives	Teaching Contents and Teaching Hours			
			Lesson Contents	Hrs	Student Activities	Hrs
7	Qualification examination	Understand the importance of qualification examination and its similarity with public tender; Understand the two forms of qualification examination; Master preliminary and delayed examination and the quality requirements of examination personnel	11. The importance of qualification examination	1		
			12. Two forms of qualification examination			
8	Bills of Quantities	Understand the interpretation structure of B. Q.; Understand the greatest characteristics of FIDIC contract, the total contract price of FIDIC; Master the importance of B. Q. for contractors	13. Definition and significance of B. Q.	1		
			14. Composition of B. Q.			
9	FIDIC contract is unit price contract	Understand the interpretation of FIDIC contract as unit price contract; Understand the characteristics of B. Q. payment mode, the main role of B. Q. in the contract, what is the quantity of labor and material; Understand that the unit price of the contract is above everything else	15. FIDIC contract is unit price contract	2		
10	Bidding Skills— Unbalanced quotation	Understand how the contractor obtains the project and the level of contractor quotation personnel; Understand the strategies of contractor's bidding quotation, the key to bidding quotation, and what is early collection of money; Master quotation skills	16. Bidding skills	1		
11	Cost Analysis and Cost Control	Understand the items included in direct fees, indirect fees and profits; Understand the cost of labor, materials, depreciation cost of machinery and equipment, project start-up costs, management fees, occupancy fees of own funds; Understand the relationship between profit and risk	17. Cost analysis and cost control	1		

序号	教学情境/任务/项目/单元	教学目标	教学内容及学时分配			
			知识点	学时	实践项目	学时
12	不可抗力	了解"不可抗力"的由来; 理解不可抗力的一般理解,不可抗力的定义,不可抗力引起的后果; 掌握不可抗力的 FIDIC 解释及包括的方面	18. 不可抗力	1		
13	工程变更令	了解工程变更令创收为什么是项目利润的中心之一; 理解变更的估价,计日工; 掌握咨询工程师确定费率的权力,变更超过 15%,估价的内容	19. 工程变更令	1		
14	暂定金额与计日工	了解直接费所包含的项目,间接费及利润; 理解暂定金额的定义,凭证的出示; 掌握暂定金额的使用	20. 暂定金额与计日工	1		
15	不可预见费	了解承包商在投标时如何进行分析; 理解不可预见费的定义	21. 不可预见费	1		
16	价格调整	了解价格调整的原因,调价争端的解决; 理解调价公式	22. 价格调整	1		
17	索赔与波纹理论	了解索赔与反索赔的情况; 理解索赔的根据及方式,索赔中波纹理论的运用; 掌握索赔的程序	23. 索赔 24. 波纹理论	1		

continued

Item	Topic	Teaching Objectives	Teaching Contents and Teaching Hours			
			Lesson Contents	Hrs	Student Activities	Hrs
12	Force Majeure	Understand the origin of Force Majeure; Understand the general interpretation of force majeure, the definition of force majeure, and the consequences of force majeure; Master the interpretation of Force Majeure in FIDIC and its aspects	18. Force Majeure	1		
13	Engineering Change Order	Understand why engineering change order revenue is one of the centers of project profits; Understand the valuation of changes and the daywork; Master the right of consultant engineer to determine the rate, change over 15%, the content of evaluation	19. Engineering change order	1		
14	Tentative sums and the daywork	Understand the items included in direct fees, indirect fees and profits; Understand the definition of Tentative sums and the presentation of vouchers; Master the Use of Tentative sums	20. Tentative sums and the daywork	1		
15	Unforeseeable expenses	Understand how the contractor conducts analysis in bidding; Understand the definition of Unforeseeable Expenses	21. Unforeseeable expenses	1		
16	Price adjustment	Understand the reasons for price adjustment and settlement of price adjustment disputes; Understand price adjustment formula	22. Price adjustment	1		
17	Claims and Ripple Theory	Understand the situation of claims and counter-claims; Understand the basis and mode of claims and the application of Ripple Theory in claims; Master the claim procedures	23. Claims 24. Ripple Theory	1		

续表

序号	教学情境/任务/项目/单元	教学目标	教学内容及学时分配			
			知识点	学时	实践项目	学时
18	索赔条款	了解国际工程中的重大风险,索赔程序; 理解风险索赔时的主要合同条款; 掌握对特殊风险不承担责任,特殊风险包括的方面	25. 风险索赔时的主要合同条款	3		
19	索赔案例分析	掌握索赔条款的应用及工程索赔的计算方法	26. 常见典型索赔案例分析	4		
20	国际仲裁	了解相关仲裁条款; 理解仲裁费用; 掌握仲裁方式解决争端的策略,律师付费的常用三种选择	27. 国际仲裁	2		
合计				32		
备注						

五、教学实施建议

(一)教学参考资料

(1)建议使用教材:张水波、陈勇强.《国际工程合同管理》,中国建筑工业出版社,2011。

(2)参考资料:

①张水波、何伯森.《FIDIC 新版合同条件导读与解析》,中国建筑工业出版社,2003。

②何伯森.《工程项目管理的国际惯例》,中国建筑工业出版社,2007。

(二)教师素质要求

任课教师应具备较强的外文(英文)阅读能力,能够搜集并阅读国外主流的合同条件文本,能够收集国外基于特定合同条件的工程实践案例,并将其翻译为中文以做课堂案例教学需要;同时任课教师应具有深厚的合同法及合同管理知识,最好曾参与工程项目合同管理实践。

(三)教学场地、设施要求

本课程可在多媒体教室或一般教室开展教学工作。

continued

Item	Topic	Teaching Objectives	Teaching Contents and Teaching Hours			
			Lesson Contents	Hrs	Student Activities	Hrs
18	Claim clause	Understand major risks in international projects and claim procedures; Understand the main contract terms in risk claims; Master the aspects of special risks that are not responsible	25. Major contract terms in risk claims	3		
19	Cases analysis of Claims	Master the application of claim clause and calculation method of Engineering claims	26. Typical cases analysis of claims	4		
20	International Arbitration	Understand the relevant arbitration clauses; Understand arbitration costs; Master approaches to settle the disputes by arbitration, and three common options used by lawyers	27. International arbitration	2		
Total				32		
Note						

5. Teaching Suggestions

5.1 Teaching Resources

(1) Suggestions on the use of textbooks and teaching materials:

Zhang Shuibo, Chen Yongqiang, *International Engineering Contract Management*. China Architecture & Building Press, 2011.

(2) References:

①Zhang Shuibo, He Bosen. *FIDIC, New Edition Conditions of Contract Reading and Analysis*. China Architecture & Building Press, 2003.

②He Bosen. *International Practice of Project Management*, China Architecture & Building Press, 2007.

5.2 Teachers' Qualification

Teachers should have strong reading ability in English, so that they can select and read the mainstream texts of contract conditions abroad, collect foreign practical engineering cases based on specific contract conditions, and translate them into Chinese for classroom teaching; at the same time, teachers should have profound knowledge of contract law and contract management, and teachers having project contract management experience are preferred.

5.3 Teaching Facilities

This course can be delivered in multimedia classrooms or common classrooms.

六、考核评价

(一) 考核方式

考试。

(二) 考核说明

(1) 考试方式:闭卷。

(2) 试卷中各种题型所占比例:单项选择题:多项选择题:简答题:案例分析题 = 20:20:40:20。

(3) 考试总成绩的构成。平时成绩和期末考试成绩所占比例:3:7。

6. Evaluation

6.1 Evaluation Method

Examination.

6.2 Evaluation Instruction

(1) Examination without reference materials.

(2) Proportion of various types of questions in examination papers: Single-choice Questions : Multiple-choice Questions: Short Answer Questions: Case Analysis = 20: 20: 40: 20。

(3) The score of the course: The daily performance accounts for 30% of the total score and the final examination accounts for 70%.

四川建筑职业技术学院

课程标准

课程名称：　　　　　建筑英语

课程代码：　　　　　210161

课程学分：　　　　　2

基本学时：　　　　　32

适用专业：　工程造价（中英合作办学项目）

执 笔 人：　　　　　邓冬至

编制单位：　　　应用英语教研室

审　　核：　国际技术教育学院（签章）

批　　准：　　　教务处（盖章）

Sichuan College of Architectural Technology

Curriculum Standard

Curriculum Name: English for Building and Construction Engineering

Curriculum Code: 210161

Credits: 2

Teaching Hours: 32

Applicable Specialty: Construction Cost

(Sino-British Cooperative Program)

Compiled by: Deng Dongzhi

On Behalf of: Teaching & Research Section of Applied English

Reviewed by: The Department of ISTE

Approved by: Teaching Affairs Office

《建筑英语》课程标准

一、课程概述

本课程是工程造价专业(中英合作办学项目)的一门主干课程,该门课程用英语讲授建筑知识,使学生既能运用英语,也能学习建筑知识。因此,学习该门课程,学生应具备较强的英语应用能力和基本的建筑知识(先修课程如《建筑工程概论》《建筑材料》等)。课程以建筑施工为主线,覆盖建筑材料、建筑工艺、项目管理、室内装饰、室外工程、安装工程、园林绿化、施工报告及施工日志写作等。

该课程的任务是培养学生"会英语,懂工程",培养学生在建筑领域使用业务英语的能力,着重培养学生对建筑工程英文资料的阅读能力和翻译能力以及涉外工程现场口译能力和交际能力。结合学生的专业知识以及学生毕业后的工作实际,力求向学生提供未来工作岗位所需要的专业英语知识和技能。

二、课程设计思路

本课程是建筑知识与语言技能的结合,是紧紧围绕该专业建设与发展的一门主干课程。从建筑管理层面,该课程涉及施工前期准备,施工过程控制,施工后期维护;从技术层面,该课程涉及基础施工、主体施工、装饰及室外工程,展现给学生一个完整的施工流程;从语言层面,该课程着重培养学生在施工现场的交流、语言的实际应用能力。建筑行业是国民经济的支柱产业之一,对国民经济的拉动作用日益显著。当前我国"一带一路"倡议的发展,既给中国建筑业带来了难得的发展机遇,又带来了不可避免的冲击和挑战。建筑业要直接面对国际承包商的竞争,要在激烈的国际工程承包市场占有一席之地,提高人才培养质量是当务之急。在此背景下,开发建筑英语课程,既解决了建筑业涉外工程"懂英语的不懂工程,懂工程的不懂英语"的尴尬,又提升了学生就业竞争力,突出了专业特色。

三、课程目标

(一)知识目标

(1)了解土木工程的概念及土木工程师的职责;

(2)了解施工过程、熟悉混凝土施工(包括混凝土基础、柱、梁、板的施工);

(3)了解预制混凝土构件及钢结构的制作与安装;

(4)了解围护结构施工、室内装饰及室外工程施工;

English for Building and Construction Engineering **Curriculum Standard**

1. Curriculum Description

This course is a compulsory content-based English course for construction cost specialty students (Sino-British Cooperative Program), which aims to help students acquire both strong English application skills and basic architectural knowledge. The prerequisite courses were *Introduction to Construction Engineering*, *Building Materials*, etc. This course covers building materials, construction technology, project management, interior decoration, outdoors engineering, installation project, landscaping, construction report and construction log writing.

The aim of this course is to train students to "speak English and understand engineering", develop students' ability to use ESP in the field of construction, with emphasis on cultivating students' reading and translation skills for English construction materials, as well as the foreign on-site interpretation and communication ability.

2. Curriculum Design Ideas

This course is designed on ideas of incorporating architectural knowledge and language skills. From the construction management level, this course involves pre-site preparation, construction process control, and post-construction maintenance; from the technical level, it involves basic construction, main building construction, decoration and outdoor project, showing students a complete construction process; from the language level, it focuses on the training of students in the construction site communication as well as the practical application of language. In the context of China's "Belt and Road" strategic development, this course solves the dilemma of "English majors do not understand construction while architects do not understand English", and enhances the competitiveness of graduates.

3. Curriculum Objectives

3.1 Knowledge Objectives

After learning this course, students will be able to

(1) understand the concept of civil engineering and the duties of a civil engineer;

(2) understand the construction process, and be familiar with the concrete construction (including the construction of concrete foundations, columns, beams, and plates);

(3) understand the production and installation of prefabricated concrete components and steel structures;

(4) understand the construction of enclosure structure, interior decoration and outdoor engineering construction;

（5）掌握基本的施工报告及施工日志的写作；

（6）掌握所学的建筑英语专业词汇及术语。

（二）素质目标

（1）培养学生良好的职业道德、坚定的爱国主义思想；

（2）培养学生科学严谨的工作态度和创造性工作能力；

（3）培养学生热爱专业、热爱本职工作的精神；

（4）培养学生一丝不苟的学习态度和工作作风。

（三）能力目标

（1）能从事涉外施工现场翻译及相关岗位的语言交流；

（2）能基本从事涉外工程项目施工及管理；

（3）能基本识读英文标注的施工图及施工文件。

四、课程内容与学时分配

教学内容与学时安排表

序号	教学情境/任务/项目/单元	教学目标	教学内容及学时分配			
			知识点	学时	实践项目	学时
1	单元1	1. 现场使用的图纸和文件； 2. 建筑物的组成部分； 3. 建筑元素	1. 工作会话 2. 现场人员，图纸和文件 3. 建筑工程的概念	4	现场使用的图纸和文件	1
2	单元2	1. 施工前活动； 2. 现场调查； 3. 现场分析中应包含的内容； 4. 初期现场工作说明中的术语	4. 工作会话 5. 初期现场工作的概念 6. 现场分析应包含的内容	4	初期现场工作	1
3	单元3	1. 基础类型； 2. 基础建设和活动； 3. 基础移动	7. 工作会话 8. 基础 9. 基础移动	4	基础建筑	1
4	单元4	1. 混凝土施工； 2. 框架结构； 3. 基本结构原则	10. 工作会话 11. 框架结构 12. 基本结构原则	4	框架结构建筑	1

(5) master the writing of basic construction reports and construction logs;

(6) master the professional vocabulary and terminology of architectural English.

3.2 Quality Objectives

After learning this course, students will have the quality of

(1) good professional ethics and firm patriotism;

(2) scientific and rigorous work attitude and creative work skills;

(3) continuous interest and love of the profession and their possible work;

(4) meticulous learning attitude and working style.

3.3 Ability Objectives

After learning this course, students will have the ability of

(1) being competent in foreign construction site translation and other job-related communication;

(2) being able to complete engineering project tasks in foreign construction and management;

(3) being able to read English construction drawings and documents.

4. Curriculum Contents and Teaching Hours

Table of Teaching Contents and Teaching Hours

Item	Topic	Teaching Objectives	Teaching Contents and Teaching Hours			
			Teaching Contents	Hrs	Student Activities	Hrs
1	Unit 1	1. Drawings and documents used on site; 2. Constituent parts of a building; 3. Building elements	1. Workplace Conversation	4	Drawings and documents used on site	1
			2. People who work on Site, Drawings and Documents			
			3. Concept of Building Construction			
2	Unit 2	1. Pre-site activities; 2. Site investigation; 3. What should be included in site analysis; 4. Terminology in the description of preliminary site work	4. Workplace Conversation	4	Preliminary Site Work	1
			5. Concept of Preliminary Site Work			
			6. What should be included in site analysis			
3	Unit 3	1. Foundation types; 2. Foundation construction and activities; 3. Foundation movement	7. Workplace Conversation	4	Foundation construction	1
			8. Foundations			
			9. Foundation Movement			
4	Unit 4	1. Concrete construction; 2. Framed structure construction; 3. Basic structural principles	10. Workplace Conversation	4	Framed structure buildings	1
			11. Framed structure			
			12. Basic structural principles			

序号	教学情境/任务/项目/单元	教学目标	教学内容及学时分配			
			知识点	学时	实践项目	学时
5	单元5	1. 解释屋顶的功能和性能要求； 2. 了解平顶和倾斜屋顶的基本概念； 3. 描述不同类型的屋顶支柱	13. 工作会话 14. 屋顶建筑 15. 屋顶支柱	4	屋顶建筑	1
6	单元6	1. 描述砖块的铺设； 2. 了解常用的各类墙体； 3. 解释承重壁	16. 工作会话 17. 墙体建筑 18. 承重墙	4	墙体建筑	1
7	总结			2		
合计				26		6
备注						

五、教学实施建议

(一)教学参考资料

(1)建议使用教材：

戴明元.《建筑工程英语》,高等教育出版社,2014。

(2)主要辅助教材：

夏唐代.《建筑工程英语》,华中科技大学出版社,2005。

教育部《土建英语》编写组.《土建英语》,高等教育出版社,2000。

(二)教师素质要求

建筑英语教学内容以建筑施工为主线,涉及现场管理、基础施工、框架结构、屋面工程、围护结构、室外工程、水电安装、室内装饰、园林绿化等。本课程要求任课教师具有扎实的专业背景,同时具备较高的英语语言能力。教师有教学热情、教学经验丰富,能根据教学需要适当灵活增补教学内容,除了课堂上的讲述及学习,教师还要指导和安排好学生的课外学习活动。

(三)教学场地、设施要求

本课程要求在多媒体语言实验室上课。

六、考核评价

本课程在课程结束后以闭卷方式对学生进行考核。平时成绩占30%,卷面成绩占70%。

continued

Item	Topic	Teaching Objectives	Teaching Contents and Teaching Hours				
			Teaching Contents	Hrs	Student Activities	Hrs
5	Unit 5	1. Explain the function and the performance requirements of roof; 2. Have basic ideas of flat roof and pitched roof; 3. Describe different types of roof strutting	13. Workplace Conversation 14. Roof Construction 15. Roof Strutting	4	Roof Construction	1
6	Unit 6	1. Describe laying of bricks; 2. Know the various types of walls commonly used; 3. Explain the load bearing wall	16. Workplace Conversation 17. Wall Constructions 18. Load Bearing Walls	4	Wall Constructions	1
7		General Review		2		
		Total		26		6
Note						

5. Teaching Suggestions

5.1 Teaching Resources

Suggestions on the use textbooks and teaching materials:

Dai Mingyuan. *Construction Engineering English*. Higher Education Press, 2014.

Recommended auxiliary materials:

Xia Tangdai. *Construction Engineering English*. Huazhong University of Science and Technology Press, 2005.

Civil English Group of Ministry of Education. *Civil English*. Higher Education Press, 2000.

5.2 Teachers'Qualification

Teachers should have a professional background anda high level of English language, with a strong motivation in teaching and rich teaching experience. Teachers should flexibly tailor the teaching content according to teaching needs. Besides the classroom instruction, teachers should also guide and arrange students'extracurricular learning activities.

5.3 Teaching Facilities

This course is required to be delivered in multimedia classrooms.

6. Evaluation

Students are required to take the closed-book final term examination. The final grade score is the total of the learning process assessment(account for 30%) and the final exam score(account for 70%).

四 川 建 筑 职 业 技 术 学 院

课 程 标 准

课程名称：　　　　　　工程结算　　　　　　

课程代码：　　　　　　150431　　　　　　

课程学分：　　　　　　2　　　　　　

基本学时：　　　　　　32　　　　　　

适用专业：　　工程造价（中英合作办学项目）　

执 笔 人：　　　　　　胡晓娟　　　　　　

编制单位：　　　　建筑工程造价教研室　　　　

审　　核：　　　工程管理系（院）（签章）　　

批　　准：　　　　教务处（盖章）

Sichuan College of Architectural Technology

Curriculum Standard

Curriculum Name: _____ Engineering Settlement _____

Curriculum Code: _____ 150431 _____

Credits: _____ 2 _____

Teaching Hours: _____ 32 _____

Applicable Specialty: _____ Construction Cost _____

_____ (Sino-British Cooperative Program) _____

Compiled by: _____ Hu Xiaojuan _____

On Behalf of: Teaching & Research Section of Construction Cost

Reviewed by: _ The Department of Engineering Management _

Approved by: _____ Teaching Affairs Office _____

《工程结算》课程标准

一、课程概述

本课程标准适用于工程造价专业(含各专业方向、中英合作办学),《工程结算》课程属于考查课程,是工程造价专业的一门综合性专业技术课程,具有总结性和归纳性。主要包括工程结算的内容和相关规定、工程结算的程序和计算方法、工程结算管理等。课程在讲解过程中,需要充分结合现行的法律法规、规范标准、合约合同、相应的专业工程案例进行讲解。

二、课程设计思路

本课程是根据工程造价专业的人才培养目标、从业人员执业要求设计的。设置依据是工程造价专业人才培养方案和实施计划;课程框架、教学目标、教学内容的设计和选取根据工程造价专业人员岗位能力、职业素质要求、造价工程师考试大纲等进行;教学资源应充分利用工程造价专业及相关专业的教学资源库。

三、课程目标

本课程的目标是要求学生在掌握各专业工程的计量与计价上,通过本课程的学习,进一步整合和提升工程造价职业能力,符合职业资格考试内容的要求。

(一)知识目标

(1)理解工程结算的原则;
(2)掌握工程结算的内容和依据;
(3)掌握工程结算的相关规定、具体程序和方法;
(4)掌握工程结算中各价款调整的规定和方法;
(5)熟悉工程结算纠纷的解决方式和应用范围;
(6)熟悉工程结算管理的方法和主要内容。

(二)素质目标

(1)具有辩证思维;
(2)具有爱岗敬业的思想,实事求是、团结合作的工作作风和创新意识;
(3)强化法律意识和合约意识;

Engineering Settlement Curriculum Standard

1. Curriculum Description

This is a comprehensive and professional course for construction cost specialty students (Sino-British Cooperative Program). It encompasses the content and related provisions of engineering settlement, the procedure and calculation method of engineering settlement, and the management of it, etc. In the course of delivery, it is necessary for teachers to fully incorporate the existing laws and regulations, standards, contracts, and corresponding professional engineering cases.

2. Curriculum Design Ideas

This course is designed according to the educational concept of construction cost specialty, vocational skills and comprehensive abilities required by the construction industry and other enterprises. The setting is based on the course description and teaching plan of construction cost specialty. The design and selection of curriculum framework, teaching objectives and teaching contents are in accordance with the qualification and skills requirements of construction cost professionals, the test syllabus of cost engineers, etc. The teaching and learning resources include the resource library for construction cost in SCAT, the construction cost information in Sichuan province and so on.

3. Curriculum Objectives

This course aims to help students, on the basis of mastering the measurement and valuation of various professional engineering, further integrate and enhance the ability of construction cost to meet the requirements of corresponding qualification certificate examination.

3. 1 Knowledge Objectives

After learning this course, students will be able to

(1) understand the principles of engineering settlement;

(2) master the contents and basis of engineering settlement;

(3) master the relevant provisions, procedures and methods of engineering settlement;

(4) master the provisions and methods for the adjustment of the prices in the sett lement of engineering;

(5) be familiar with the solutions and application range for engineering settlement disputes;

(6) be familiar with the methods and main contents of engineering settlement management.

3. 2 Quality Objectives

After learning this course, students will have the quality of

(1) preliminary critical thinking ability;

(2) devoted, responsible, practical and cooperative work style and innovative consciousness;8

(3) strengthened legal awareness and contract awareness;

(4)遵循客观、公平、公正原则,具有遵纪守法、诚实守信、公平竞争、廉洁自律的职业道德意识。

(三)能力目标

(1)培养学生综合应用工程造价相关知识的能力;

(2)熟悉、理解并应用施工合同的能力;

(3)根据施工合同处理工程各种变更和价款调整的能力;

(4)初步具体沟通及结算谈判的能力;

(5)编制结算文件的能力。

四、课程内容与学时分配

教学内容与学时安排表

序号	教学情境/任务/项目/单元	教学目标	教学内容及学时分配			
			知识点	学时	实践项目	学时
1	工程结算概论	掌握工程结算的内容,理解工程结算的作用,熟悉工程结算的依据	1. 工程结算的概念、原则、种类; 2. 工程结算的作用; 3. 工程结算的依据	2		
2	工程结算程序	熟悉工程结算的相关规定;掌握工程结算程序及计算方法	4. 工程预付款的计算、支付和扣回(★); 5. 工程进度款的计算与支付; 6. 竣工结算款的计算与支付; 7. 工程尾款的计算与支付	8	案例分析、计算练习	2

(4) abiding principles of objectivity, fairness and impartiality, and fostering their professional ethics of law observing, honesty and trustworthiness, fair play, integrity and self-discipline.

3.3 Ability Objectives

After learning this course, students will have the ability of

(1) using knowledge of construction cost comprehensively;

(2) being familiar with construction contracts and can apply the knowledge into practice;

(3) handling projects changes and price adjustments according to the construction contract;

(4) acquiring the ability of preliminary communication for specific tasks and settlement negotiations;

(5) preparing settlement documents.

4. Curriculum Contents and Teaching Hours

Table of Teaching Contents and Teaching Hours

| Item | Topic | Teaching Objectives | Teaching Contents and Teaching Hours | | | |
			Lesson Contents	Hrs	Student Activities	Hrs
1	Introduction to the engineering settlement	Master the contents of engineering settlement; understand the role of engineering settlement; familiar with the basis of engineering settlement.	1. Concept, principle and types; 2. The role of engineering settlement; 3. The basis of engineering settlement	2		
2	Procedure of engineering settlement	Be familiar with the relevant provisions of engineering settlement; master engineering settlement procedures and calculation methods	4. The calculation, payment and withholding of the advance payment of the project(★); 5. Calculation and payment of progress funds for the project; 6. The calculation and payment of the completed settlement; 7. Calculation and payment of the final payment of the project	8	Case study and calculation exercise	2

序号	教学情境/任务/项目/单元	教学目标	教学内容及学时分配			
			知识点	学时	实践项目	学时
3	工程价款调整	1. 理解价款调整的原则、依据； 2. 实习价款调整的相关规定； 3. 掌握各种价款调整的计算方法和处理程序	8. 政策法规变化引起的价款调整； 9. 工程变更引起的价款调整； 10. 物价变化引起的价款调整； 11. 工程索赔引起的价款调整； 12. 其他情况引起的价款调整	10	案例分析	2
4	工程结算纠纷处理	1. 理解工程价款纠纷处理的原则； 2. 熟悉工程价款纠纷处理的方式和适应范围		2		
5	工程结算管理	1. 理解工程结算管理的重要性； 2. 了解政府工程结算管理的方式和内容； 3. 了解企业工程结算管理的方式和内容	13. 政府结算管理 14. 各方企业（建设单位、施工单位、咨询单位）工程结算管理的方法和内容	2		
6	工程结算综合案例	通过案例整合工程结算的程序、方法	15. 分析案例的可取之处； 16. 分析案例的不足之处	2	案例讨论	2
合计				26		6

备注：教学过程中持续更新案例，收集最新的法律法规，培养学生的综合能力。根据学生的接受情况可以对授课计划进行微调

continued

Item	Topic	Teaching Objectives	Teaching Contents and Teaching Hours			
			Lesson Contents	Hrs	Student Activities	Hrs
3	Adjustment of the price of the project	1. Understand the principle and basis of price adjustment; 2. The relevant provisions of the adjustment of the internship price; 3. Master the calculation methods and procedures for various price adjustments	8. Price adjustment caused by changes in policies and regulations; 9. Caused by changes of engineering; 10. Caused by commodity fluctuation; 11. Caused by engineering claims; 12. Caused by other circumstances	10	Case study	2
4	Dealing with the disputes of engineering settlement	1. Understand the principle of handling disputes over the price of the project; 2. Familiar with solutions and the applicable range for disputes over project prices		2		
5	The management of engineering settlement	1. Understand the importance of engineering settlement management; 2. Understand the way and content of government engineering settlement management; 3. Understand the way and content of enterprise engineering settlement management	13. Government settlement management; 14. The methods and contents of the project settlement management of the enterprises (construction units, construction enterprises and consulting units) of all parties	2		
6	Comprehensive case study of engineering settlement	Integrating procedures and methods for engineering settlement through cases	15. The strengths of the case; 16. The weaknesses of the case	2	Case discussion	2
	Total			26		6

Note: Teachers should keep updating cases, collect the latest laws and regulations, encourage more discussions and inspirations, and cultivate students'comprehensive abilities. The teaching content can be tailored depending on students'feedback

五、教学实施建议

(一)教学参考资料

胡晓娟.《工程结算》,重庆大学出版社,2015。

(二)教师素质要求

任课教师应具备教师资格证,应具有工程造价专业或相关专业执业资格和一定的社会实践和专业实践能力,有一定的专业教学经验。

(三)教学场地、设施要求

实施该课程教学,无须特殊环境,一般教室即可。

六、考核评价

(一)考核方式

闭卷。考核依据:《四川建筑职业技术学院学生学业考核办法》。

(二)考核说明

总评成绩的构成比例按:平时:期末 = 3:7(原则上)。

考核重点应覆盖的范围:工程结算程序和各阶段工程价款的计算,工程价款变更的规定和计算方法,工程结算纠纷的解决方式,工程结算管理的现行规定等。

考核的原则要求:工程造价职业资格制度、工程结算的相关规定、工程价款计算的方法为考核重点,突出动手能力的培养。

5. Teaching Suggestions

5.1 Teaching Resources

Hu Xiaojuan. *Engineering Settlement*. Chongqing University Press, 2015.

5.2 Teachers'Qualification

Teachers must have teacher qualification certificate, with construction cost major background and relevant professional qualification certificate and practical working and teaching experience.

5.3 Teaching Facilities

This course can be delivered in common classrooms.

6. Evaluation

6.1 The Method of Assessment

Examination without reference material in the examination room is suggested in accordance with the *Regulation of Academic Evaluation for Students in SCAT*.

6.2 Assessment Requirement

The final grade score is the total of the learning process assessment(including the usual attendance, in-class performance and homework, accounting for 30%)and the final term examination(accounting for 70%).

Key points of the assessment include the calculation of the engineering settlement procedure and the engineering price at each stage, the provisions and calculation methods for the change of the project price, solutions of settlement dispute, the available provisions of engineering settlement management, etc.

The assessment principle stresses the core methods of calculating the engineering price in accordance with the professional qualification system of construction cost and the relevant provisions of the engineering settlement, highlighting the cultivation of hands-on ability.

四川建筑职业技术学院

课程标准

课程名称： **工程造价职业资格实务培训**

课程代码： 150347

课程学分： 2

基本学时： 32

适用专业： **工程造价**

执笔人： 胡晓娟

编制单位： 建筑工程造价教研室

审核： 工程管理系（院）（签章）

批准： 教务处（盖章）

Sichuan College of Architectural Technology

Curriculum Standard

Curriculum Name : Professional Qualification Training of Construction Cost

Curriculum Code : 150347

Credits : 2

Teaching Hours : 32

Applicable Specialty : Construction Cost

Compiled by : Hu Xiaojuan

On Behalf of : Teaching & Research Section of Construction Cost

Reviewed by : The Department of Engineering Management

Approved by : Teaching Affairs Office

《工程造价职业资格实务培训》课程标准

一、课程概述

本课程标准适用于工程造价专业(中英合作办学项目),《工程造价职业资格实务培训》课程属于考查课程,是工程造价专业的一门综合性专业技术课程,具有总结性和归纳性。主要内容应包括工程造价职业资格制度、职业资格考试内容和要求、职业岗位及能力分析,由于工程造价专业前行课程已经对工程材料、结构与构造、施工技术、造价原理、计量与计价、造价控制、项目管理、工程经济等内容进行了详细介绍,本课程除了对职业资格管理和考试进行整体介绍外,应注重对工程结算的内容予以介绍。

课程任务包括介绍工程造价职业资格考试制度、职业资格考试内容和要求、职业岗位及能力要求、工程结算的内容和相关规定、工程结算的程序和计算方法、工程结算管理等。课程在讲解过程中,需要充分结合现行的法律法规、规范标准、合约合同、相应的专业工程案例进行讲解。

二、课程设计思路

本课程是根据工程造价专业的人才培养目标、从业人员执业要求设计的。设置依据是工程造价专业人才培养方案和实施计划;课程框架、教学目标、教学内容的设计和选取根据工程造价专业人员岗位能力、职业素质要求、造价工程师考试大纲等进行设计;教学资源应充分利用工程造价专业及相关专业的教学资源库。

三、课程目标

本课程的目标是要求学生在掌握各专业工程的计量与计价基础上,通过本课程的学习,进一步整合和提升工程造价职业能力,符合职业资格考试内容的要求。

Professional Qualification Training of Construction Cost **Curriculum Standard**

1. Curriculum Description

This curriculum standard is applicable to the specialty of construction cost(Sino-British cooperative program). *Professional Qualification Training of Construction Cost* is a required course for this specialty. It is a comprehensive professional technical course and it is quite inductive. The main contents of the course are about the professional qualification system, the contents and requirements of the examinations of professional qualification of construction cost and professional post demand and capacity analysis. According to the training program of this specialty, courses about construction materials, construction structure and principles, construction technology, cost measurement and valuation, cost control, project management, engineering economy, etc have been delivered, so this course will be mainly about professional qualification management and examination, and engineering settlement should also be covered in this course.

The main contents of the course are about the professional qualification system, the contents and requirements of the examinations of professional qualification of construction cost and professional post demand and capacity analysis, engineering settlement and relevant regulations, engineering settlement procedures and calculation methods, engineering settlement management, etc. In the teaching process, the current laws and regulations, norms and standards, contracts and corresponding professional engineering cases should be introduced or explained.

2. Curriculum Design Ideas

This course is designed according to the objectives in the professional training program and professional requirements for graduates of the specialty of construction cost. The basis for the curriculum standard is the professional training program and the implementing plan for the specialty of construction cost. The curriculum plan, curriculum objectives, the selecting and organizing of the teaching contents are based on the analysis of the posts and requirements for the specialty of construction cost and the requirements of examinations of professional qualification, etc. The teaching and learning resources include the resource library of construction cost and other related specialties.

3. Curriculum Objectives

In this course, students can master the pricing and estimating of different construction projects; furthermore, they can integrate and improve the professional ability of construction cost through the study of this course, and meet the requirements of professional qualification examination.

(一)知识目标

(1)熟悉工程造价职业资格考试制度的规定、考试内容和要求,工程造价岗位设置及能力要求;

(2)理解工程结算的原则;

(3)掌握工程结算的内容和依据;

(4)掌握工程结算的相关规定、具体程序和方法;

(5)掌握工程结算中各价款调整的规定和方法;

(6)熟悉工程结算纠纷的解决方式和应用范围;

(7)熟悉工程结算管理的方法和主要内容。

(二)素质目标

(1)初步具备辩证思维的能力;

(2)具有爱岗敬业的思想,实事求是、团结合作的工作作风和创新意识;

(3)强化法律意识和合约意识;

(4)遵循客观、公平、公正原则,具有遵纪守法、诚实守信、公平竞争、廉洁自律的职业道德意识。

(三)能力目标

(1)职业规划能力;

(2)综合应用工程造价相关知识的能力;

(3)熟悉、理解并应用施工合同的能力;

(4)根据施工合同处理工程各种变更和价款调整的能力;

(5)初步具体沟通及结算谈判的能力;

(6)编制结算文件的能力。

3. 1　Knowledge Objectives

After learning this course, students will be able to

(1) be familiar with the regulations, examination contents and requirements of the professional qualification examination system of construction cost, as well as the setting and ability requirements of posts related with construction const;

(2) understand the principles of engineering settlement;

(3) master the content and basis of engineering settlement;

(4) master relevant regulations, specific procedures and methods of engineering settlement;

(5) master the regulations and methods of adjustment of price in engineering settlement;

(6) be familiar with settlement methods and application scope of engineering settlement disputes;

(7) be familiar with the methods and main contents of engineering settlement management.

3. 2　Quality Objectives

After learning this course, students will have the quality of

(1) critical thinking;

(2) loving work, seeking truth from facts, unity and cooperation and the consciousness of innovation;

(3) strengthened legal awareness and contract awareness;

(4) objectivity, fairness and justice, and professional ethics consciousness of abiding discipline and law, being honest and trustworthy, fair competition, honesty and self-discipline.

3. 3　Ability Objectives

After learning this course, students will have the ability of

(1) planning career;

(2) applying knowledge related with construction cost comprehensively;

(3) understanding and applying construction contracts;

(4) handling all kinds of engineering changes and price adjustments according to the construction contract;

(5) basic communication and settlement negotiation;

(6) preparing settlement documents.

四、课程内容与学时分配

教学内容与学时安排表

序号	教学情境/任务/项目/单元	教学目标	教学内容及学时分配			
			知识点	学时	实践项目	学时
1	工程造价职业资格管理制度	了解工程造价现行的职业资格制度、考试内容和要求,清楚本课程与前行课程的关系,了解本课程的主要内容和学习目标	1. 职业资格考试制度; 2. 本专业的课程设置情况; 3. 本课程的主要内容和学习目标	2		
2	工程结算概论	掌握工程结算的内容,理解工程结算的作用,熟悉工程结算的依据	4. 工程结算的概念、原则、种类; 5. 工程结算的作用; 6. 工程结算的依据	2		
3	工程结算程序	熟悉工程结算的相关规定;掌握工程结算程序及计算方法	7. 工程预付款的计算、支付和扣回(★); 8. 工程进度款的计算与支付; 9. 竣工结算款的计算与支付; 10. 工程尾款的计算与支付	6	案例分析、计算练习	2
4	工程价款调整	理解价款调整的原则、依据;实习价款调整的相关规定;掌握各种价款调整的计算方法和处理程序	11. 政策法规变化引起的价款调整; 12. 工程变更引起的价款调整; 13. 物价变化引起的价款调整; 14. 工程索赔引起的价款调整; 15. 其他情况引起的价款调整	10	案例分析	2

4. Curriculum Contents and Teaching Hours

Table of Teaching Contents and Teaching Hours

Item	Topic	Teaching Objectives	Teaching Contents and Teaching Hours			
			Lesson Contents	Hrs	Student Activities	Hrs
1	Construction Cost professional qualification management system	Understand the current professional qualification system of construction cost, examination contents and requirements; Understand the relationship between this course and the previous courses, and understand the main content and learning objectives of this course	1. Professional qualification examination system; 2. Courses setting of this specialty; 3. Main contents and learning objectives of this course	2		
2	Introduction to engineering settlement	Master the contents of engineering settlement; Understand the role of engineering settlement, and be familiar with the basis of engineering settlement	4. Concepts, principles and types of engineering settlement; 5. Role of engineering settlement; 6. Basis for engineering settlement	2		
3	Engineering settlement procedure	Familiar with relevant regulations of engineering settlement; Master the procedure and calculation method of engineering settlement	7. Calculation, payment and deduction of advance payment(★); 8. Calculation and payment of project progress payment; 9. Calculation and payment of completion settlement amount; 10. Calculation and payment of pro-ject balance payment	6	Case analysis, calculation exercises	2
4	Adjustment of project price	Understand the principle and basis of price adjustment; Understand relevant regulations on the adjustment of internship price; Master the calculation methods and processing procedures of various price adjustments	11. Price adjustment caused by changes in policies and regulations; 12. Price adjustment caused by engineering change; 13. Price adjustment caused by price changes; 14. Price adjustment caused by engineering claim; 15. Price adjustment caused by other circumstances	10	Case analysis	2

序号	教学情境/任务/项目/单元	教学目标	教学内容及学时分配			
			知识点	学时	实践项目	学时
5	工程结算纠纷处理	理解工程价款纠纷处理的原则;熟悉工程价款纠纷处理的方式和适应范围		2		
6	工程结算管理	理解工程结算管理的重要性; 了解政府工程结算管理的方式和内容; 了解企业工程结算管理的方式和内容	16. 政府结算管理; 17. 各方企业(建设单位、施工单位、咨询单位)工程结算管理的方法和内容	2		
7	工程结算综合案例	通过案例整合工程结算的程序、方法	18. 分析案例的可取之处; 19. 分析案例的不足之处	2	案例讨论	2
	合计			26		6
备注:重要知识点标注★						

五、教学实施建议

(一)教学参考资料

胡晓娟.《工程结算》,重庆大学出版社,2015。

(二)教师素质要求

任课教师应具备教师资格证,应具有工程造价专业或相关专业执业资格和一定的社会实践和专业实践能力,有一定的专业教学经验。

(三)教学场地、设施要求

实施该课程教学,无须特殊环境,一般教室即可。

continued

Item	Topic	Teaching Objectives	Teaching Contents and Teaching Hours			
			Lesson Contents	Hrs	Student Activities	Hrs
5	Settlement of project disputes	Understand the principles of dispute settlement of project price; Be familiar with the way and scope of settlement of project price disputes		2		
6	Project settlement management	Understand the importance of engineering settlement management; Understand the methods and contents of government engineering settlement management; Understand the method and content of enterprise project settlement	16. Government settlement management; 17. Methods and contents of project settlement management for enterprises of all parties (development units, construction units and consulting units)	2		
7	Project settlement comprehensive case	Through case integration project settlement procedures, methods	18. Analyzing the merits of the case; 19. Analyzing the shortcomings of the case	2	Case discussion	2
Total				26		6
Note: Mark the important points with ★						

5. Teaching Suggestions

5.1 Teaching Resources

Hu Xiaojuan. *Engineering Settlement.* Chongqing University Press, 2015.

5.2 Teachers'Qualification

The teachers for this course should have the qualification certificate of teachers, the professional qualification of construction cost or related professional or social practice, and certain professional teaching experience is preferred.

5.3 Teaching Facilities

There is no special environment requirement for the implementation of this course, so it can be delivered in the common classrooms.

六、考核评价

(一)考核方式

闭卷。

(二)考核说明

总评成绩的构成比例原则上按《四川建筑职业技术学院学生学业考核办法》执行:平时:期末=3:7。

考核重点应覆盖的范围:工程造价职业资格考试制度、考试要求,工程结算的原则、种类、依据、作用等,工程结算程序和各阶段工程价款的计算,工程价款变更的规定和计算方法,工程结算纠纷的解决方式,工程结算管理的现行规定等。

考核的原则要求:工程造价职业资格制度、工程结算的相关规定,工程价款计算的方法为考核重点,突出动手能力的培养。

6. Evaluation

6.1   Assessment method

Examination without taking reference books.

6.2   Assessment statement

The score of the course is based on the *Regulation of Academic Evaluation for Students in SCAT*, and the proportion for the final score based on the principle that the daily performance accounts for 30% and the final examination accounts for 70%.

The assessment should cover the contents and requirements of professional qualification examination of the construction cost, the principles, categories, basis and function of the engineering settlement, engineering settlement procedure and the calculation of construction cost on each stage, construction cost change regulation and calculation methods, dispute solution in engineering settlement, the regulations for engineering settlement management, etc.

Assessment principles: the focus of the assessment is about professional qualification system of construction cost, relevant regulations of engineering settlement, and calculation method of construction cost. The assessment should promote students'practical ability.